PRINCIPLES OF FOOD SCIENCE

PART II
PHYSICAL PRINCIPLES OF FOOD PRESERVATION

FOOD SCIENCE

A Series of Monographs

Series Editor

OWEN R. FENNEMA
DEPARTMENT OF FOOD SCIENCE
UNIVERSITY OF WISCONSIN
COLLEGE OF AGRICULTURE
MADISON, WISCONSIN

Volume 1:
FLAVOR RESEARCH: Principles and Techniques
by A. Teranishi, I. Hornstein, P. Issenberg, and E. L. Wick
(Out of Print)

Volume 2:
PRINCIPLES OF ENZYMOLOGY FOR
THE FOOD SCIENCES *by John R. Whitaker*

Volume 3:
LOW-TEMPERATURE PRESERVATION OF FOODS AND LIVING MATTER
by Owen R. Fennema, William D. Powrie, and Elmer H. Marth

Volume 4:
PRINCIPLES OF FOOD SCIENCE *Edited by Owen R. Fennema*

Part I: FOOD CHEMISTRY *Edited by Owen R. Fennema*
Part II: PHYSICAL PRINCIPLES OF FOOD PRESERVATION
by Marcus Karel, Owen R. Fennema, and Daryl B. Lund
Part III: FOOD FERMENTATIONS *by E. H. Marth, N. F. Olson, and C. H. Amundson (In Preparation)*

OTHER VOLUMES IN PREPARATION

PRINCIPLES OF FOOD SCIENCE

Edited by Owen R. Fennema

PART II
PHYSICAL PRINCIPLES OF FOOD PRESERVATION

Marcus Karel
Department of Nutrition and Food Science
Massachusetts Institute of Technology
Cambridge, Massachusetts

Owen R. Fennema
Department of Food Science
University of Wisconsin-Madison
Madison, Wisconsin

Daryl B. Lund
Department of Food Science
University of Wisconsin-Madison
Madison, Wisconsin

MARCEL DEKKER, INC. New York and Basel

MARCEL DEKKER, INC.

270 Madison Avenue, New York, New York 10016

LIBRARY OF CONGRESS CATALOG CARD NUMBER: 75-29694

ISBN: 0-8247-6322-X

Current printing (last digit):
10 9 8 7

PRINTED IN THE UNITED STATES OF AMERICA

PREFACE

The subject of food processing occupies a position of major importance in the food science curriculum and this emphasis is likely to continue. Thus it is both puzzling and reprehensible that a university-level textbook tailored to the needs of typical students in food science does not exist. In the United States, the two largest categories of students represented in food processing courses are (1) upper-level undergraduates in food science, and (2) graduate students that are entering food science from other disciplines. These students generally have solid backgrounds in chemistry, bacteriology, biology, mathematics, and physics and minimal backgrounds in engineering. This book on Physical Principles of Food Preservation is intended primarily for these students, and secondarily for persons in the food industry and for scientists in food-related groups of the government.

In the past, knowledge of food preservation has been used to develop new or improved products or more economical (labor-saving) processes. In the future, knowledge of food preservation techniques will no doubt be applied increasingly to matters of a more critical nature, i.e., to reducing wastage of the world's food supply and to devising processes that optimize the factors of cost, nutritional quality, environmental impact, and consumption of resources and energy.

In this book, a qualitative and semiquantitative approach is used, and stress is given to long-enduring principles rather than to detail. This approach is necessary if one is to adapt successfully to the changing objectives of food preservation, as discussed above.

In the introductory chapter, attention is drawn to the fact that food wastage is a major problem on a world-wide basis, that much of this wastage can be prevented by proper application of food-preservation techniques, and that processing advances must be accomplished with due consideration given to the amounts of energy and resources consumed and to the environmental effects. The remainder of the book is divided into four sections: the first deals with preservation by means of temporary increases in the product's energy content (heat processing, irradiation); the second deals with preservation by controlled reduction of the product's temperature (chilling, freezing); the third deals with preservation by controlled reduction of the product's water content (concentration, air — dehydration, freeze drying); and the last deals with packaging — an important means of protecting foods during storage. Chapters are also included at appropriate places to familiarize the reader with the principles of phase equilibria, heat transfer, mass transfer, and water activity.

It is hoped that this book shall enlighten readers with respect to the principles of physical methods of food preservation and thereby encourage use of methods that are most suitable to the needs of society.

Marcus Karel
Owen Fennema
Daryl Lund

ACKNOWLEDGMENTS

The number of individuals to whom we are indebted for help in preparing this book is too large to mention by name. I would like, however, to express my very special appreciation to the following:

My wife, Cal, for her loving encouragement of my work,
My children, Debra, Karen, Steven, and especially Linda for "daring" me to write "a book,"
My editor, co-author, and friend, Owen Fennema, whose critical reviews of my chapters were simply invaluable,
Our co-author, Daryl Lund, who joined the team when we arrived at a critical juncture and whose fine chapters allowed us to complete this work,
And perhaps most of all to my students in courses 20.13 and 20.52 at M.I.T. whose persistent questions and critical comments helped to shape the lectures from which my chapters are derived.

Marcus Karel

CONTENTS

PRINCIPLES OF FOOD SCIENCE

PART II
PHYSICAL PRINCIPLES OF FOOD PRESERVATION

Chapter 1

INTRODUCTION TO FOOD PRESERVATION

Owen Fennema

CONTENTS

I. WORLD FOOD PROBLEM

A great need exists for more effective and more widely used methods of food preservation. This need can be readily substantiated by documenting the grave problems, both current and future, with respect to providing an adequate diet for the world's expanding population and by drawing attention to the staggering amount of food that is unnecessarily wasted because of inadequate preservation.

A. Dietary Inadequacies

In underdeveloped countries, where two-thirds of the world's population reside, there is overwhelming evidence that substantial numbers of people suffer from undernutrition or malnutrition [5, 8, 14]. Much, if not all, of this problem could be at least temporarily remedied if existing food supplies were adequately preserved, properly distributed on the basis of need, and willingly consumed. Unfortunately, this has not been possible, because of technical, economic, political, cultural, and religious reasons, along with a large measure of unconcern. Shown in Figure 1-1 are actual changes in population and food production in underdeveloped countries during the period 1957-1970, as well as estimates for the

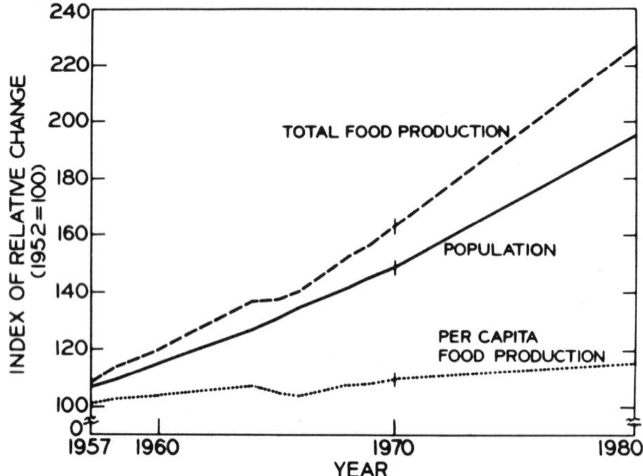

FIG. 1-1. Trends in food production and population in underdeveloped countries. Data after 1970 are estimates [10].

period 1970-1980. It is evident that per capita food production has remained essentially constant from 1959 to 1970 (i.e., dietary inadequacies persisted with little change), and that only small improvements are predicted up to 1980. Although no one can determine with accuracy how well future supplies of food can fulfill future needs, it is abundantly clear that the predicted doubling of the world's population by the year 2000 must severely tax man's technological capabilities just to maintain current per capita diets in underdeveloped countries [2, 19]. Solutions to this dilemma are deserving of the highest priority.

B. Long-term Solutions

Adequate feeding of the world's population can be achieved only by a multi-faceted program involving all of the following approaches [1, 3, 5, 6, 8, 14, 15, 17, 19].

1. Control of Population Growth

It is of the utmost importance that the growth rate of the world's population be drastically reduced in the near future if present per capita diets are to be maintained or improved. It must be realized that resources and space are finite [12].

2. Increased Supply of Food

Vastly increased supplies of food will be needed to provide minimal diets for a world population growing at even the slowest anticipated rate. Increased supplies of food can be obtained by devoting a greater portion of the earth's surface to food production (obvious limitations), by increasing the yield per acre (improved crop varieties; soil conservation; greater use of conventional plants as a source of protein; harvesting more nearly at optimum maturity; and greater use of fertilizers, pesticides, and irrigation), by developing new sources of food (leaves and unicellular organisms as sources of protein, artificial meat from soybean proteins, etc.), and by greater exploitation of sea foods.

3. Decreased Food Wastage

The importance of food wastage as a factor affecting the world's supply of food is clearly indicated by the following statement from the Food and Agricultural Organization of the United Nations [9]:

> Nobody knows how much food men labor every year to produce, only to see it taken off by rats, insects or spoiled in a hundred different ways. The figures and estimates fly around — 10 percent here, 50 percent there, 20 percent in the world as a whole — and almost every figure has its firm supporters and its equally resolute opponents. About the only thing that is generally agreed upon is that the losses are enormous and that, if modern storage and pest control techniques were available on a wider scale, we would have many millions of tons of extra food available for people to eat each year without planting an extra hectare (Preface, p. V).

Data and opinions in support of these remarks are readily available from several sources [2, 4, 7, 13, 15, 17, 20].

Urbanization has greatly intensified the problem of food wastage in some areas of the world. The rapid increase in the nonfarm population of developed countries is well recognized. However, this occurrence has had little effect on food wastage. Less well known is that urbanization in diet-deficient regions of the world is fully as intensive as in developed countries. "As early as 1950, more than one-third of the world's cities with populations exceeding 100 thousand were in Asia and the exodus from rural areas has accelerated each year since" [14]. The cause lies in the desire of people to leave poverty-stricken rural areas in the hope of finding better living conditions in the cities. Urbanization in diet-deficient regions of the world generally has had many undesirable effects, and increased food wastage is one of them. Urbanization moves the consumer further away from areas of food production and increases the time between harvest and consumption. Under these circumstances, increased wastage of food can be avoided only if an extensive, efficient distribution system is developed and if effective methods of food preservation are widely employed [16]. Unfortunately, the systems for distributing and preserving food in diet-deficient regions of the world generally have not improved at a pace equal to urbanization, thus establishing urbanization as a significant cause of food wastage in these areas. Thus, it is abundantly clear that food wastage is currently a major problem and is likely to remain so in the foreseeable future.

Food wastage occurs through contamination, consumption by pests, inefficient utilization, and spoilage. The last two points deserve further consideration.

Wastage through inefficient utilization occurs when nutritious portions of food are not fully utilized for human consumption and when food tissues are harvested or slaughtered at maturities that provide a less than optimum combination of yield and quality. Cheese whey and bran are important examples of nutritious food constituents that frequently are not used for human consumption. If fruits are harvested when excessively immature or overripe or if vegetables are harvested when excessively overmature, wastage can occur because the consumer frequently rejects these products. Inefficient utilization of foods is unquestionably a major cause of food wastage.

Spoilage is also a major cause of food wastage. Much of the fresh food supply is perishable because of its moderate to high water content and its nutritious nature. In Table 1-1, essentially unmodified foods are grouped according to degree of perishability. Among these foods, only mature, dry plant tissues can resist spoilage for extended periods at room temperature, and these materials, of course, remain subject to attack by numerous pests.

The causes of deterioration and spoilage of foods include growth of microorganisms (by far the most common cause of spoilage of foods with moderate to high water contents), contamination (filth, absorption of off flavors and odors), normal respiration (plant tissues), loss of water, sprouting, autolysis (especially fish), various chemical reactions (e.g., oxidation), physiological disorders (e.g., scald of apples, cold shortening of muscle, chilling injury and anaerobic respiration of plant tissues), and mechanical damage.

Spoilage of perishable foods can be averted only by prompt consumption, which often is not possible, or by prompt, effective preservation. Effective preservation not only greatly retards spoilage but also helps reduce both contamination of food

TABLE 1-1

Perishability of Essentially Unmodified Foods

Degree of perishability (storage life at room temperature)	Water content	Type of product
Highly perishable (1-7 days)	Medium to high	Animal tissues: meats, poultry, seafoods
		Plant tissues: soft, juicy fruits and vegetables with rapid respiration, e.g., berries, peaches, spinach, immature or leafy vegetables
		Milk
Moderately perishable (1 to several weeks)	Medium	Plant tissues: root crops, mature apples, and pears
		Eggs
Slightly perishable (1 year or more)	Low	Plant tissues: food grains, beans (dry), peas (dry), nuts

and consumption of food by pests. Thus control of three of the four major causes of food wastage is achieved through effective food preservation (the efficiency of food utilization is not necessarily improved through effective food preservation).

II. FOOD PRESERVATION

A. General Comments

The aim of commercial food preservation is to prevent undesirable changes in the wholesomeness, nutritive value, or sensory quality of food by economical methods which control growth of microorganisms; reduce chemical, physical, and physiological changes of an undesirable nature; and obviate contamination.

Preservation of food can be accomplished by chemical, biological, or physical means. Chemical preservation involves the addition to food of such substances as sugars, salts, or acids or exposure of food to chemicals, such as smoke or fumigants. Biological preservation involves alcoholic or acidic fermentations. Physical approaches to preserving food include temporary increases in the product's energy level (heating, irradiation), controlled reduction of the product's temperature (chilling, freezing), controlled reduction in the product's water content (concentration, air dehydration, freeze drying), and the use of protective packages. Physical methods of preservation are used extensively in developed countries of the world and they are likely to become more common world-wide.

B. Physical Methods of Food Preservation

During preservation of moderately or highly perishable foods, the greatest concern is with microorganisms. Physical methods of preservation result either in death of microorganisms (by temporarily increasing the energy level of a food which is suitably packaged to avoid recontamination), or suppression of their growth (by maintaining the food at subambient temperatures or by removing water followed by packaging to avoid readsorption of water).

Although certain physical methods of food preservation completely stop growth of microorganisms and greatly retard the rates of many chemical reactions, it is important to recognize that none of these methods can completely stop chemical and physical changes. For example, in frozen foods stored at a recommended temperature of -18°, microorganisms cannot grow, but degradation of vitamin C, insolubilization of protein, oxidation of lipids, and recrystallization can occur at significant rates.

It also should be noted that those methods of preservation which successfully stop growth of microorganisms generally have some undesirable consequences on the sensory and nutritive attributes of food. For example, thermal sterilization softens food tissues, degrades chlorophylls and anthocyanins, alters flavor, and results in loss or degradation of substantial amounts of some vitamins. Thus preservation techniques for food are by no means perfect and although the advantages almost always outweigh the disadvantages, there is an unending need to decrease the disadvantages while maintaining the advantages.

C. Constraints

Industrialized nations of the world have until recent years been able to develop food systems (production, processing, distribution, utilization) with little regard for consumption of energy and resources or for impact on the environment. These factors must now receive major consideration. Concern with quality of the environment has already begun to drastically alter practices in food-processing plants, and this trend is likely to intensify. One important effect is that discharges from food plants now must be monitored with care, and control of both the quantity and quality of these discharges is frequently necessary.

The fact that natural resources are finite and some are scarce or nearing positions of scarcity is having an increasing influence on the food industry. For example, increasing costs and increasing intervention by governmental agencies provide food processors with strong incentives to reduce the amount of materials used for food packaging and to use materials that are recyclable and/or easily disposable in a fashion that does not have detrimental consequences on the environment.

The energy problem is directly related to the problem of natural resources since most of our energy is derived from nonreplenishable natural resources, mainly coal and oil. Since natural resources used for energy are no longer abundant, costs of energy are rapidly increasing. This affects the United States food supply system significantly since this system consumes about 12% of the total energy generated in the United States [11]. The food-processing portion of the food supply system consumes about 4.5% of the total energy generated in the United States [18]. The severity of the situation in the United States is further aggravated by the long-term pattern of large annual increases in energy consumption by the food supply system. Energy consumption by this system increased from about 300 x 10^{12} kcal in 1940 to 850 x 10^{12} kcal in 1970 [18]. Part of this increase is attributable to increases in the population but most is attributable to markedly increased energy input per calorie of food output. In 1910, about 1 cal of energy input was needed to produce 1 cal of food output. Since then, this ratio has steadily increased, until in 1970 9 cal of energy input were required per calorie of food output [18]. Although this pattern of increase has continued unabated through 1970, the much higher energy costs that now prevail are almost certain to have a moderating effect.

Despite the success with which the United States food supply system has functioned, it is not likely to be adopted on a world-wide basis, since to do so necessitates consumption of almost 80% of the world's annual energy supply [18].

In summary, it is clear that food-processing industries in industrialized nations of the world face formidable new constraints and that these constraints often differ in degree or kind from those prevailing in underdeveloped regions of the world. Thus food scientists are confronted with unprecedented challenges — challenges that cannot be met without a thorough knowledge of the principles underlying the various methods of food preservation.

REFERENCES

1. J. L. Bennett, World Rev. Nutr. Diet., 11, 1 (1969).
2. G. Borgstrom, Principles of Food Science, 2 vols., Macmillan, New York, 1968.
3. G. Borgstrom, Too Many, Macmillan, New York, 1969.
4. J. E. Clayton, in The Yearbook of Agriculture 1966, U. S. Department of Agriculture, U. S. Government Printing Office, Washington, D. C., 1966, pp. 118-122.
5. W. W. Cochrane, The World Food Problem, T. Y. Crowell, New York, 1969.
6. P. Ehrlich, The Population Bomb, Ballantine Books, New York, 1968.
7. B. H. Farmer, in Population and Food Supply (J. Hutchinson, ed.), Cambridge University Press, Cambridge, 1969, Chapter 6, pp. 75-95.
8. Food and Agricultural Organization, World Food Problems No. 4, 6 Billion to Feed, Food and Agricultural Organization of the United Nations, Rome, 1962.
9. Food and Agricultural Organization, World Food Problems No. 9, Bigger Crops and Better Storage, Food and Agricultural Organization of the United Nations, Rome, 1969.
10. Food and Agricultural Organization, Agricultural Commodity Projection 1970-1980, Vol. 1, Food and Agricultural Organization of the United Nations, Rome, 1971.
11. E. Hirst, Energy Use for Food in the United States, ORNL-NSF-E9-57, Oak Ridge National Laboratories, Oak Ridge, Tenn., 1973.
12. T. Malthus, Essay on the Principle of Population, in The House We Live In (S. D. Blau and J. von B. Rodenbeck, eds.), Macmillan, New York, 1971, pp. 209-213.
13. S. V. Pingale, Pesticides, 1(11), 66 (1968).
14. President's Science Advisory Committee, The World Food Problem, Vol. 1, U. S. Government Printing Office, Washington, D. C., 1967.
15. President's Science Advisory Committee, The World Food Problem, Vol. 2, U. S. Government Printing Office, Washington, D. C., 1967.
16. J. C. Somogyi, in The Influence of Industrial and Household Handling on the Composition of Food (J. C. Somogyi, ed.), S. Karger, Basel, 1965, pp. 1-35.

17. J. J. Spengler, World Rev. Nutr. Diet., 9, 1 (1968).

18. J. S. Steinhart and C. E. Steinhart, Science, 184, 307 (1974).

19. J. M. Thoday, in Population and Food Supply (J. Hutchinson, ed.), Cambridge University Press, Cambridge, 1969, Chapter 1, pp. 1-10.

20. United States Department of Agriculture, Losses in Agriculture, Agricultural Handbook No. 291, U. S. Government Printing Office, Washington, D. C., 1965.

PRESERVATION THROUGH ENERGY INPUT

Chapter 2

HEAT TRANSFER IN FOODS

Daryl Lund

CONTENTS

I. INTRODUCTION

Heat transfer is one of the most important unit operations in the food industry. Nearly every process requires either heat input or heat removal to alter the physical, chemical, and storage characteristics of the product. For example, in the storage of fresh fruits, vegetables, meats, and dairy products heat is removed to decrease temperature and thereby increase storage life. On the other hand, heat is temporarily added to achieve preservation by thermal means. Examples of heat input to change the characteristics of products include cooking to alter the texture of foods and heating to develop color and flavor in candy. Removal of heat is used to produce desirable physical changes (crystallization) in frozen desserts and some candies.

In all of these applications, transfer of heat is governed by physical laws. An understanding of these physical principles enables one to predict heating phenomenon and to determine optimum operating conditions. In determining optimum operating conditions consideration must of course be given not only to the rate of energy transfer but also to food quality. This necessitates a working knowledge of both engineering and food science.

In this chapter the basic physical principles of heat transfer are presented with particular attention given to the nature of the material being heated.

II. MECHANISMS OF HEAT TRANSFER

A. Introduction

Heat transfer is an energy-transfer phenomenon. Increased heat energy causes a product's molecules to move more rapidly. That is, the kinetic energy of a molecule increases as heat energy is absorbed. Heat is transferred when a fast moving molecule collides with a slow moving molecule, causing the fast moving molecule to lose kinetic energy and the slow moving molecule to gain kinetic energy. From this it is apparent that heat transfer involves a molecular mechanism [5].

One manifestation of the level of heat energy in a group of molecules is temperature, a substance's relative hotness or coldness. Several arbitrary "yardsticks" or scales have been developed for measuring temperature but the two most common ones are the Celsius and Fahrenheit scales. However, if the energy level of atoms and molecules are compared to zero kinetic energy (i.e., no motion of the particles and consequently zero on an absolute temperature scale) the temperature is expressed in degrees Kelvin or degrees Rankine. The relationships among these temperature scales are shown in Table 2-1.

The amount of heat energy is also measured in arbitrary units. In the cgs system the calorie is the unit of heat energy; whereas in the fps system the Btu (British thermal unit) is the unit of heat energy. A calorie is the amount of heat necessary to raise the temperature of 1 g of water 1° C (from 14.5° to 15.5° C). A Btu is the amount of heat necessary to raise 1 lb of water 1°F (usually from 61° to 62°F).

"Heat capacity" (C_p; the p indicating constant pressure) is a term used to define the amount of heat energy necessary to cause a given mass of a substance to change 1°. In the cgs system heat capacity is the calories required to raise 1 g of substance 1° C; in the fps system, it is the Btus required to raise 1 lb of substance 1° F.

Since the definition of the calorie and of the Btu is based on water, the heat capacity for water is 1 cal/g $^\circ$C or 1 Btu/lb $^\circ$F. If the heat capacity of a substance is compared to that of water the resulting ratio is called "specific heat." Specific heat is a dimensionless ratio that is numerically equal to the heat capacity of the substance.

$$\text{Specific heat (sh)} = \frac{C_p \text{ substance}}{C_p \text{ water}}$$

The heat capacities of a variety of common foods are given in Table 2-2 [2].

When heat energy is added to or taken from a substance, either the temperature or the physical state of the substance changes. Heat input which causes a rise in temperature is called "sensible heat" since the heat can be "sensed." Added heat that does not contribute to a temperature increase is called "latent heat." Latent heat is the heat that is used to bring about a change of state of a substance without a change in temperature. If the substance goes from a solid to a liquid, the latent heat is called "heat of fusion." If the change is from liquid to a gas, it is called

TABLE 2-1

Relationships among Various Temperature Scales

Reference points for water	Celsius (°C)[a]	Fahrenheit (°F)	Kelvin (°K)[b]	Rankine (°R)[c]
Boiling point (1 atm)	100	212	373	672
Freezing point (1 atm)	0	32	273	492
Zero kinetic energy	-273	-460	0	0

[a] °C = (5/9) (°F - 32).

[b] °K = °C + 273.

[c] °R = °F + 460.

"heat of vaporization." For the reverse process of gas to liquid or liquid to solid the latent heat is usually called "heat of condensation" and "heat of solidification" (heat of crystallization), respectively. Energy involved in a change of state is added or emitted as a result of changes in various forces of attraction among molecules.

For many of the heating and cooling applications in food, a change in the physical state of water is the primary objective. For example, changing liquid water to vapor is the primary objective in dehydration and evaporation, and changing liquid water to a solid is the objective of freezing foods that are to be consumed in a frozen state.

The latent heat of vaporization is dependent on the pressure and boiling point of the aqueous phase. For pure water at 212° F (100° C) the latent heat of vaporization is 540 cal/g or 971 Btu/lb. Values of latent heat of vaporization as a function of pressure and temperature can be found in most handbooks of the physical sciences (<u>Handbook of Chemistry and Physics</u>). The latent heat of fusion for pure water at 32° F (0° C) is 80 cal/g or 144 Btu/lb.

In many heat-transfer processes it is necessary to know the total quantity of heat that must be transferred. Knowing this, the rate of the process can be estimated from process parameters of the heat-transfer system. In assessing the total quantity of heat transferred, sensible heat, heat of vaporization, and heat of fusion all must be considered whenever appropriate. Sensible heat can be calculated from the relationship

$$Q_s = MC_p \Delta T \tag{2-1}$$

where

Q is sensible heat transferred (Btu)
M is mass of material (lb)
C_p is heat capacity (Btu/lb °F), and
ΔT is temperature change of the product (°F)

TABLE 2-2

Mean Heat Capacities of Foodstuffs Between 32° and 212° F [a]

	Water content (%)	Heat capacity (Btu/lb °F)
Soups		
Broth		0.71
Cabbage soup		0.90
Pea soup		0.98
Potato soup	88	0.94
Fish		
Fried	60	0.72
Fresh	80	0.86
Dried, salt	16-20	0.41-0.44
Fats		
Butter	14-15.5	0.49-0.51
Margarine	9-15	0.42-0.50
Vegetable oils		0.35-0.45
Beverages		
Cacao		0.44
Cream, 45-60% fat	57-73	0.73-0.78
Rich (cow) milk	87.5	0.92
Skim (cow) milk	91	0.95-0.96
Vegetables		
Artichokes	90	0.93
Carrots, fresh	86-90	0.91-0.94
Carrots, boiled	92	0.90
Cucumber	97	0.98
Leek	92	0.95
Lentils	12	0.44
Mushrooms, fresh	90	0.94
Mushrooms, dried	30	0.56
Onions	80-90	0.86-0.93
Parsley	65-95	0.76-0.97
Peas, air dried	14	0.44
Potatoes	75	0.81
Potatoes, boiled	80	0.87
Sorrel	92	0.96
Spinach	85-90	0.90-0.94
White cabbage, fresh	90-92	0.98
White cabbage, boiled	97	0.98
Fruit		
Apples	75-85	0.89-0.96
Berries, fresh	84-90	0.89-0.98

TABLE 2-2 (cont'd)

	Water content (%)	Heat capacity (Btu/lb °F)
Berries, dried	30	0.51
Fruits, fresh	75-92	0.80-0.90
Fruits, dried	30	0.50
Plums, fresh	75-78	0.84
Plums, dried	28-35	0.53-0.59
Meat		
Bacon	50	0.48
Beef, fat	51	0.69
Beef, lean	72	0.82
Beef, mincemeat		0.81
Beef, boiled	57	0.73
Bones		0.40-0.60
Goose (eviscerated)	52	0.70
Kidneys		0.86
Mutton	90	0.93
Pork, fat	39	0.62
Pork, lean	57	0.73
Sausages, fresh	72	0.82
Veal	63	0.77
Veal cutlet	72	0.82
Veal cutlet, fried	58	0.71
Venison	70	0.81
Miscellaneous		
Bread, white	44-45	0.65-0.68
Bread, brown	18.5	0.68
Curd (cottage cheese)	60-70	0.78
Dough		0.45-0.52
Flour	12-13.5	0.43-0.45
Grains	15-20	0.45-0.48
Salt	15-20	0.27-0.32
Macaroni	12.5-13.5	0.44-0.45
Pearl barley		0.67-0.68
Porridge (buckwheat)		0.77-0.90
Raisins	24.5	0.47
Rice	10.5-13.5	0.42-0.44
Sugar		0.30
White of egg	87	0.92
Yolk of egg	48	0.67
Water		1.00

[a] From Ref. [2].

The heat of fusion can be calculated from the relationship

$$Q_f = M_f \lambda_f \tag{2-2}$$

where

Q_f is latent heat transferred (Btu)
M_f is mass of material solidified or melted (lb)
λ_f is heat of solidification or melting (Btu/lb)

Likewise, the heat removed or supplied for condensation or vaporization can be calculated from the relationship

$$Q_v = M_v \lambda_v \tag{2-3}$$

where

Q_v is latent heat transferred (Btu)
M_v is mass of material evaporated or condensed (lb)
λ_v is heat of vaporization or condensation (Btu/lb)

The total heat required (or released) is then the sum of these three, i.e.,

$$Q_{total} = Q_s + Q_f + Q_v \tag{2-4}$$

Equation (2-4) can be used to calculate the total quantity of heat transferred but not the rate of heat transfer. To arrive at the rate of heat transfer the process must be expressed on a time basis. If it is known that the mass of material must undergo a change in sensible heat or a change of state in a given length of time, then the rate of heat transfer can be calculated. The rate of heat transfer is obtained by dividing Eq. (2-4) by the time involved. For example, if the process is to be completed in 1 hr then Q_{total} would be in units of Btu/hr. On a rate basis the change in sensible heat becomes

$$Q_s = WC_p \Delta T \tag{2-5}$$

where

W is mass flow rate (lb/hr)
Q_s is heat transfer rate (Btu/hr)
$C_p \Delta T$ is as defined in Eq. (2-1)

Likewise, Eqs. (2-2) and (2-3) can be expressed on a time basis as follows:

$$Q_f = W_f \lambda_f \tag{2-6}$$

where W_f is the mass of material solidified or melted per unit time (lb/hr) and

$$Q_v = W_v \lambda_v \tag{2-7}$$

where W_v is the mass of material evaporated or condensed per unit time (lb/hr). The total rate of heat transfer is then

$$Q_{total} = WC_p \Delta T + W_f \lambda_f + W_v \lambda_v \quad (2\text{-}8)$$

Heat transfer is a dynamic process wherein heat is transferred from a hotter body to a colder body, the rate being dependent on the temperature difference between the two bodies. A temperature difference is therefore necessary for heat transfer, and this temperature difference is called the "driving force" for heat transfer. Without a temperature difference there is no heat transfer. For example, water cannot be boiled at 212°F by heating with steam at 212°F.

In transferring heat energy from one body to another resistance may be encountered. This resistance to heat flow can occur within or at the surface of the material. Expressions for these resistances are given in Sections II, B and C. Like all rate processes, the rate of heat transfer is inversely proportional to resistance and directly proportional to driving force. Thus,

$$\text{Rate of transfer} = \frac{\text{driving force}}{\text{resistance}}$$

There are three basic mechanisms by which heat transfer can occur: conduction, convection, and radiation. Conduction heat transfer occurs when thermal energy is transferred from one molecule or atom to an adjacent molecule or atom without gross change in the relative positions of the molecules. Molecular movement is limited to oscillation about a fixed position, allowing any given molecule to make contact with only its immediate neighbors.

In heat transfer by convection the molecules are free to move about, resulting in mixing of warmer and cooler portions of the same material. Convection is thus restricted to flow of heat in fluids, either gases or liquids.

Transfer of heat by radiation occurs by means of an electromagnetic mechanism. Electromagnetic radiation passes unimpeded through space and is not converted to heat or any other form of energy until it collides with matter. Upon collision the radiation can be transmitted, absorbed, or reflected. Only the absorbed energy can (but need not) appear as heat.

One final introductory concept should be discussed, namely the differences between steady-state and unsteady-state processes. The criterion of a steady state is that conditions at all points in the system do not change with time. If one of the conditions at any given point changes with time an unsteady state exists. An example will help illustrate the difference. Consider a pipe through which water is flowing at a constant inlet temperature and a constant rate while surrounded by a jacket in which steam is circulating at a constant temperature. Conditions in the water vary from one location to another throughout the length of the heat exchanger but at any given point the conditions are constant with time. This system is said to be in steady state. Obviously there is a flow of energy from the hot fluid (steam) to the cold fluid (water) but the equations which describe the conditions at any point in the exchanger are independent of time. Now consider a tank of cold water in which a steam coil is immersed. In this situation, the mean temperature of the water increases and there is no point in the mass of water where conditions are constant with time. This system is said to be in unsteady state. Generally, steady state is characteristic of continuous processes, whereas unsteady state is characteristic of batch or discontinuous operation.

B. Conduction

The basic law of heat transfer by conduction is given by Fourier's law. Fourier's law states that the rate of heat transfer through a uniform material is directly proportional to the area for heat transfer, directly proportional to the temperature drop through the material and inversely proportional to the thickness of the material. Stated mathematically, Fourier's law is

$$\frac{dQ}{dt} = -kA\frac{dT}{dL} \tag{2-9}$$

where

dQ/dt is rate of heat transfer (Btu/hr)
A is area for heat transfer (ft^2)
dT/dL is temperature drop per unit thickness ($^{\circ}F/ft$)
k is a proportionality constant

The principles of Fourier's law are illustrated schematically in Figure 2-1. In Eq. (2-9) the negative sign is included on the right-hand side of the equation because the term dT/dL is negative and therefore must be multiplied by the minus sign to yield a positive heat-transfer rate, dQ/dt. If the temperature gradient does not vary with time, then the rate of heat flow is constant and Eq. (2-9) can be written

$$q = -kA\frac{dT}{dL} \tag{2-10}$$

where q is the rate of heat flow (Btu/hr). For constant conditions of A, L, and k, Eq. (2-10) can be separated and integrated, resulting in

$$q = k\frac{A\Delta T}{L} \tag{2-11}$$

The proportionality constant k is called the "thermal conductivity" and it is a physical property of the material through which heat is transferred. The units of k may be derived from Eq. (2-11) as

$$k\ (=)\ \frac{\text{Btu ft}}{\text{hr ft}^2\ {}^{\circ}\text{F}} \quad \text{or} \quad \frac{\text{cal cm}}{\text{sec cm}^2\ {}^{\circ}\text{C}}$$

Like most physical properties, k is dependent on temperature and the k in Eq. (2-11) is actually k between T_1 and T_2. Thermal conductivity values of various materials are available in the literature [7] and a few representive values are presented in Table 2-3. The most notable feature of food products is their extremely low values of thermal conductivity compared to metals. This difference in thermal conductivity is due to differences in abundance of free electrons. In metals the electrons transmit most of the heat energy, whereas in foods, where water is the main constituent, the free electron concentration is low and

TABLE 2-3

Thermal Conductivity of Various Materials

Material	Temperature ($^\circ$F)	Thermal conductivity (Btu hr^{-1} ft^{-1} $^\circ F^{-1}$)
Ice	32	1.28
Water	32	0.314
Water	199.5	0.391
Air	68	0.0148
Peanut oil	39.2	0.0973
Meats (average)	32	0.290
Fruits (average)	32-81	0.22-0.29
Vegetables (average)	32-81	0.22-0.29
Stainless steel		9.4
Copper		230
Pyrex glass		0.05

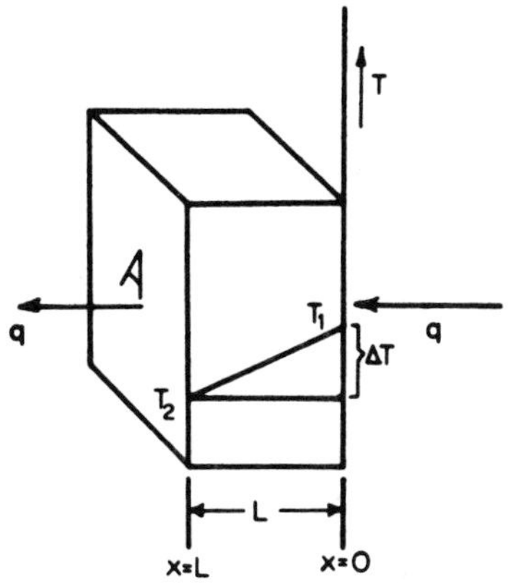

FIG. 2-1. Heat transfer by conduction.

the transfer mechanism involves primarily vibration of atoms and molecules. Therefore good electrical conductors are also good thermal conductors. Similarly, those materials which are good electrical insulators, such as air and glass, are also good thermal insulators (see Table 2-3).

Another striking feature in Table 2-3 is that the thermal conductivity of ice is nearly four times greater than that of water. This difference in thermal conductivity partly accounts for the difference in freezing and thawing rates of food tissues. During freezing, heat is conducted through an ice layer, a layer which has a higher conductivity than that of water and therefore offers comparatively little resistance to heat flow. During thawing, on the other hand, the frozen material becomes surrounded by a continually expanding layer of immobilized water and heat flow through this layer is comparatively difficult. In essence, the water is acting as an insulator compared to ice. Consequently the thawing process for foods that are not fluid in nature is inherently slower than the freezing process.

C. Convection

As pointed out earlier, convective heat transfer occurs as a result of bulk movement in a fluid. Obviously, heat transfer by conduction occurs simultaneously but it is generally negligible compared to convective heat transfer. There are two types of convective heat transfer: natural, or free, convection and forced convection. In natural convection, bulk movement of the fluid occurs as a result of density gradients established during heating or cooling the fluid. With forced convection an external source of energy, such as a pump, stirrer, or fan, is used to move the fluid. In most processing applications with foods, forced-convection heat transfer is used since it is inherently much faster than free convection.

The rate of convective heat transfer is governed by Newton's law of cooling. This law states that the rate of heat transfer by convection is directly proportional to the heat transfer area and the temperature difference between the hot and cold fluid. The quantitative relationship is

$$q = hA\Delta T \tag{2-12}$$

where

q is rate of heat transfer (Btu/hr)
A is heat transfer area (ft^2)
ΔT is temperature difference ($^\circ F$)
h is a proportionality constant (Btu hr^{-1} ft^{-2} $^\circ F^{-1}$)

The proportionality constant h is called the "heat transfer coefficient."

The magnitude of h is dependent on properties of the fluid, nature of the surface, and velocity of flow past the heat-transfer surface. It can be regarded as the conductance of heat through a thin stagnant layer of fluid of thickness X_f and is often referred to as the film heat-transfer coefficient. Thus, $h = k/X_f$ where k is the thermal conductivity of the fluid. Some typical values of h are shown in Table 2-4.

TABLE 2-4

Typical Film Heat-transfer Coefficients

Condition	h ($Btu\ hr^{-1} ft^{-2}\ {}^{\circ}F^{-1}$)
Gases, natural convection	0.5-5
Gases, forced convection	2-20
Viscous liquids, forced convection	10-100
Water, forced convection	100-1000
Boiling water	300-4000
Condensing steam	1000-3000

A method for calculating individual film heat-transfer coefficients must take into consideration properties of the fluid and the conditions of flow that affect heat transfer. Since there are so many factors affecting h, the most useful method for developing equations involves dimensional analysis. This method shows the relationship between variables in the form of dimensionless groups. Actual data are then used to establish the constants and exponents in a power function relating these dimensionless groups. This results in empirical relationships for calculating film heat-transfer coefficients. The technique and equations are beyond the scope of this discussion; however, the equations are readily available in handbooks and texts on heat transfer.

Most food heating applications involve indirect heating (i.e., heat is transmitted from a hot fluid, through a wall, into a cold fluid). Direct heating occurs when hot steam and cold fluids are mixed together.

The process of indirect heating is shown schematically in Figure 2-2. In steady state, the rate of heat transfer across each physical boundary is constant. There is a heat-transfer coefficient characterizing transfer of heat from the hot fluid to the wall, a thermal conductivity characterizing heat transfer through the wall, and finally a heat-transfer coefficient for transfer of heat from the wall to the cold fluid. The resistance to heat transfer in each of these sections can be determined by rearranging Eqs. (2-11) and (2-12) as follows:

$$q = \frac{\Delta T_{wall}}{(L/kA)} = \frac{\Delta T_{hot\ fluid}}{(1/hA)_{hot\ fluid}} = \frac{\Delta T_{cold\ fluid}}{(1/hA)_{cold\ fluid}} \qquad (2\text{-}13)$$

The term L/kA then represents the resistance of the wall to heat transfer, $(1/hA)_{hot\ fluid}$ is the resistance in the hot fluid, and $(1/hA)_{cold\ fluid}$ is the resistance in the cold fluid.

It can be shown that the overall resistance to heat transfer is the sum of the individual resistances. Thus,

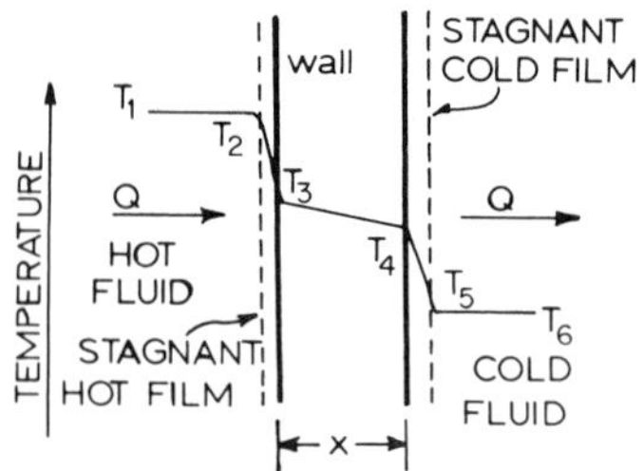

FIG. 2-2. Indirect heat transfer.

$$R_{total} = (1/hA)_{hot\ fluid} + (L/kA)_{wall} + (1/hA)_{cold\ fluid} \tag{2-14}$$

The rate of heat transfer is then directly proportional to the overall driving force and inversely proportional to the total resistance.

$$q = \frac{\Delta T_{overall}}{R_{total}} = \frac{(T_1 - T_6)}{(1/hA)_{hot\ fluid} + (L/kA)_{wall} + (1/hA)_{cold\ fluid}} \tag{2-15}$$

Since the total resistance to heat transfer is inversely proportional to the overall conductance, Eq. (2-15) can be expressed as

$$q = \frac{\Delta T_{overall}}{(1/UA)} = UA\Delta T_{overall} \tag{2-16}$$

where U is the overall heat-transfer coefficient (Btu hr^{-1} ft^{-2} $^{\circ}F^{-1}$). Equation (2-16) is similar to Eq. (2-12), except U is substituted for h. Equation (2-16) indicates that the rate of heat transfer is the product of three factors: overall heat-transfer coefficient, temperature drop, and area of heating surface.

In every heating application it is necessary to establish the magnitude of U so that the equipment can be properly sized. By knowing the rate of heat transfer desired, the overall heat-transfer coefficient, and the driving force for heat transfer, the required surface area can be calculated. In food applications, generally the major resistance to heat transfer is encountered in the food. Thus, U values are generally of the same order of magnitude as the h value for the food side of the heat exchanger.

Equation (2-16) actually applied only when the temperature drop is constant for all parts of the heating surface. In reality the temperature of the hot fluid decreases as the fluid moves through the heat exchange. Thus, ΔT is not constant over the length of the heat exchanger. The ΔT used in Eq. (2-16) must, therefore, represent a mean driving force for heat transfer. It can be shown that the mean ΔT should be a logarithmic mean. To illustrate this point, consider the heat-exchange system shown in Figure 2-3. As the hot fluid moves through the heat exchanger its temperature changes from T_{1H} to T_{2H} and the cold fluid temperature changes from T_{1C} to T_{2C}. Obviously the ΔT between the two streams is constantly changing and heat is transferred more rapidly at the entrance than at the exit. Analysis of this system by means of heat energy balances shows that

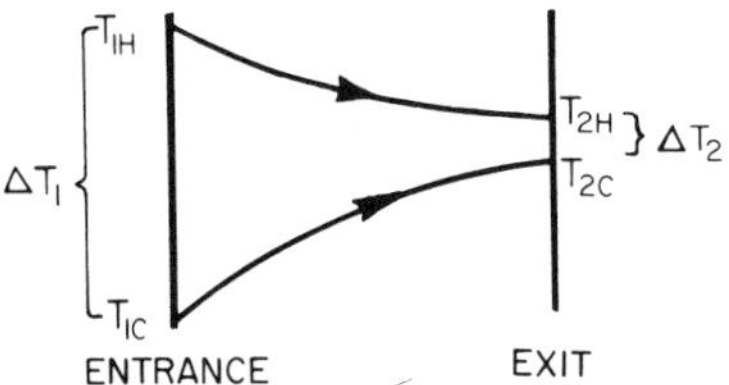

FIG. 2-3. Temperatures in a parallel-flow heat exchanger.

$$q = UA \frac{\Delta T_1 - \Delta T_2}{\ln(\Delta T_1/\Delta T_2)} = UA\Delta T_{lm} \tag{2-17}$$

where ΔT_{lm} is log mean temperature difference.

Equation (2-17) can be used to calculate the size of various heat-transfer systems, such as tube-in-shell heat exchangers used for continuous heating of food slurries (confectionery manufacture, tomato paste processing, aseptic canning), and for plate-type heat exchangers used for milk and beer pasteurization.

Before the subject of convection heating is abandoned, there is one other important point to make. When most biological fluids are heated there is a tendency for some components to deposit on the heat-transfer surface. This phenomenon is known as "fouling" or "burn-on" and can contribute significantly to heat-transfer resistance. Generally burn-on is due to destabilization and/or polymerization of components. Fouling is generally compensated for by oversizing the heat exchanger, the degree of oversizing being dependent on the known fouling characteristics of the fluid. Unfortunately, fouling characteristics of fluids and factors affecting fouling are not well understood.

D. Radiation

As stated in Section II. A, energy can be transferred in the form of electromagnetic waves emitted from one body and absorbed by another. Electromagnetic radiation travels at the speed of light and is characterized by its wavelength and frequency. The relationship between wavelength and frequency is

$$c = \lambda f \tag{2-18}$$

where

- c is velocity of light (3×10^{10} cm/sec)
- λ is wavelength (cm)
- f is frequency (cps)

The large ranges of wavelengths and frequencies for various types of electromagnetic radiation are shown in Table 2-5.

Radiation of wavelength 0.8-400 μm (infrared) is referred to as "thermal" radiation or heat rays since electromagnetic radiation with this wavelength is most readily absorbed and converted to heat energy. Since infrared radiation is

TABLE 2-5

Electromagnetic Radiation

Type of radiation	Major effect on matter	Wavelength (μm)[a]	Frequency (cps)
Cosmic rays	Transmitted	1×10^{-6}	3×10^{20}
γ-rays	Absorbed, reflected, or transmitted	$1 \times 10^{-6} - 140 \times 10^{-6}$	$3 \times 10^{20} - 2 \times 10^{18}$
X-rays	Strongly absorbed	$6 \times 10^{-6} - 100{,}000 \times 10^{-6}$	$5 \times 10^{19} - 1 \times 10^{15}$
Ultraviolet rays	Absorbed, reflected, or transmitted	0.014 - 0.4	$2 \times 10^{16} - 8 \times 10^{14}$
Visible or light rays	Absorbed, reflected, or transmitted	0.4 - 0.8	$8 \times 10^{14} - 4 \times 10^{14}$
Infrared or heat rays	Strongly absorbed	0.8 - 400	$4 \times 10^{14} - 8 \times 10^{11}$
Radio	Absorbed, reflected, or transmitted	$10 \times 10^{6} - 30{,}000 \times 10^{6}$	$3 \times 10^{8} - 1 \times 10^{4}$

[a] $1\ \mu m = 1 \times 10^{-6}\ m = 0.001\ mm$.

transmitted rapidly to the surface of the material, it is used primarily for surface heating. Examples of use include [4] (1) dehydration of fruits and vegetables, (2) roasting of cocoa beans and nut kernels, (3) dehydration of grain, (4) dehydration of tea, (5) freeze drying, and (6) baking.

All bodies emit electromagnetic radiation, the amount and wavelength being dependent on the physical state and temperature of the body. For comparative purposes it is desirable to define a body that emits the maximum possible amount of energy at a given temperature. This theoretical substance is called a "black body." The energy emitted by a black body is given by the Stefan-Boltzmann law,

$$q = \sigma T^4 \tag{2-19}$$

where

q is energy radiated (Btu/hr)
σ is the Stefan-Boltzmann constant (0.17×10^{-8} Btu hr^{-1} ft^{-2} $^{\circ}R^{-4}$)
T is absolute temperature of the black body ($^{\circ}$R)

Examination of Eq. (2-19) shows that the quantity of heat emitted is highly dependent on temperature, varying as the fourth power of absolute temperature.

Since no real bodies behave as black bodies, Eq. (2-19) must be corrected. This is easily done by defining the emissivity of a body. "Emissivity" is the ratio of the energy emitted by a real body to that emitted by a black body at the same temperature, and its numerical value is always less than 1.0. Emissivity values for representative materials are given in Table 2-6. The radiation by any actual body is then

$$q = \epsilon\sigma T^4 \tag{2-20}$$

Bodies obeying Eq. (2-20) are called "grey bodies."

If a body is in thermal equilibrium with respect to radiant energy then it is absorbing as much radiation as it is emitting. For grey bodies the amount of energy absorbed is expressed by an equation similar to Eq. (2-20) except the term absorptivity (α) is used instead of emissivity. Thus,

$$q = \alpha\sigma T^4 \tag{2-21}$$

Absorptivity is defined analogously to emissivity in that it is the fraction of radiation absorbed by a real substance compared to a black body (a perfect absorber with $\alpha = 1.0$). For grey bodies it can be shown that $\alpha \approx \varepsilon$.

The radiant energy transferred between two surfaces depends on the differences in surface temperatures, geometric arrangements, and emissivities. Although the effect of geometric variables on heat transfer by radiation is too complex for treatment here, there are several applications in the food industry where simplifications can be made. In the case of a relatively small body completely surrounded by a surface at a uniform temperature, the net exchange of heat by radiation is given by,

$$q = A\varepsilon\sigma\,[T_1^4 - T_2^4] \tag{2-22}$$

TABLE 2-6

Emissivities of Solid Surfaces

Surface	Emissivity
Polished iron	0.14 - 0.38
Rough steel plate	0.94 - 0.97
Asbestos board	0.96
Water	0.95 - 0.96
Foods	0.50 - 0.97

where

A is surface area of the body (ft^2)
ϵ is emissivity of the body
T_1 is absolute temperature of the body ($^\circ R$)
T_2 is absolute temperature of the surroundings ($^\circ R$)

Equation (2-22) can be used to calculate rates of heat transfer in infrared ovens, radiation heat transfer in tunnels for freezing or drying, and in other similar situations.

E. Dielectric and Microwave Heating

Electromagnetic radiation in the radio frequency range can be used to generate heat in substances which efficiently absorb these wavelengths. Substances containing water or other polar molecules absorb radio frequency (rf) energy efficiently; therefore rf heating is applicable to foods.

There are two types of radio frequency heating. In one case heating is accomplished by placing the object between parallel electrodes (dielectric heating). In the other, the product is placed in a resonance cavity (microwave heating). In dielectric heating an electric field is established between two electrodes and the polarity is reversed millions of times per second. Dipolar molecules contained in the object to be heated attempt to align themselves in this field and this molecular movement results in generation of heat. This may be thought of as capacitance heating since a relatively poor electrical conductor (the food) is situated between the electrodes. The heat developed in the product is a function of frequency, potential across the product (field strength), and the dielectric properties of the product [3].

Since the frequencies used in rf heating are also used in communications, the Federal Communications Commission has ruled that only certain frequencies may be used for industrial, scientific, and medical (ISM) purposes. The four frequencies commonly used are: 915, 2450, 5800 and 22,150 Mc/sec. Of these, dielectric heating is generally accomplished at 915 Mc/sec (1 Mc/sec = 1 MHz).

Microwave heating is a truly radiative process. Electromagnetic radiation is generated in special oscillator tubes, such as magnetrons or klystrons, and is radiated into a closed chamber containing the product. The chamber is constructed of highly reflective walls so the radiation reflects back and forth until it is absorbed by the product. There is usually a wave deflector located at the point where the radiation enters the oven (cavity) to distribute the radiation uniformly in the oven. Microwave radiation generally has a frequency of 2450 Mc/sec.

The heat generated in microwave and dielectric heating is given by

$$P = 55.61 \times 10^{-14} E^2 f \varepsilon'' \qquad (2\text{-}23)$$

where

P is power generated (W/cm^3)
E is electric field strength (V/cm)
f is frequency (cps)
ε'' is dielectric loss factor

It can be seen that the power generated (this can be converted to thermal units because 1 W = 14.4 cal/min) is directly proportional to the square of the field strength, the frequency, and the loss factor. The loss factor is an intrinsic property of the object and is a measure of the efficiency with which electromagnetic radiation is converted to heat. Thus a large loss factor indicates that the product would heat readily with dielectric or microwave heating.

From Eq. (2-23) it would appear that increasing the field strength would be the most effective way to increase the rate of heating. However, this approach is limited by the voltage breakdown strength of the product. If the breakdown strength is exceeded an electrical discharge occurs resulting in localized burning.

The other means of increasing the rate of heating is to increase the frequency. However, since nearly 100% of the electromagnetic radiation is absorbed at frequencies between 1000 and 3000 Mc/sec, commercial rf heating systems are designed to operate in this range. Since dielectric heating is usually accomplished at 915 Mc/sec and microwave at 2450 Mc/sec, the power generated from microwave heating is more than twice that of dielectric heating at constant field strength.

Radio frequency heating has been referred to as volume heating because the energy is absorbed within the product, in contrast to convection, conduction, and infrared heating, wherein the heat energy is absorbed at the surface of the product. In general, the degree of penetration increases as the frequency decreases. Thus dielectric heating penetrates better than microwave heating. Consequently microwave heating is generally less uniform, particularly in large objects, than is dielectric heating.

Another important consideration in microwave heating is the behavioral differences between ice and water. Liquid water absorbs radio frequency energy much more efficiently than does ice. Thus if ice and pockets of liquid water exist together (as in frozen foods) the water pockets absorb energy much more rapidly than the adjacent frozen areas. Pulsed applications of rf energy can help overcome this problem.

Dielectric and microwave heating have been successfully applied in a few situations in the food industry, e.g., to bake crackers, to accomplish the final drying of potato chips, and to precook chicken pieces [6]. The high initial cost of dielectric and microwave equipment and the failure to heat nonhomogeneous products uniformly are factors which tend to limit their use.

III. UNSTEADY-STATE HEAT TRANSFER

In food processing, there are many situations where temperature at some point in a product is a function of time; i.e., unsteady-state heat transfer is occurring. The most notable examples of unsteady-state heat transfer are heating or cooling of particulate materials, such as in blanching and aseptic thermal processing, or cooling and heating of products in containers (canning). Calculations of unsteady-state heat transfer are extremely complicated and basically involve use of the Fourier equation, Eq. (2-9), written in terms of partial differentials in three dimensions. There are some instances where simplification is possible, and charts have been prepared to enable calculation of temperature as a function of time [1].

As with heat-transfer coefficients for convection, it is most convenient to express the variables in unsteady-state heat transfer as dimensionless groups. Factors affecting unsteady-state heat transfer are:

1. Initial temperature of the body (T_i)
2. Temperature of the heating medium (T_o)
3. Surface heat-transfer coefficient (h)
4. Thermal conductivity of the body (k)
5. Radius on half-thickness of the body (R or L)
6. Heat capacity of the body (C_p)
7. Density of the body (ρ)
8. Time of heating (t)

SYMBOLS

A	heat-transfer area (ft^2)
c	velocity of light (cm/sec)
C_p	heat capacity (Btu/lb/$^\circ$F)
E	electric field strength (V/cm)
f	frequency (cps)
h	film heat-transfer coefficient (Btu hr^{-1} ft^{-2} $^\circ F^{-1}$)
k	thermal conductivity (Btu ft hr^{-1} ft^{-2} $^\circ F^{-1}$)
L	characteristic length or half-thickness (ft)
M	mass (lb)
P	power (W/cm^3)
Q or q	heat transferred (Btu or Btu/hr)
R	thermal resistance (hr $^\circ$F/Btu) or radius (ft)
t	time (hr)
T	temperature ($^\circ$C, $^\circ$F, $^\circ$R, $^\circ$K)
ΔT	temperature difference ($^\circ$F)
ΔT_{lm}	logarithmic mean temperature difference ($^\circ$F)
U	overall heat-transfer coefficient (Btu hr^{-1} ft^{-2} $^\circ F^{-1}$)
W	mass flow rate (lb/hr)

X_f ficticious film thickness (ft)
α absorptivity
ε emissivity
ε'' dielectric loss factor
λ latent heat (Btu/lb) or wavelength (cm)
σ Stefan-Boltzmann constant (Btu hr^{-1} ft^{-2} $^{\circ}R^{-4}$)
ρ density (lb/ft^3)

Subscripts

f solidification or melting process
s sensible heat
v vaporization or condensation process

REFERENCES

1. H. S. Carslaw and J. C. Jaeger, Conduction of Heat in Solids, 2nd ed., Clarendon Press, Oxford, 1959.

2. S. E. Charm, Fundamentals of Food Engineering, 2nd ed., AVI Publ Co., Westport, Conn., 1971.

3. D. A. Copson, Microwave Heating in Freeze-drying, Electronic Ovens and Other Applications, AVI Publ. Co., Westport, Conn., 1962.

4. A. S. Ginzberg, Application of Infra-red Radiation in Food Processing (translated by A. Grochowski), CRC Press, Cleveland, Ohio, 1969.

5. R. L. Sproul, Sci. Amer., 207, 92 (1962).

6. N. W. Tape, Can. Inst. Food Technol. J., 3, 39 (1970).

7. E. E. Woodams and J. E. Nowrey, Food Technol., 22, 194 (1968).

GENERAL REFERENCES

ASHRAE Guide and Data Book -- Applications. American Society of Heating, Refrigerating and Air-conditioning Engineers, New York, 1971.

ASHRAE Guide and Data Book Handbook of Fundamentals. American Society of Heating, Refrigerating and Air-conditioning Engineers, New York, 1972.

R. W. Dickerson, Jr., in The Freezing Preservation of Foods, Vol. 2, 4th ed. (D. K. Tressler, W. B. Van Arsdale, and M. J. Copley, eds.), AVI Publ. Co., Westport, Conn., 1968, Chapter 2, pp. 26-51.

C. G. Harrel and R. J. Thelen, Conversion Factors and Technical Data for the Food Industry, 6th ed., Burgess, Minneapolis, Minn., 1959.

J. E. Hill, J. D. Leitman, and J. E. Sunderland, Food Technol., 21, 1143 (1967).

A. C. Jason and R. Jowitt, Dechema Monographien., 62, 21 (1969).

N. N. Mohsenin, Physical Properties of Plant and Animal Materials, Department Agricultural Engineering, Pennsylvania State University, University Park, Penn., 1966.

J. H. Perry (ed.), Chemical Engineers Handbook, 5th ed., McGraw Hill, New York, 1973.

G. A. Reidy and A. L. Rippen, Trans. ASAE, 14, 248 (1971).

R. D. Weast (ed.), Handbook of Chemistry and Physics, 54th ed., CRC Press, Cleveland, Ohio, 1973.

Chapter 3

HEAT PROCESSING

Daryl Lund

CONTENTS

I. INTRODUCTION

Preservation of food has occupied and continues to occupy a large portion of man's time and effort. Early man preserved his food supply by applying methods of preservation available to him in his environment. Sun drying, salting, and fermentation were used to provide food in periods when fresh foods were not available. As civilization developed, demand for greater quantities of better quality processed food also increased. Hence there has developed a large food-preservation industry which attempts to supply food that is satisfying, nutritious, and economical. Thermally processed foods constitute a large part of this industry.

II. APPLICATION OF HEAT ENERGY TO FOODS

It is appropriate to introduce this subject on the basis of objectives of the heat process. General principles governing heat transfer and response of foods to heat energy can be applied to all heat processes but each type of thermal process has specific objectives. Severity of the thermal process is dependent on these objectives.

A. Cooking

Cooking is a heating process the primary object of which, is to produce a more palatable food. The word "cooking" is a broad term embodying at least six forms of heating, including baking, broiling, roasting, boiling, frying, and stewing. The method of applying heat energy and the duration differs somewhat for each of these processes. Baking, broiling, and roasting usually require dry heat at relatively high temperatures (greater than 212°F), boiling and stewing are done by placing the product in boiling water, and frying involves cooking oil and temperatures much greater than 212° F.

Cooking can be thought of as a preservation technique since cooked foods generally can be stored for longer periods than uncooked foods provided recontamination of the food with spoilage organisms is minimized. Refrigerated storage following cooking is a common household method of preservation.

Two important preservative changes occur in food as a result of cooking: destruction or reduction of microorganisms and inactivation of undesirable enzymes.

Other desirable changes that may occur during cooking include (1) destruction of potentially hazardous toxins present naturally or through microorganisms; (2) alteration of color, flavor, and texture; and (3) improved digestibility of food components. Undesirable changes also may occur, such as degradation of nutritive components and sensory attributes. Principles which can be applied to describe the effect of cooking on the reduction of microorganisms and enzymes also can be applied to other changes occurring during cooking. Application of these principles is dealt with in Section III. F.

B. Blanching

Blanching is a heat treatment commonly applied to tissue systems prior to freezing, drying, or canning. The objectives of blanching depend on the process which follows it. Blanching prior to freezing or dehydration is done primarily to inactivate enzymes. Unblanched frozen or dried foods exhibit relatively rapid changes in such properties as color, flavor, and nutritive value as a result of enzyme activity.

Two of the more heat-resistant and widely distributed enzymes in plant tissues are peroxidase and catalase. Activity of these enzymes, therefore, can be used to evaluate the effectiveness of a blanching treatment. If both are inactivated then it can be assumed that other significant enzymes also are inactivated. The heating time necessary to destroy catalase or peroxidase depends on the type of fruit or vegetable, the method of heating, the size of the fruit or vegetable, and the temperature of the heating medium. For commercial blanching, typical times at 212°F are given in Table 3-1. Heating media other than water (e.g., steam, hot air, microwave) and at temperatures other than 212°F can be used. Although blanching of vegetables is most often done in hot water or steam, blanching of fruits is often done in calcium brines to firm the fruit through the formation of calcium pectates. Colloidal thickeners, such as pectin, carboxymethyl cellulose, and alginates, also can be used to aid fruit firmness following blanching.

For frozen fruits which are not to be further heated after thawing, blanching is not used because it results in undesirable changes in texture and flavor. Instead, other preservative techniques are used to combat undesirable enzymatic changes (mainly oxidative browning and oxidation of ascorbic acid). This is accomplished by chemical inactivation of enzymes, avoidance of contact with oxygen (sugar syrups), and addition of antioxidants (ascorbic acid).

Blanching prior to conventional canning fulfills several important objectives: to remove tissue gases, to increase the temperature of the tissue, to cleanse the tissue, to wilt the tissue to facilitate packing, and to activate or inactivate enzymes. Since the product generally receives a thermal process severe enough to inactivate enzymes, enzyme inactivation is not necessarily a primary objective of blanching. Removal of tissue gases and preheating the product prior to filling, however, are important objectives of precanning operations since they have a great influence on the final level of oxygen in the container and therefore directly influence storage life. Accomplishment of these two objectives results in high can vacuum, which necessarily implies low gas concentration in the product. Figure 3-1 shows the relationship between can vacuum and closing temperature.

In at least one instance blanching prior to canning is used to activate an enzyme. With some green beans, the epidermis is loosened upon heating and this results in a quality defect known as "sloughing." A moderate heat treatment (less than 175° F) followed by a holding period prior to thermal processing can prevent sloughing through heat activation of an enzyme, pectin methylesterase, which demethylates the pectin molecule allowing crosslinking with calcium ions. This crosslinking bonds the outer layer of tissue to the underlying structure, thus preventing sloughing.

C. Pasteurization

Pasteurization is a heat treatment that kills part but not all of the vegetative microorganisms present in the food and consequently it is used for foods which are to be further handled and stored under conditions which minimize microbial growth. In many cases, the primary objective of pasteurization is to kill

TABLE 3-1

Blanching Times for Vegetables Prior to Freezing[a]

Vegetable	Blanching time (min) in water at 212° F
Asparagus	
< 5/16 in. per butt	2
5/16 - 9/16 in. per butt	3
> 10/16 in. per butt	4
Beans, green and wax	
Small	$1\text{-}1\frac{1}{2}$
Medium	2-3
Large	3-4
Beets	
Small, whole	3-5
Diced	3
Broccoli	2-3
Corn	2-3
Peas	$1\text{-}1\frac{1}{2}$
Spinach	$1\frac{1}{2}$

[a]Reprinted from Ref. [4], p. 63.

pathogenic microorganisms (e.g., milk). Some vegetative spoilage organisms can survive this heat treatment, and thus more severe preservation methods are needed if microbial spoilage is to be prevented. In other cases (e.g., beer), pasteurization serves primarily to kill vegetative spoilage organisms. Other preservation methods used in conjunction with pasteurization include (1) refrigeration, (2) chemical additives which result in an undesirable environment for microbial growth (e.g., sugar in sweetened condensed milk, food acids in pickles and fruit juices), (3) packaging (e.g., maintenance of anaerobic conditions in bottled beer), and (4) fermentation with desirable organisms.

The time-temperature treatment used in pasteurization depends on (1) the heat resistance of the particular vegetative or pathogenic microorganism the process is designed to destroy and (2) the sensitivity of product quality to heat. The high-temperature and short-time (HTST) method involves a comparatively high temperature for a short time (e.g., 161°F for 15 sec for milk), whereas the low-temperature and long-time procedure involves relatively low temperatures for longer times (e.g., 145° F for 30 min for milk). Optimization of the pasteurization

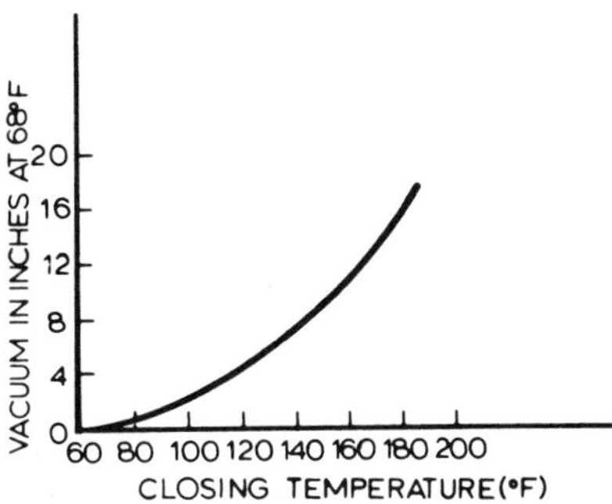

FIG. 3-1. Relation of can vacuum to closing temperature. Reprinted from Ref. [45], p. 5, by courtesy of Continental Can Company.

process depends on the relative destruction rates of organisms as compared to quality factors, but generally the HTST process results in maximum product quality.

For market milk, pasteurization conditions are based on thermal destruction of Coxiella burnetti, the rickettsia organism responsible for Q fever. The thermal resistance parameters for this organism are given in Table 3-6. For high-acid fruits, such as cherries, the pasteurization process is based on successful destruction of yeast or molds. For fermented beverages, such as wine or beer, destruction of wild yeasts is the pasteurization criterion.

D. Sterilization

A sterile product is one in which no viable microorganisms are present — a viable organism being one that is able to reproduce when exposed to conditions optimum for its growth. Temperatures slightly above the maximum for bacterial growth result in the death of vegetative bacterial cells, whereas bacterial spores can survive much higher temperatures. Since bacterial spores are far more heat resistant than are vegetative cells, they are of primary concern in most sterilization processes.

"Sterilization" is not a good term to apply to the thermal processing of foods since the criterion of success is the inability of microorganisms and their spores to grow under conditions normally encountered in storage. This means that there could be some dormant nonpathogenic microorganisms in the food (any pathogenic microorganisms are dead) but that environmental conditions are such that the organisms do not reproduce. Foods which have been thermally processed with this as the criterion are referred to as "commercially sterile," "bacterially inactive," or "partially sterile."

The thermal conditions needed to produce commercial sterility depend on many factors, including (1) nature of the food (e.g., pH); (2) storage conditions of the food following the thermal process; (3) heat resistance of the microorganisms or spore; (4) heat-transfer characteristics of the food, its container, and the heating medium; and (5) initial load of microorganisms.

The food being heated affects not only the type of organism that the process should be based on but also the resistance of the organism to heat energy. Food that is commercially sterile is generally present in hermetically (gas-tight) sealed containers to prevent recontamination. Since low oxygen levels are purposely achieved in most of these containers, microorganisms that require oxygen (obligate aerobes) are unable to grow sufficiently to pose spoilage or health problems. Also spores of most obligate aerobes are less heat resistant than are spores of organisms that are capable of growing under anaerobic conditions (facultative or obligate anaerobes). However, in canned, cured, meat products where oxygen is not completely removed and where a mild heat treatment is used in conjunction with other preservative methods (curing agents and/or refrigeration), such aerobes as Bacillus subtilis and Bacillus mycoides have been found to cause spoilage [32].

For processes based on inactivation of facultative or obligate anaerobes, pH of the food is a critical factor. There are some extremely heat-resistant spores that might survive a commercial process but due to the low pH of the food they do not constitute a health or spoilage hazard. For thermal process design, foods are divided into three pH groups: (1) high-acid foods with pH values less than 3.7, (2) acid foods with pH values between 3.7 and 4.5, and (3) low-acid foods with pH values greater than 4.5. Tables 3-2 and 3-3 contain examples of foods in the acid and low-acid groups along with their pH values. Since spore-forming bacteria do not grow at pH values less than 3.7, heat processes for high-acid foods are generally based on inactivation of yeasts or molds. The pH value 4.5, which represents the dividing line between acid and low-acid foods, was carefully chosen to be slightly lower than the pH at which strains of Clostridium botulinum can grow and produce toxin. Clostridium botulinium is an obligate anaerobe which is widely distributed in nature and is assumed to be present on all products intended for canning. For low-acid canned foods the anaerobic conditions that prevail are ideal for growth of and toxin production by C. botulinum. This organism is also the most heat-resistant, anaerobic, spore-forming pathogen that can grow in low-acid canned foods, and consequently its destruction is the criterion for successful heat processing of this type of product. There are several important strains of C. botulinum which produce toxin, the most heat-resistant spores being types A and B. The toxin is extremely potent (one millionth of a gram will kill man) but can be destroyed by exposing it to moist heat for 10 min at 212°F.

There exists a nontoxic obligate anaerobic spore-former that is significantly more resistant to heat than C. botulinum and it is used to determine safe thermal processes for low-acid foods. This mesophilic organism, identified only as putrefactive anaerobe (PA) 3679, resembles Clostridium sporogenes. Fortunately contamination of food with the more heat-resistant strains of PA 3679 is minimal. PA 3679 instead of C. botulinum is used to determine process adequacy for low-acid foods because it is nontoxic, is easy to assay, and has suitable heat resistance.

In low-acid foods, processing could be based on inactivation of spores more heat resistant than those of C. botulinum since some such spores do exist. Of greatest importance are such facultative anaerobes as Bacillus stearothermophilus. Spores of this organism are extremely heat resistant (up to 20 times more resistant than C. botulinum spores). Growth of these spores results in "flat sour" spoilage because acid is produced but little or no gas. Bacillus stearothermophilus is sometimes referred to as FS (flat sour) 1518. Optimum growth temperatures for these flat sour thermophiles vary from about 49° to 55° C (120°-131° F). Fortunately, most do not grow at temperatures below about 38° C (100° F). In the

TABLE 3-2

Canned Foods with pH Values of Less than 4.5 (Acid Foods)[a]

Canned product	pH Value		
	Average	Minimum	Maximum
Apples	3.4	3.2	3.7
Applesauce	3.6	3.2	4.2
Apricots	3.9	3.4	4.4
Blackberries	3.5	3.1	4.0
Blueberries	3.4	3.3	3.5
Cherries, black	4.0	3.8	4.2
Cherries, red sour	3.5	3.3	3.8
Cherries, Royal Ann	3.8	3.6	4.0
Cranberry sauce	2.6	2.4	2.8
Grape juice	3.2	2.9	3.7
Grapefruit juice	3.2	2.8	3.4
Lemon juice	2.4	2.3	2.8
Loganberries	2.9	2.7	3.3
Orange juice	3.7	3.5	4.0
Peaches	3.8	3.6	4.0
Pears, Bartlett	4.1	3.6	4.4
Pickles, fresh cucumber	3.9	3.5	4.3
Pickles, sour dill	3.1	2.6	3.5
Pickels, sweet	2.7	2.5	3.0
Pineapple juice	3.5	3.4	3.5
Plums, Green Gage	3.8	3.6	4.0
Plums, Victoria	3.0	2.8	3.1
Prunes, fresh prune plums	3.7	2.5	4.2
Raspberries, black	3.7	3.2	4.1
Raspberries, red	3.1	2.8	3.5
Sauerkraut	3.5	3.4	3.7
Strawberries	3.4	3.0	3.9
Tomatoes	4.3	4.1	4.6
Tomato juice	4.3	4.0	4.4
Tomato puree	4.4	4.2	4.6

[a]Reprinted from Ref. [46], p. 263, by courtesy of AVI Publ. Co.

commercial processing of canned foods these types of microorganisms can be ignored provided the product is quickly cooled and stored below 110°F. This causes these organisms to remain in a dormant state.

For acid foods, thermal processes are usually based on facultative anaerobes, such as *Bacillus coagulans* (*B. thermoacidurans*), *B. mascerans*, and *B. polymyxa*. There are several obligate anaerobes that are important in food spoilage but they are less heat resistant than the *Bacillus* organisms. For tomatoes and tomato

TABLE 3-3

Canned Foods with pH Values Greater than 4.5 (Low-acid Foods)[a]

Canned product	pH Value		
	Average	Minimum	Maximum
Asparagus, green	5.5	5.4	5.6
Asparagus, white	5.5	5.4	5.7
Beans, baked	5.9	5.6	5.9
Beans, green	5.4	5.2	5.7
Beans, lima	6.2	6.0	6.3
Beans and pork	5.6	5.0	6.0
Beans, wax	5.3	5.2	5.5
Beets	5.4	5.0	5.8
Carrots	5.2	5.0	5.4
Corn, whole grain, brine packed	6.3	6.1	6.8
Corn, cream style	6.1	5.9	6.3
Figs	5.0	5.0	5.0
Mushrooms	5.8	5.8	5.9
Olives, ripe	6.9	5.9	8.0
Peas, Alaska	6.2	6.0	6.3
Peas, sweet wrinkled	6.2	5.9	6.5
Potatoes, sweet	5.2	5.1	5.4
Potatoes, white	5.5	5.4	5.6
Pumpkin	5.1	4.8	5.2
Spinach	5.4	5.1	5.9

[a]Reprinted from Ref. [46], p. 264, by courtesy of AVI Publ. Co.

products, destruction of B. coagulans is the basis of the thermal process. Summarized in Table 3-4 are spore-forming spoilage organisms that are important at various storage conditions and food pH values.

II. INTERACTION OF HEAT ENERGY AND FOOD COMPONENTS

A. Introduction

In order to assess whether objectives of a thermal process have been accomplished, the food product must be tested. The test may be for surviving microorganisms, for active nutrients, or for such quality factors as color, texture, or flavor. If the outcome of the test indicates that the objective has been met, the process can be adopted. This was the procedure used to develop thermal processes before mathematical approaches and prediction of events were available, i.e., before reaction rates and their dependence on temperature were incorporated into procedures for establishing thermal processes. Use of these techniques has not

TABLE 3-4

Spore-forming Bacteria Important in Spoilage of Canned Food[a]

Approximate temperature range for vigorous growth ($^\circ$C)	Acidity of food	
	Acid $3.7 < pH < 4.5$	Low-acid $pH > 4.5$
Thermophilic (55°-35°)	B. coagulans	C. thermosaccharolyticum C. nigrificans B. stearothermophilus
Mesophilic (40°-10°)	C. butyricum C. pasteurianum B. mascerans B. polymyxa	C. botulinum, A and B C. sporagenes B. licheniformis B. subtilis
Psychrotrophic (35°-$<5^\circ$)		C. botulinum E

[a] Reprinted from Ref. [29], p. 567, by courtesy of Academic Press.

totally replaced the experimental technique since validity of the calculated process must be verified by actual testing. When the test procedure involves microorganisms, it is referred to as an "inoculated pack."

Mathematical methods allow a process to be calculated based on known behavior of the system. In this instance, the system is the food and its components, and the response is the effect of heat on the components (including microorganisms). Two types of information are needed: (1) reaction rate constants for the particular component under conditions that prevail in practice, and (2) dependence of the rate constants on temperature.

B. Reaction Kinetics

Fortunately a majority of reactions occurring in foods obey well-established kinetics. The thermal destruction of microorganisms, most nutrients, quality factors (texture, color, and flavor) and enzymes generally obey first-order reaction kinetics. That is, the destruction rate of each of these components is dependent on the concentration of the component. This relationship is frequently referred to as a "logarithmic order of inactivation or destruction" since, regardless of the initial concentration, a constant fraction of the remaining substrate reacts per unit time. In other words, if at a given temperature the concentration reduces by 0.9 in 1 min, then 0.9 of the remaining concentration reacts in the next minute, 0.9 the third minute, and so on.

Since microorganisms are so important and are the basis of thermal process calculations, microbial inactivation is used here to illustrate first-order kinetics.

Expressing the first-order response mathematically gives:

$$\frac{-dc}{dt} = kc \qquad (3\text{-}1)$$

where

-dc/dt is the rate at which concentration decreases
c is the concentration of viable microorganisms
k is the first-order reaction rate constant

Integrating Eq. (3-1) between limits, c_1 at time $t_1 = 0$, and c at time t results in:

$$-\int_{c_1}^{c} \frac{dc}{c} = k \int_{t_1}^{t} dt$$

$$-\ln c + \ln c_1 = k(t - t_1)$$

or

$$\log c = \log c_1 - \frac{kt}{2.303} \qquad (3\text{-}2)$$

The graphical expression of Eq. (3-2) is shown in Figure 3-2. The initial concentration is $c_1 = 10^5$ survivors per unit volume. The number of survivors for any heating time can be obtained directly from the curve. The slope of the line is -k/2.303.

From Figure 3-2, it is evident that the time required for the survivor curve to transverse one log cycle corresponds to a 90% reduction in the number of survivors. This, it turns out, is a convenient way to characterize the reaction rate constant k. The time required to reduce the population by 90% is called the "decimal reduction time" or D value [33] and the slope of the survivor curve can be expressed as:

$$\frac{(-\log c_1 - \log c)}{D} = \frac{-\log c_1/c}{D} = \frac{-\log 100/10}{D} = \frac{-1}{D}$$

Since the slope is also -k/2.303,

$$\frac{-k}{2.303} = \frac{-1}{D} \quad \text{or} \quad D = \frac{2.303}{k} \qquad (3\text{-}3)$$

C. Temperature Dependence of Kinetics

Figure 3-2 is the survivor curve for a particular organism being heated in a particular medium at a constant temperature. Since real food products do not heat instantaneously but go through a time-dependent temperature treatment, the rate

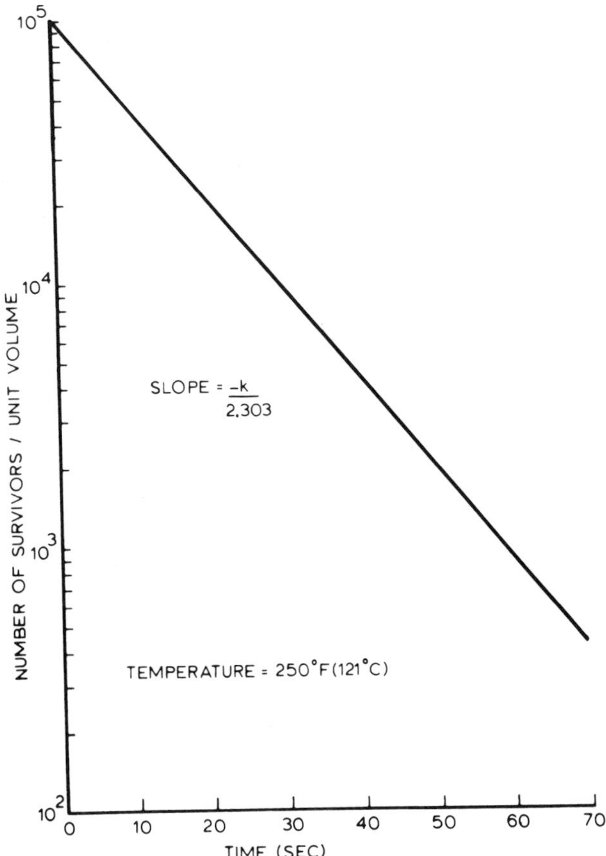

FIG. 3-2. First-order destruction of microorganisms.

inactivation at several lethal temperatures must be known. Dependence of reaction rate on temperature is also required to enable treatments at different temperatures to be compared with respect to thermal severity. There are two principal methods of describing the dependence of the reaction rate constant on temperature: (1) the Arrhenius equation, and (2) thermal death time curves.

The dependence of reaction rate constants on temperature as described by the Arrhenius equation is:

$$k = s \exp\left(-E_a/RT\right) \tag{3-4}$$

where

k is the reaction rate constant (min^{-1})
s is a constant, the frequency factor (min^{-1})
E_a is the activation energy (cal/mole)
R is the gas constant (1.987 cal/K mole)
T is the absolute temperature (K)

The activation energy is the energy required to get the molecules into an active state, and the term may be determined in the following manner. Taking logarithms of both sides of Eq. (3-4) results in:

$$\ell n\ k = \ell n\ s - (E_a/RT) \tag{3-5}$$

Therefore, a plot of ln k vs 1/T is a straight line of slope $-E_a/R$.

The frequency factor s can be evaluated by letting the reaction rate constant be k_1 at temperature T_1. Then,

$$\ln s = \ln k_1 + (E_a/RT_1) \tag{3-6}$$

Substitution of Eq. (3-6) into Eq. (3-5) yields:

$$\log \frac{k}{k_1} = \frac{-E_a}{2.303R}\left(\frac{1}{T} - \frac{1}{T_1}\right) = \frac{-E_a}{2.303R}\left[(T_1 - T)/T_1 T\right] \tag{3-7}$$

The thermal death time (TDT) method was introduced by Bigelow [9], who observed that when the thermal death time (minimum time to accomplish a total destruction) was plotted vs temperature in °F a straight line resulted. Figure 3-3 is a typical thermal death time curve.

The following equation describes a TDT curve:

$$\log (TDT_1/TDT_2) = (-1/z)(T_1 - T_2) = (T_2 - T_1)/z \tag{3-8}$$

where

TDT_1 is the thermal death time at temperature T_1 (min)
TDT_2 is the thermal death time at temperature T_2 (min)
T_1 is temperature 1 (°F)
T_2 is temperature 2 (°F)
z is the °F temperature change required to change the TDT by a factor of 10

The slope of a TDT curve, -1/z, is used to characterize the dependence of the reaction rate constant on temperature and it is therefore related to E_a. The significance of z with respect to relative effects of temperature on reaction rates is readily illustrated. A small z value (say 10° F) indicates that the destruction time decreases by a factor of 10 for a 10° F temperature increase, whereas a large z value (say 50° F) indicates that the temperature must increase 50° F to reduce the destruction time by a factor of 10. Therefore, reactions that have small z values are highly temperature dependent; whereas reactions with large z values are less influenced by temperature.

For microbial inactivation the thermal death time is given the symbol F. Since the F value is dependent on temperature and is specific for one organism it is usually identified with a superscript denoting the z value of the organism and a subscript denoting the temperature (° F). Thus F_{250}^{18} denotes the thermal death time for a microorganism at 250° F that is characterized by z = 18° F. Substituting the F notation into Eq. (3-8) and using 250° F as a reference temperature yields:

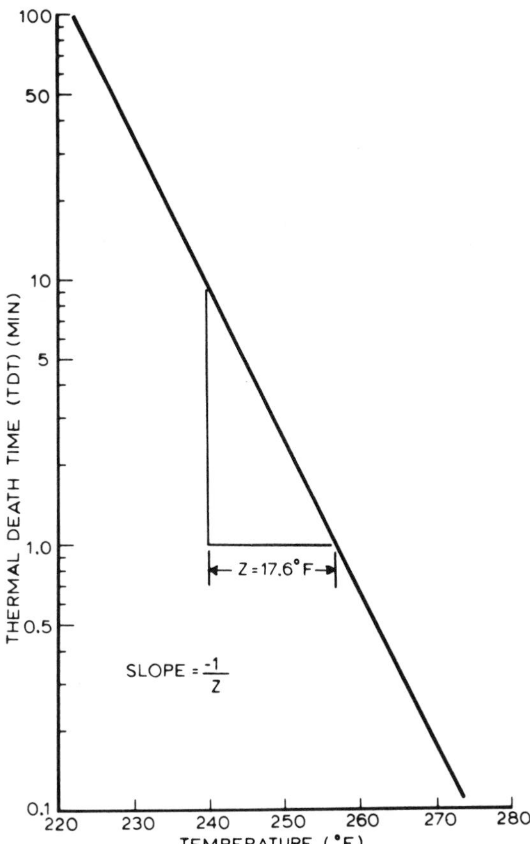

FIG. 3-3. Thermal death time (TDT) curve for Clostridium botulinum in phosphate buffer [17].

$$\log (F_T^z / F_{250}^z) = (250 - T)/z \tag{3-9}$$

An equation similar to Eq. (3-9) also can be developed for the decimal reduction value of a food component. If this is done, the procedure is referred to as the thermal resistance method and the resulting curve is called a "thermal resistance curve." This curve describes the effect temperature has on the time to reduce the concentration of the component by 90%. The thermal resistance equation is:

$$\log (D/D_1) = (T_1 - T)/z \tag{3-10}$$

Since k and D are related by Eq. (3-3), substitution into Eq. (3-10) results in:

$$\log (k_1/k) = (T_1 - T)/z \tag{3-11}$$

or

$$\log (k/k_1) = (1/z) (T - T_1) \tag{3-12}$$

Equation (3-12) states that the logarithm of the reaction rate constant is directly proportional to temperature. This appears to be in direct contradiction to Eq. (3-7), which states that the logarithm of the reaction rate constant is inversely proportional to temperature. This apparent contradiction has concerned investigators for many years. The basis of thermal processes was traditionally the TDT method but it could not be reconciled with reaction kinetic theory (Arrhenius method). However, if it is recognized that over small temperature ranges $T \propto 1/T$ then the systems are reconcilable.

The relationship between E_a and z can be obtained by equating Eqs. (3-7) and (3-12) and changing °F in Eq. (3-12) to K. Then

$$\frac{-E_a}{2.303R}\left[\frac{T_1 - T}{T_1 T}\right] = \frac{(T - T_1)}{z}(9/5) = \frac{-(T_1 - T)}{z}(9/5) \tag{3-13}$$

Rearrangement yields:

$$E_a = \frac{2.303R\ T\ T_1}{z}(9/5) \tag{3-14}$$

where

T_1 is the reference temperature (K)
T is the temperature (K)
1/z is the slope of the TDT curve (° F)
9/5 is the conversion of °F to K
R is the gas constant (1.987 cal/mole K)
E_a is the activation energy (cal/mole)

It is noteworthy that Eq. (3-14) is dependent on a reference temperature even though both z and E_a are independent of temperature. This dependence on reference temperature results from the fact that the relationship between reciprocal absolute temperature (K^{-1}) and temperature (° F) is defined for a small temperature range about a reference temperature. The relationship between E_a and z for reference temperatures of 210° F and 250° F is shown in Figure 3-4. The value of T was chosen z°F less than T_1.

Another term used to describe the response of biological systems to temperature change is the Q value, a quotient indicating how much more rapidly the reaction proceeds at temperature T_2 than at a lower temperature T_1. If Q reflects the change in rate for a 10° C rise in temperature, it is then called Q_{10}. The relationship between z and Q_{10} is [49]:

$$z = \frac{18}{\log Q_{10}} \tag{3-15}$$

In summary the three sets of interconvertible terms used to describe the response of a system's kinetics to temperature change are listed in Table 3-5. The most

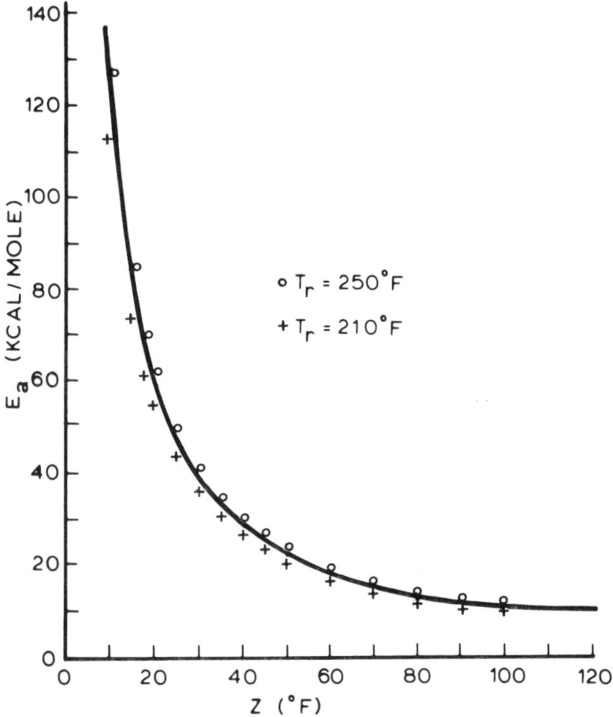

FIG. 3-4. Relationship between activation energy (E_a) and z value.

useful set of terms is D and z since they are the basis of thermal process calculations and are commonly used in industry.

D. Thermal Destruction of Microorganisms

As mentioned, when microorganisms are subjected to moist heat at temperatures slightly above the maximum growth temperature, reduction of viable cells or spores generally follows first-order kinetics. There are several factors which can cause deviation from the logarithmic order of death but deviations usually involve factors other than a change in the order of destruction. For a discussion of these factors, see Stumbo [58].

Many explanations have been suggested to account for the logarithmic order of death of bacteria. One of the most popular is that offered by Rahn [49] who has suggested that loss of reproductive ability is due to the denaturation of one gene or a protein essential to reproduction. This is consistent with the first-order kinetics observed when microorganisms are inactivated by heat, chemicals, or radiation. Regardless of the cause for first-order response to thermal energy, it does occur and is a great aid to thermal process calculations. Since all microorganisms obey the same kinetics this simplifies comparison of the heat resistances of different species at a given temperature and, more importantly, to different temperatures.

TABLE 3-5

Important Terms in Thermal Processing

Method	Reaction rates	Temperature dependence
Arrhenius equation	k	E_a
Thermal death time	D or F	z
Q_{10}	k	Q_{10}

There are two important concepts to gain from the logarithmic relationship for inactivation of bacteria: (1) the smaller the initial concentration of bacteria, the less the heating time needed to achieve a given final concentration; (2) the survivor curve never reaches zero population. The first concept is extremely important since it implies that a standard heat treatment given a product may or may not be sufficient depending on initial population. It is therefore necessary at each stage of food preparation prior to heat treatment to minimize contamination. The attitude that a standard thermal process can take care of any microbial population is erroneous but difficult to eradicate.

The second concept, that the microbial population may approach but never attain zero, can be easily illustrated. If the initial concentration of a suspension is 10^4 spores per milliliter, then for each D minutes the suspension is heated the population is reduced 90% as shown below.

Duration of heating (min)	Number of spores present during heating
0	10^4
D	10^3
2D	10^2
3D	10^1
4D	10^0
5D	10^{-1}
6D	10^{-2}

The meaning of the negative powers becomes readily apparent when interpreted in terms of probability of survival. Suppose, for example, that 100 experiments are run under the same conditions. Then 1/10 survival means that ten out of 100 samples heated for 5D min have one surviving spore per sample. Similarly 1/100 survival obtained after 6D min means that one sample out of 100 contains a surviving spore. The concept of probability of survival is extremely important when one considers the number of cans of thermally processed products that reach the consumer yearly.

1. Factors Affecting Thermal Resistance

It has been pointed out in Section II. D that spores are more heat resistant than vegetative cells and that spores of different species exhibit a wide range of heat resistance. For thermal process calculations in nonacid foods, the heat process is based on the resistance of spores. Many factors affect the thermal resistance of microorganisms and failure to consider all of these factors has contributed to conflicting results. The three general types of factors affecting thermal resistance of microorganisms are (1) inherent resistance, (2) environmental influences active during growth and formation of cells or spores, and (3) environmental influences active during the time of heating cells or spores.

Inherent resistance is illustrated by the fact that different strains of the same species prepared under the same conditions exhibit widely different degrees of resistance. Those spores or cells with the greatest resistance are believed to contain proteins with the greatest stability.

Conditions present during sporulation or growth which affect heat resistance are: (1) temperature, (2) ionic environment, (3) organic compounds other than lipids, (4) lipids, and (5) age or phase of growth of the culture [47, 53].

1. Temperature. Effect of growth temperature on resistance of microorganisms has been studied extensively. Generally spores produced at higher temperatures are more heat resistant than those produced at lower temperatures.
2. Ionic environment. Although the ionic environment affects spore resistance, there is no general trend. Ions shown to influence spore resistance are calcium, magnesium, iron, phosphates, manganese, sodium, and chloride. Since salts can influence resistance, the exact composition of the growth medium should be specified.
3. Organic compounds other than lipids. The presence of various organic compounds has been shown to influence the resistance of microorganisms to heat. Again, no general inference can be drawn except to reemphasize that specification of conditions for producing the spores or cell suspension is a prerequisite for obtaining reliable, interpretable data.
4. Lipids. Sugiyama [60] reported that *C. botulinum* spores exhibited increased heat resistance when low concentrations of certain saturated or unsaturated fatty acids were present during sporulation.
5. Age or phase of growth of the culture. Although Esty and Meyer [17] have reported that young moist spores are more heat resistant than old moist spores, there has been little corroborative data. Thus a general conclusion should not be drawn at this time. For vegetative cells, it has been generally observed that growth phase affects thermal resistance but again no rule of thumb can be applied. Some investigators have found that young cells are more susceptible to heat destruction than older, more mature cells [57]; others have found that resistance increases during the early logarithmic phase [2]; and still others have reported that resistance increases during the initial stationary phase, decreases when reproduction

begins, and arrives at a minimum during the logarithmic growth phase [15]. Thus, for vegetative cells, there is sufficient evidence to suggest that the phase of growth should be specified when thermal resistance is reported.

The third group of factors are the environmental influences active during the heat treatment: (1) pH and buffer components, (2) ionic environment, (3) water activity, and (4) composition of the medium.

1. pH and buffer components. Generally spores are less heat resistant at pH extremes than near neutrality. Optimal resistance may be exhibited at slightly alkaline or acidic pH depending on the species, but most species of significance are more resistant in slightly acidic conditions. As pointed out in Section II. D, pH of the food product dictates the microorganism to be used for the basis of the thermal process. Buffer constituents have been shown to influence heat resistance of microorganisms, but generalizations are not possible.
2. Ionic environment. Presence of ionic species has been shown to influence spore resistance but reports are conflicting and generalities are difficult to make. In phosphate buffer, low concentrations of magnesium or calcium reduced spore resistance, as did the presence of chelating agents, such as EDTA (ethylene diamine tetraacetic acid) or glycylglycine. Early reports claimed that sodium chloride increased spore resistance but the results could not be duplicated. At levels present in foods it is doubtful that sodium chloride has any effect on resistance.
3. Water activity. Spores are much more resistant to dry heat than to moist heat. This difference is related to the mechanism of spore destruction in the two environments: denaturation of protein in moist heat and oxidation in dry heat. Since oxidation requires greater energy input than denaturation, equivalent time-temperature treatments for dry and moist heat result in greater destruction with moist heat. For example, dry spores of *C. botulinum* can be inactivated in 120 min at 248°F in dry gas heat, but inactivation is achieved in only 4-10 min at 250°F in moist heat.
4. Composition of the medium. The carbohydrate, protein, and lipid contents of the substrate; the presence of colloidial systems, such as emulsions or foams; and the presence of other organic or inorganic compounds can influence the apparent thermal resistance of microorganisms. The response of the microorganism to heat while present in a specified medium depends on the microorganism, and generalities are once again subject to exceptions. However, at the risk of oversimplification, some trends are apparent. High concentrations of soluble carbohydrates generally result in increased thermal resistance. Although the mechanism of the protecting action is obscure, some have suggested that the increased resistance is due to passive dehydration of the protoplasm. This apparently is not the only protective mechanism since protection is not proportional to molarity of the carbohydrate [60]. Proteins likewise have protective effects, as do lipids. In dry fat, organisms may show a pronounced increase in resistance presumably due to the localized absence of moisture. With increased salt concentration and in the presence of germicidal or antibiotic substances, thermal resistance generally decreases.

Although all of the factors previously mentioned affect the thermal resistance of microorganisms directly, there is still one other critical factor — the outgrowth medium used after thermal processing. Usually the microorganisms are incubated in the heating medium to determine whether there are survivors, and this automatically establishes the culture environment. However, agents which inhibit germination may be present as a result of heating, and this may produce misleading information about thermal resistance. Thus the supposed resistance may be due to inhibition of germination after heating rather than the direct action of heat energy. To overcome this problem, the culture is sometimes transferred to a new medium (subculture) following completion of heating. If a subculture is used, its nature should be reported in detail.

In summary, the thermal resistance of microorganisms is dependent on many factors, each of which must be considered and reported. To increase the reliability and usefulness of thermal destruction data, the National Canners Association has published procedures for preparation of spore and cell suspensions [43].

2. Methods of Measuring Resistance

In the 50-odd years that food microbiologists have been measuring thermal resistance of microorganisms several techniques have been developed for obtaining data. Generally the techniques consist of putting the inoculated menstruum into containers and then subjecting the containers to a thermal treatment. Since the object is to describe the response of the microorganism to a particular time-temperature treatment, primary emphasis has been devoted to the design of heating vessels where thermal lags in heating and cooling are minimized. The glass tube method (7 mm i.d. Pyrex), the capillary tube method (0.8 mm i.d.), and the can method (208 x 006) are the most popular systems. Pflug and Schmidt [47] and Stumbo [58] have discussed the advantages and disadvantages of the various methods for obtaining thermal resistance data.

Interpretation and reliability of thermal-resistance data are dependent on the method of obtaining the data. Although the first impression may be that determination of thermal resistance is a relatively easy task, this is not true. Considerable effort is required to generate a standard spore or cell suspension, to perform the series of heat treatments (which requires replication at several temperatures), and to analyze the data.

As stated in Section II. D, the thermal process applied to a food is based on the most heat resistant microorganism which could present a health hazard or cause spoilage if it should grow following the thermal process. Summarized in Table 3-6 are the approximate resistances of bacterial spores of importance in foods.

Although the D value can be used for comparative purposes, a parameter defining the thermal process should describe the total reduction in population that must be accomplished. Since thermal destruction is logarithmic, total destruction theoretically cannot be accomplished. However, the probability of survival can be reduced to some small nonzero level. The question is what should that level be? For nonacid foods the thermal process is based on the reduction of _C. botulinum_ and on the classic work of Esty and Meyer [17]. Esty and Meyer experimented with the largest concentration of spores of _C. botulinum_ they could effectively manipulate in buffer and it was assumed that this was equivalent to 10^{12} spores per milliliter. The indicated thermal death time at 250° F for reduction to

TABLE 3-6

Thermal Resistance Characteristics of Spore-forming Microorganisms Used as a Basis of Thermal Processing[a]

Organisms	z value (°F)	D_{250} value (min)
B. stearothermophilus	12.6	4.0
B. subtilis	13.3-23.4	0.48-0.76
B. cereus	17.5	0.0065
B. megaterium	15.8	0.04
C. perfringens	18.0	—
C. sporogenes	23.4	0.15
C. sporogenes (PA 3679)	19.1	0.48-1.4
C. botulinum	17.8	0.21
Coxiella burnetti[b]	8	—
C. thermosaccharolyticum	16-22	3.0-4.0

[a]Ref. [55].
[b]Ref. [16]; Coxiella burnetti is a rickettsia organism.

10^0 (= 1) survivors in phosphate buffer was 2.78 min. Later Townsend et al. [66] corrected these data for thermal lag and reported a heating time of 2.45 min at 250°F to reduce the population by a factor of 10^{12}. Use of a heating time at 250°F to reduce the population by a given factor is a very convenient way of comparing heat treatments which a product may receive. While the D value gives the time for the reduction of 90% of the population, the F value gives the time to reduce the population by a multiple of the D value. The F value is the thermal death time as defined in Eq. (3-9). For C. botulinum the F value represents the heating time necessary to reduce the population by a factor of 10^{12} (12 D values) at 250°F. When z = 18°F (C. botulinum) and the temperature is 250°F, the symbol F_o is used. Since the F_o value describes the time to reduce the population by a factor of 10^{12} and the D value describes the time to reduce the population by a factor of 10^1, then:

$$F_o = D \log a/b \qquad (3\text{-}16)$$

where a is initial number of microorganisms and b is final number of microorganisms. With C. botulinum, F_o = 12D.

A thermal process equivalent to the F_o of C. botulinum is referred to as the "bot cook" and the destruction of 10^{12} spores as the "12D concept." To restate the 12D concept, if 1000 C. botulinum spores (estimated maximum) were in each container of low-acid food then the incidence of survival of C. botulinum following a 12D process would be less than one in 10^9 containers. At first this may seem to be gross overprocessing and that the probability of survival can be increased. However, when it is realized that United States' production alone amounts to billions of cans of low-acid foods, the 12D concept is justified.

Sometimes the situation exists wherein products which receive a thermal process sufficient to inactivate spore-forming pathogens exhibit high spoilage rates as a result of growth of mesophilic spore-forming bacteria. In this case, such microorganisms as those listed in Table 3-4 are used as the basis of the process. They exhibit D_{250} values on the order of 1 min. For a 12D process, this would result in an F_o of 12 min. A thermal process based on this F_o value would result in a product of low quality due to the many heat-induced physical and chemical changes. Consequently, the process is based on keeping the frequency of spoilage at a level economically tolerable to the processor. Most processors would like to keep spoilage rates at less than one container per thousand. To assure that this is not exceeded, a 5D value can be used for the process. This results in:

$$F_o = 1.0(5) = 5.0 \text{ min}$$

When the 5D process based on a mesophilic spore-forming bacterium is used, it is necessary to verify that the F value is at least equivalent in lethality to a 12D process based on the most heat-resistant spore-forming pathogenic microorganism presumed present in the food.

When canned foods are stored at high temperatures where thermophilic anaerobes can cause spoilage, the thermal process must be extremely severe. From Table 3-6 the D_{250} value for C. thermosaccharolyticum is on the order of 3-4 min, and thus for a $5D_{250}$ process:

$$F_o = 4(5) = 20 \text{ min}$$

Under these circumstances it is imperative to use preventive measures for minimizing microbial load on the product since a process of this severity must be very damaging to product quality.

The multiple of D used in thermal processing has been termed the "order of the process" factor. A factor of 12 is used for commercial processing of low-acid foods whereas a factor of 5 or less is used when mesophilic spore-forming bacteria or thermophilic spore-forming bacteria are the most heat-resistant organisms of concern.

E. Thermal Destruction of Enzymes

Before considering methods of calculating thermal processes for foods, the effect of time-temperature treatments on enzymes and other food components should be discussed. With enzymes, the objective of the thermal process generally is inactivation, whereas with nutrients and other quality attributes, such as flavor, color, or texture, the objective is maximum retention. In order to compare the total effect of equivalent heat processes it is necessary to know the response of each component of the food to thermal energy.

The thermal stabilities of enzymes are, like those of microorganisms, dependent on many factors. In fact the same factors that affect microbial inactivation can be expected to affect enzyme inactivation since the most popular theory dealing with microbial inactivation holds that denaturation of a critical gene or protein causes loss of reproductive ability. A complete review of enzyme inactivation by heat is not necessary here; however, knowledge of the relative dependence of enzyme inactivation on temperature is essential. Whitaker [68] has discussed the effect of heat on enzymes in great detail.

The dependence of enzyme inactivation on temperature can be expressed in the same ways that have been used for microorganisms, i.e., with D and z values. For most enzymes the D and z values are the same order of magnitude as those for microorganisms. However, there are some heat-resistant isozymes causing off flavor and off color which exhibit unusually great thermal resistance. For these heat-resistant isozymes the z value may be as much as two to four times greater than those for the heat-labile isozymes. For example Adams and Yawger [1] have reported that peroxidase in peas has a z value of 55°F (E_a = 20,000 cal/mole). A denaturation reaction characterized by such a low activation energy indicates that the active site of the protein is dependent on few bonds, perhaps a disulfide bond [e.g., the activation energy for H_2S formation from corn is on the order of 30,000 cal/mole [67], or four to five H bonds (E_a = 5000 cal/mole each)].

In thermal process calculations based on enzyme inactivation, the most heat-resistant enzyme which can alter product quality during storage is selected as the basis of the process, just as the most heat-resistant pathogenic or spoilage organism is used for microbial inactivation. To reiterate the point made in the discussion of blanching, generally peroxidase or catalase is selected as the basis of the process.

F. Thermal Destruction of Nutrients and Quality Factors

It is evident from a TDT curve of microorganisms that there are an infinite number of time-temperature combinations that can produce commercial sterility. Part of the responsibility of the food processor is not only to produce a commercially sterile product but also to provide a product that is wholesome and retains nutritional and quality attributes to a maximum degree. To accomplish this objective, the processor must be aware of the relative changes in nutrients and quality factors as a function of the time-temperature treatment.

Many of the same factors that accelerate or inhibit microbial destruction by heat also affect nutrient destruction. Harris and von Loesecke [24] have reviewed the effect of heat processing on nutrients, and it is evident that numerous variables affect the rate of nutrient destruction. If one of the requirements of a thermal process is retention of a particular nutrient, it is necessary to establish the kinetics of its degradation in the system and the dependence of the rate constant on temperature.

Presented in Table 3-7 is a summary of the reaction rate-temperature relationship for microorganisms, enzymes, nutrients, and a few quality attributes. Thiamine is the most thoroughly investigated nutrient in food systems, and activation energies and relative reaction rates in foods are fairly abundant. Response of kinetics for nutrient destruction to temperature is one area where more research is needed. As is to be expected, the activation energies characterizing the degradation of nutrient and quality factors are similar to those for chemical reactions.

A striking pattern emerging from Table 3-7 is that the z values for nutrient and quality-factor destruction are larger than for microorganisms or heat-labile enzymes. That is, a given increase in temperature causes a larger increase in the rate of destruction of microorganisms than in the rate of destruction of nutrients or quality factors. This situation is exploited to good advantage in high-temperature and short-time processing (e.g., 270°F for a few seconds). At 270°F the rate of destruction of microorganisms would be at least tenfold ($z = 18^\circ$F) greater, and the rate of destruction of nutrients and quality factors would be only two to threefold greater than at a standard temperature of 250°F.

The D values for thermal destruction of various nutrients and quality factors are more dependent on the environmental conditions than are the z values. Therefore, such conditions as pH and composition of medium must be specified when D values are reported. From Table 3-7 it can be seen that D values for nutrients and quality factors are generally 100-1000 times greater than those for microorganisms. For example, for C. botulinum the D_{250} value is 0.1-0.2 min, whereas for thiamine the D_{250} value is on the order of 150 min. The fact that microorganisms have much smaller D values than do quality factors and nutrients permits foods to be sterilized by heat without total destruction of quality and wholesomeness factors. If the D values for nutrients and quality factors were essentially equal to those for microorganisms and a 12D process were used, then a product containing N milligrams nutrient at the start of the process would contain $N \times 10^{-12}$ mg at the end of the process and would be nutritionally worthless. The z and D_{250} values of nutrients and quality factors also can be used to optimize thermal processes for nutrient or quality-factor retention, as is shown in Section VI.A.

IV. HEAT PENETRATION INTO FOODS

A. Introduction

In order to calculate the effectiveness of a thermal process two items of information are needed: (1) the thermal resistance characteristics of the microorganism used as the basis of the process (z and F value) and (2) the temperature history of the product. Thermal processing of food is accomplished in one of two ways. Either the food is heated to accomplish commercial sterility and then placed in a sterile container and sealed, or the food is placed in a container, sealed, and then heated. The first method is referred to as aseptic processing, whereas the second corresponds to conventional canning. Since the conventional canning process is older and more common, it is used for illustrative purposes here. It should be noted, however, that the principles apply equally well for aseptic processing.

If a container of product at low temperature is placed in a vessel and the vessel is filled with steam at high temperature, the steam condenses on the container and the latent heat of condensation (see Chapter 2) is transferred through the wall of the container and into the product. The process involves unsteady-state heat transfer since there is no point in the container where the temperature is constant. In this situation, the temperature at any point in the container is a function of

TABLE 3-7

Kinetic Parameters for the Thermal Degradation of Food Components

Reference	Component	Medium	pH[a]	Temperature range (°F)	z (°F)	E_a (kcal/mole)	D_{250}[b]	Other
Bendix et al. [8]	Thiamine	Whole peas	Nat.	220-270	47	21.2	164 min	
Feliciotti and Esselen [19]	Thiamine	Carrot puree	5.9	228-300	45	27	158 min	
		Green bean puree	5.8	228-300	45	27	145 min	
		Green pea puree	6.6	228-300	45	27	163 min	
		Spinach puree	6.5	228-300	45	27	134 min	
		Beef heart puree	6.1	228-300	45	27	115 min	
		Beef liver puree	6.1	228-300	45	27	124 min	
		Lamb puree	6.2	228-300	45	27	120 min	
		Pork puree	6.2	228-300	45	27	157 min	
Mulley et al. [42]	Thiamine	Phosphate buffer	6.0	250-280	45	29.4	156.8 min	
		Pea puree	Nat.	250-280	48	27.5	246.9 min	
		Beef puree	Nat.	250-280	48	27.4	254.2 min	
		Peas-in-brine puree	Nat.	250-280	49	27.0	226.7 min	
Gillespy [21]	Thiamine	—	—	—	56	20.0	—	
	Riboflavin	—	—	—	50	23.0	—	
Garrett [20]	$B_1 \cdot HCl$	Liquid multi-vitamin prep.	3.2	39-158	44.5	26	1.35 days	
	d-Pantothenic acid	Liquid multi-vitamin prep.	3.2	39-158	56	21	4.46 days	

	C	Liquid multi-vitamin prep.	3.2	39-158	50	23.1	1.12 days	
	B_{12}	Liquid multi-vitamin prep.	3.2	39-158	50	23.1	1.94 days	
	Folic acid	Vitamin prep.	3.2	39-158	66	16.8	1.95 days	
	A	Vitamin prep.	3.2	39-158	72	14.6	12.4 days	
Davidek et al. [13]	Inosinic acid	Buffer soln.	3	140-208	34	34.0	—	
	IMP	Buffer soln.	4	140-208	38.5	30.4	—	
	IMP	Buffer soln.	5	140-208	41	28.1	—	
Gupte et al. [23]	Chlorophyll a	Spinach	6.5	260-300	92	15.5	13.0 min	
	Chlorophyll b	Spinach	5.5	260-300	143	7.5	14.7 min	
Gold and Weckel [22]	Chlorophyll	Pea puree	Nat.	240-280	66	16.1	14.0 min	Blanched
	Chlorophyll	Pea puree	Nat.	240-280	81	12.6	13.9 min	Unblanched
Lenz and Lund [37]	Chlorophyll	Pea puree	6.5	175-280	52	22	113 min	
	Chlorophyll	Spinach puree	6.5	175-280	59	19	116 min	
Mackinney and Joslyn [40]	Chlorophyll a	Buffered soln.	—	32-122	143	7.5	—	
	Chlorophyll b	Buffered soln.	—	32-122	125	9.0	—	
Dietrich et al. [14]	Chlorophyll	Green beans	Nat.	190-212	84	12.0	10.1 min	
Timbers [65]	Color ($^{-}$a/b)	Peas	Nat.	175-300	71	15.0	25.0 min	
		Asparagus	Nat.	175-300	75	14.0	17.0 min	
		Green beans	Nat.	175-300	70	15.0	21.0 min	
	Quality by taste panel	w.k. corn	Nat.	175-300	57	19.5	6.0 min	Time for taste panel to judge 4.0/9.0 compared to frozen control
		whole peas	Nat.	175-300	51	22.5	2.3 min	
		whole green beans	Nat.	175-300	52	22.0	4.0 min	

TABLE 3-7 (cont'd)

Kinetic Parameters for the Thermal Degradation of Food Components

Reference	Component	Medium	pH^a	Temperature range (° F)	z (° F)	E_a (kcal/mole)	$D_{250}{}^b$	Other
Herrmann [26]	Chlorophyll a	Spinach	Nat.	212-266	81	12.5	34.1 min	
	Chlorophyll b	Spinach	Nat.	212-266	106	10.0	48.3 min	
	Maillard Rxn.	Apple juice	Nat.	100-266	45	27.0	4.52 hr	
	Nonenz. brn.	Apple juice	Nat.	100-266	55	20.7	4.75 hr	
	B_i	Pork		? -250	57.6	19.5	6.03 hr	
Ponting et al. [48]	Anthocyanin	Grape juice	Nat.	68-250	41.7	28.0	17.8 min	
		Boysenberry juice	Nat.	68-250	55.7	20.0	102.5 min	
		Strawberry juice	Nat.	68-250	59.8	19.0	110.3 min	
Tanchev and Joncheva [62]	Cyanidin-3-rutinoside	Citrate buffer	4.5	170-225	49	23.7	28.5 min	
		Plum juice	4.5	170-225	50	23.1	41.6 min	
	Peonidin-3-rutinoside	Citrate buffer	4.5	170-225	44.5	26.2	21.6 min	
		Plum juice	4.5	170-225	46	22.1	27.7 min	
Ramakrishnan and Francis [51]	Carotensids	Paprika	Nat.	125-150	34	34.0	0.038 min	
von Elbe et al. [67]	Betanin	Buffer	5.0	122-212	81	12.5	19.5 min	
		Beet juice	5.0	122-212	106	10.0	46.6 min	
Burton [11]	Browning	Goat's milk	6.5-6.6	200-250	45	27.0	1.08 min	Homogenized
		Goat's milk	6.5-6.6	200-250	45	27.0	0.91 min	Unhomogenized

Williams and Nelson [69]	Methyl methionine sulfonium bromide → DMS	Sodium citrate buffer	6.0	178-212	33.5	34.8	8.4 min	
		Sweet corn	6.9	178-212	37	31.6	4.5 min	
		Tomato	4.4	178-212	45	27.2	23.2 min	
Taira et al. [61]	Lysine	Soybean meal	—	212-260	38	30.0	13.1 hr	
Mansfield [41]	Texture and overall cook quality	Peas	Nat.	170-200	58	19.5	1.4 min	Time to acceptable product
		Peas	Nat.	210-240	58	19.5	2.5 min	
		Beets	Nat.	180-210	34	34.0	2.0 min	
		Whole kernel corn	Nat.	212-250	66	16.0	2.4 min	
		Broccoli	Nat.	212-250	80	13.0	4.4 min	
		Squash	Nat.	182-240	46	25.0	1.5 min	
		Carrots	Nat.	176-240	30	38.0	1.4 min	
		Green beans	Nat.	182-240	28	41.0	1.0 min	
		Potato	Nat.	161-240	42	27.5	1.2 min	
Hackler et al. [28]	Trypsin inhibitor	Soybean milk	—	200-250	60	18.5	13.3 min	
Adams and Yawger [1]	Peroxidase	Whole peas	Nat.	230-280	67	16.0	3.0 min	
Licciardello et al. [38]	<u>C. botulinum</u> toxin-type E	Growth	6.2	125-135	44	26.4	3.0 min at 60°C	Temperature-dependent activation energy
		Media	6.2	135-140	22	57.4		
		Media	6.2	140-145	6	153.5		
Read and Bradshaw [52]	Staphylococcus milk enterotoxin B		6.4-6.6	210-260	44	25.9	9.4 min	

TABLE 3-7 (cont'd)

Kinetic Parameters for the Thermal Degradation of Food Components

Reference	Component	Medium	pH[a]	Temperature range (°F)	z (°F)	E_a (kcal/mole)	D_{250}[b]	Other
Stumbo [58]	*C. botulinum* spores (types A and B)	Variety	>4.5	220- ?	12-19	64-82	0.1-0.2 min	
	B. stearothermophilus	Variety	>4.5	230- ?	12-22	53-82	4.0-5.0 min	

[a] Nat. indicates the natural pH of the system.

[b] D value at 250° F.

(1) the surface heat-transfer coefficient, (2) the physical properties of the product and container, (3) the difference between the steam temperature and the initial temperature of the product, and (4) the size of the container. For thermal processing in condensing steam it is reasonable to assume that the surface heat-transfer coefficient is very large compared to the thermal conductivity of the product and, therefore, the only resistance to heat transfer is the product.

The rate of heat penetration into the container is dependent on the mechanism of heat transfer within the product. Some food products are relatively nonviscous or contain small particles in brine. These products exhibit relatively rapid heat penetration because natural convection currents occur in the container. The natural convection currents can be aided by rotating or agitating the can, thus increasing the heating rate. This approach is used in continuous retorts (Section VI.B).

Extremely viscous or solid food products heat primarily by conduction. For these products, the charts developed for unsteady-state heat transfer can be used to predict the temperature-time heating curve. Examples include cream-style corn, pumpkin, most thick pureed vegetables, potato salad, baked beans, and most intact food tissues.

Finally, there are products in which the mode of heat transfer changes from convection to conduction during heating. Since convection heating is much faster than conduction heating, these products exhibit a change in the rate of heating, and a plot of temperature vs heating time shows an abrupt change in temperature rise. This is referred to as a broken heating curve. Foods containing significant quantities of starch that gels upon heating exhibit this behavior. Examples of products exhibiting broken heating curves are cream-style corn (when the starch has not been gelatinized prior to initiating the thermal process), soups, and noodle products.

B. Determination of Time-Temperature Profile for Thermal Process Calculations

Since commercial sterility must be assured throughout the container, process evaluation is based on the slowest heating point in the container. It is then assumed that if the slowest heating point receives a heat process sufficient to accomplish commercial sterility then all other points have received an equal or greater thermal process and are also commercially sterile. For conduction heating foods the center is the slowest heating point in the container since it is the farthest point from the heat source. For convection heating products, however, the center is not the slowest heating point. The slowest heating point (cold point) for convection heating products in cans is located on the vertical axis near the bottom of the container. These differences are illustrated in Figure 3-5. For those products in which the mode of heat transfer changes, causing broken heating curves during heating, it is difficult to predict the point at which heating occurs most slowly and, therefore, this point must be determined experimentally. If a thermal process is being designed for a new product, it is advisable to determine the cold point in the container. This is done by placing a series of thermocouples along the vertical axis at appropriate intervals (depending on the size of the container) and conducting a heat penetration test.

To determine the temperature changes at the slowest heating point, a temperature-sensing device is inserted into the container at this point. A heating

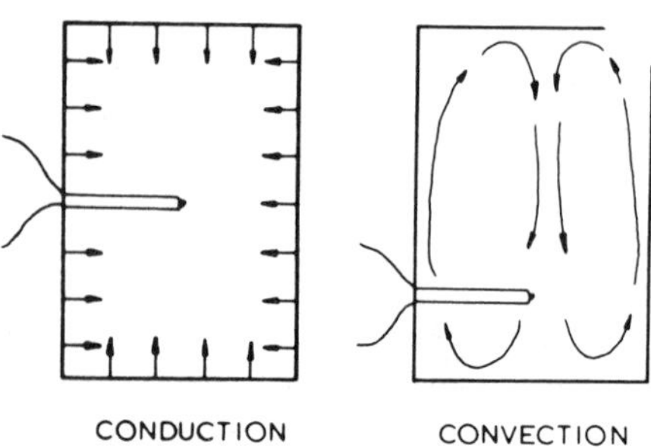

FIG. 3-5. Heat penetration into food in cans.

process is then started and the temperature is recorded as a function of time. The most popular means of measuring temperature in heat penetration tests involves a thermocouple. Recently, however, use of radiotelemetry has increased, primarily for monitoring temperature in continuous, agitated retorts where thermocouples are not appropriate. The radiotelemetry system, which is based on the change in resistance of a thermistor with temperature, requires no electrical hookup to the recording or readout device. This system, however, is very expensive.

C. Evaluation of Heat Penetration Data

Heat penetration data are most easily evaluated and utilized by plotting heat penetration curves. For calculating the lethality of a thermal process by the general method (Section V. B), heat penetration data need not be plotted unless one desires to know the temperature at a time when a measurement has not been taken. Once the data are plotted, the desired information can be easily obtained by interpolation. For the mathematical method of calculating lethality of thermal processes, some parameters which characterize the rate of heating are needed and, therefore, heat penetration data must be plotted.

Factors affecting unsteady-state heat transfer were given in Chapter 2; however, methods of plotting heat penetration data were not considered. In order to characterize heat penetration into foods, it is most convenient to plot the data so that there is a linear relationship between product temperature and heating time. It can be shown that the logarithm of the thermal driving force ($T_s - T_c$, where T_s is steam temperature and T_c is temperature at the center or the cold point) is a linear function of heating time. Thus, a plot of log ($T_s - T_c$) vs time is linear (Fig. 3-6). Product temperature (PT) at the slowest heating point is often substituted for T_c, and T_s is often called retort temperature (RT). Therefore ($T_s - T_c$) can be replaced by (RT - PT). Heat penetration curves which are linear throughout the heating process are referred to as simple heating curves; whereas heating curves exhibiting an abrupt change in heat transfer are referred to as broken heating curves.

A simple way of plotting heat penetration data also can be developed from examination of Figure 3-6. It is evident that (RT - PT) = 1 on the left-hand axis corresponds to PT = RT - 1 on the right-hand axis. By means of this relationship,

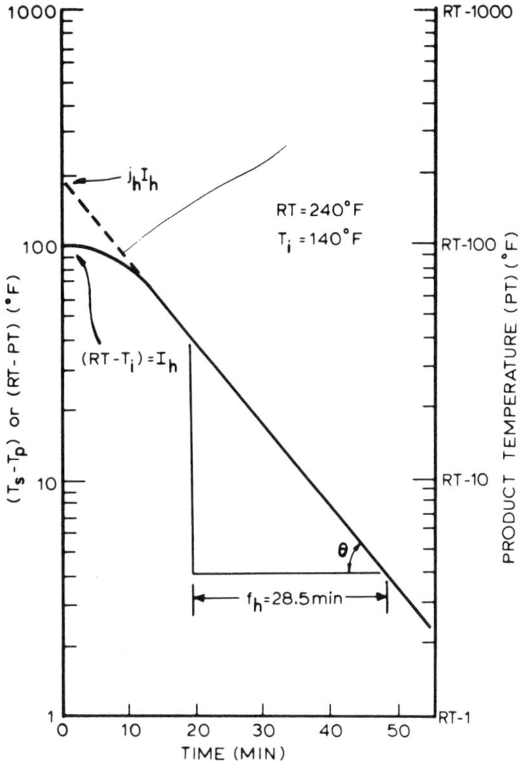

FIG. 3-6. Heat penetration curve.

product temperature can be plotted directly on semilog paper and the mechanical step of subtracting PT from RT need not be done. This is accomplished by inverting the semilog paper 180° and labeling the top line with a numeral equivalent to RT - 1. The next log cycle as labeled with a numeral equivalent to RT - 10, and the third with a numeral equivalent to RT - 100. This method of plotting heat penetration data is shown in Figure 3-7.

There are several important characteristics of a heat penetration curve. First, in the development of a mathematical method for calculating sterility of a thermal process it is necessary to express the temperature of the slowest heating point as a function of time. This is easily done on a linear heating curve by giving the slope and a characteristic intercept. Second, it should be recognized that theoretically the product cold point never reaches retort temperature. The cold point may get very, very close to RT but RT - PT can never be zero. This concept is important in the formula method of calculating thermal processes.

The curve shown in Figure 3-6 can be characterized in part by its slope. It is convenient to define the slope of the heat penetration curve in a fashion similar to that used to define the slope of a TDT curve. Thus, just as the z value is the °F temperature change required for the thermal death time to change by a factor of 10 (go through one log cycle on a TDT curve), so f_h is defined as the time in minutes required for the heat penetration curve to traverse one log cycle. Thus:

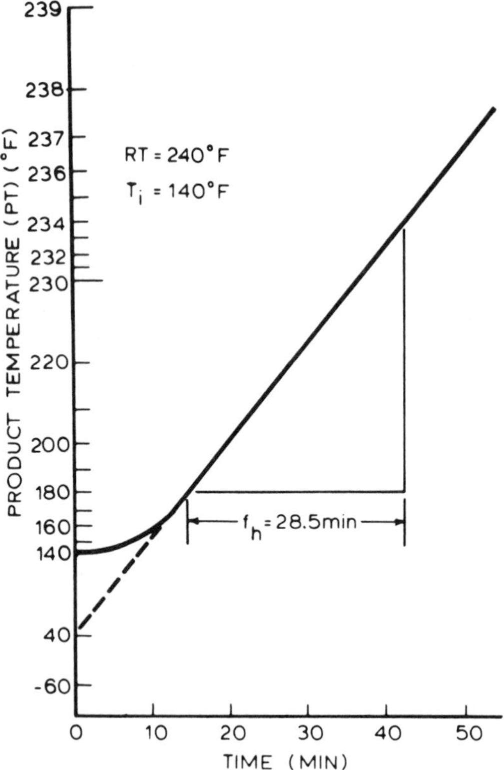

FIG. 3-7. Plotting heat penetration data.

$$\text{slope} = \tan\theta = \frac{\Delta y}{\Delta x} = \frac{-(\log 100 - \log 10)^*}{\Delta\,\text{time}} = \frac{-1}{28.5\ \text{min}} = \frac{-1}{f_h} \tag{3-17}$$

The subscript h indicates that the f value applies to a heating process. In the cooling process a similar value, f_c, is defined for the cooling curve.

The other necessary characteristic for defining the heat penetration curve is an intercept. The intercept is obtained by linearly extrapolating the heat penetration curve back to zero time. The pseudo-initial product temperature is defined as T_{pih} and the intercept is then $(RT - T_{pih})$. The equation for the heat penetration curve thus becomes:

$$\log (RT - PT) = (-t/f_h) + \log (RT - T_{pih}) \tag{3-18}$$

Although Eq. (3-18) completely describes the linear portion of the heat penetration curve it is not suitable for thermal process calculations since it does not identify the time at which the product starts to exhibit logarithmic heating. That is, it does not identify the lag period. The lag period can be characterized by considering

---

*Any full log cycle may be used.

the initial driving force for heat transfer, $RT - T_{ih}$, where T_{ih} is the initial product temperature. Then the ratio $(RT - T_{pih})/(RT - T_{ih})$ is a measure of the thermal lag. This ratio is called the j_h factor (h identifies the j factor for heating since there is also a j factor for cooling, j_c). If the retort temperature minus initial product temperature is called I_h, notice that

$$j_h I_h = \left(\frac{RT - T_{pih}}{RT - T_{ih}}\right)(RT - T_{ih}) = (RT - T_{pih})$$

Thus $j_h I_h$ can be substituted into Eq. (3-18) resulting in:

$$\log (RT - PT) = (-t/f_h) + \log j_h I_h \tag{3-19}$$

At the end of the heating cycle (time B) in a thermal process, the cold point will have reached some finite maximum temperature. If the temperature difference (RT - PT) existing at this time is defined as "g," then Eq. (3-19) can be written:

$$\log g = -(B/f_h) + \log j_h I_h$$

or

$$B = f_h \log (j_h I_h / g) \tag{3-20}$$

where

B is thermal process time (min)
f_h is the time in minutes for heat penetration curve to traverse one log cycle
j_h is $(RT - T_{pih})/(RT - T_{ih})$
I_h is $(RT - T_{ih})$

Equation (3-20) can be used to calculate heating times if the value of g is known. In the formula method for process calculations, means for determining the g value have been developed.

As pointed out earlier, heat penetration into a product is dependent on many factors. There are methods available for adjusting f_h and j_h for changes in these variables; however, discussion of these methods is beyond the scope of this chapter. Suffice it to say that a change in process conditions, container size, or fill weight generally should be accompanied by a new series of heat penetration tests.

Another important factor to consider in heat penetration tests is that the test should be conducted under conditions that duplicate those that are encountered in a commercial retort. If a batch-type retort is used then the slowest heating point must be determined in that container which receives the least thermal energy. The packing characteristics of the retort can influence the accessibility of steam to can surfaces.

Finally, in thermal process calculations temperatures at the cold point also must be known for the cooling phase. In many processes wherein the product heats by conduction the cooling portion of the process can contribute significantly to the overall lethality of the process. The cooling curve can be characterized in a manner analogous to that used for heating (parameters j_c and f_c). In plotting cooling data the ordinate is labeled log (PT - CT) (CT = cooling water temperature) or log PT if the scale is labeled CT - 100 on the line corresponding to PT - CT = 100 and is thereafter numbered in the same manner as for the heating curve.

V. METHODS OF DETERMINING LETHALITY OF THERMAL PROCESSES

A. Introduction

Three procedures have been developed for evaluating lethality of thermal processes: (1) improved general method [5, 56], (2) formula method [6], and (3) nomogram method [44].

The general method was the first scientific approach to calculating the adequacy of thermal processes. Basically, the method involves graphical integration of the total lethal effect for the time-temperature combinations which the cold point experiences in a thermal process. The underlying principles form the basis of both the formula and nomogram methods. The general method has undergone modification resulting in the "improved general method," which is accurate and does not rely on assumptions about the nature of the heat penetration curve. However, its broad applicability and accuracy are counter-balanced by the fact that it is laborious.

The formula method is more rapid and as accurate as the improved general method. Its use with products exhibiting simple heating curves (linear on semilogarithmic plots) is very straightforward and requires a minimum of calculations. Stumbo [58] presents methods of applying the formula method to products exhibiting broken heating curves.

The nomogram method is the quickest and least complicated method available for thermal process calculations. The parameters from heat penetration and thermal death time curves must be known before the nomogram method can be used. Ball and Olson [7] caution that this method should be used only by those thoroughly familiar with the assumptions inherent in establishing the nomogram. Determination of whether or not the particular process meets the assumptions must be done before the method is applied.

B. Improved General Method

In the improved general method, use is made of the fact that different combinations of time-temperature conditions can achieve the same lethal effect on microorganisms. Recall Eq. (3-9), which is the equation for a TDT curve:

$$\log \frac{F_T^z}{F_{250}^z} = \frac{250 - T}{z}$$

or

$$\frac{F_T^z}{F_{250}^z} = 10^{(250-T)/z} = \frac{TDT_T}{1} \qquad (3\text{-}21)$$

For convenience in thermal process calculations, the right-hand term in Eq. (3-21) ($TDT_T/1$) is used. This term, calculated from $10^{(250-T)/z}$, is equal to the thermal death time at temperature $T(TDT_T)$ equivalent to 1 min at 250° F (F = 1). This approach, while absolutely sound, arbitrarily establishes an F value of 1.0 as the starting basis for thermal process calculations. This is done to simplify calculations, and conversion to the actual F value for the process is easily accomplished as the last step in the calculation. Using Eq. (3-21), for example, if the product temperature is 232° F and z = 18° F, then,

$$10^{(250 - T)/z} = 10^{(250 - 232)/18} = 10^1 = 10 = TDT_T/1$$

This means that 10 min at 232°F is equivalent in lethality to 1 min at 250° F. For purposes of determining the total lethal effect during a heat process it is convenient to express the lethality in terms of minutes at some reference temperature. Obviously, 250°F is a convenient reference temperature for low-acid food processing since the TDT curve is defined with 250°F as the reference temperature. In order to express the lethal effect at temperature T in terms of the lethal effect at 250°F, it is convenient to use the reciprocal of the term on the right-hand side of Eq. (3-21):

$$\frac{1}{TDT_T} = \frac{1}{10^{(250 - T)/z}} = 10^{(T - 250)/z} \tag{3-22}$$

In terms of the example just cited,

$$\frac{1}{TDT_T} = \frac{1}{10^{(250 - T)/z}} = \frac{1}{10^{(250 - 232)/18}} = \frac{1}{10} = 0.1$$

This means that 1 min at 232° F is equivalent to 0.1 min at 250° F. In other words, if the F^z_{250} value for the process if 1.0 min, then ($1/TDT_T$) = 0.1 is the fraction of unit sterility (unit sterility is a thermal process resulting in inactivation of microorganisms characterized by F^z_{250} = 1.0 min) accomplished in 1 min at temperature T. The term $1/TDT_T$ is called "lethal rate" and is used in a graphical integration procedure to determine thermal process time. Since lethal rate is a function only of product temperature and z, it is possible to prepare a series of tables of lethal rates; Table 3-8 gives lethal rates for a z value of 18° F. The z value for the microorganism upon which the process is based is used in the lethal rate calculations.

The procedure used in applying the improved general method requires heat penetration data and the conversion of product temperature to lethal rates. To obtain the temperature profile, the temperature-sensing device is located at the cold point and the can is sealed and placed in the retort. Steam is brought in at the desired retort temperature. The time required to bring the retort up to operating temperature must be noted. The temperature at the cold point is recorded initially and at frequent intervals thereafter. The interval is dependent on the rate at which the product heats. When the can has been processed until the cold point temperature is within 1° or 2° of retort temperature, the steam is shut off, air pressure is maintained if required, and water is added for cooling. During

TABLE 3-8

Lethal Rates[a] (Minutes at 250° F per Minute at T) for a z Value of 18° F[b]

T, Temp. (° F)	Minutes at 250°F/min at T	T, Temp. (° F)	Minutes at 250°F/min at T	T, Temp. (° F)	Minutes at 250°F/min at T
214	0.010	232	0.100	250	1.000
215	0.011	233	0.114	251	1.136
216	0.013	234	0.129	252	1.292
217	0.015	235	0.147	253	1.468
218	0.017	236	0.167	254	1.668
219	0.019	237	0.190	255	1.896
220	0.022	238	0.215	256	2.154
221	0.024	239	0.245	257	2.448
222	0.028	240	0.278	258	2.783
223	0.032	241	0.316	259	3.162
224	0.036	242	0.359	260	3.594
225	0.041	243	0.408	261	4.084
226	0.046	244	0.464	262	4.642
227	0.053	245	0.527	263	5.275
228	0.060	246	0.600	264	5.995
229	0.068	247	0.681	265	6.813
230	0.077	248	0.774	266	7.743
231	0.088	249	0.880	267	8.800
				268	10.000

[a] Lethal rate $= 1/10^{(250 - T)/z}$.

[b] Reprinted from Ref. [46], p. 269, by courtesy of AVI Publ Co.

cooling, temperatures also should be recorded. The same cooling procedure should be followed in the test as is to be used in the actual process. The data obtained consist of a series of cold point temperatures at various times. From the product temperatures lethal rate values can be calculated by Eq. (3-22) or obtained from such tables as Table 3-8.

The lethal rate values are now plotted on arithmetic graph paper vs process time. A typical curve known as a lethal rate curve is shown in Figure 3-8. The

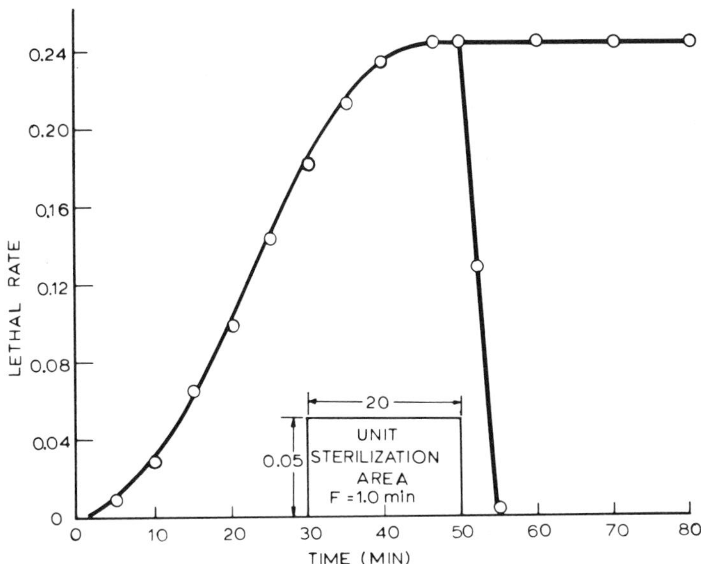

FIG. 3-8. Lethal rate curve.

area under the curve corresponding to a value of one is called a "unit sterility area" and it represents a process equivalent to 1 min at 250°F (F_{250}^{z} = 1 min). Since most processes based on microbial inactivation have F_{250}^{z} values greater than 1 min, the total thermal process must result in an area F times the unit sterility area. The graphical integration procedure enables us to readily find the area and calculate process time. The area under the curve can be estimated either with a planimeter or by counting squares.

Since it is improbable that the steam would be shut off and cooling would be started to yield exactly a process equivalent to commercial sterility, it is usually necessary to interpolate to other steam-off times. For convection heating products this is done by drawing in a cooling curve parallel to the determined cooling curve. Since the lethality accomplished in cooling is counted in the process, a simple procedure for determining steam-off time is to draw two or three cooling curves parallel to the measured cooling curve. Then by counting the total area under each heating/cooling process and by knowing steam shut-off time, a plot can be made of total area versus steam shut-off time. The appropriate steam shut-off time can be easily determined from this figure.

For conduction heating products, this procedure for determining steam shut-off time should not be used. In these products, significant thermal lag usually occurs at the start of cooling, and if the center of the product is not near retort temperature at the time of steam off, the center temperature continues to rise. This additional temperature rise can contribute significantly to lethality since the lethal rates are large at high temperatures (lethal rate = 1.0 at 250° F). Since the additional temperature rise is not easily predicted it is necessary to run cooling curves at several different steam shut-off times for conduction heating products.

Thermal processes are always reported based on steam shut-off time. Since cooling lethality is counted in the total lethality of the process it is necessary to state conditions for cooling and to use cooling procedures in the test that actually are to be used in practice.

Since the improved general method is so popular, a special graph paper has been developed for plotting lethal rate curves. Stumbo [58] has described its use.

C. Formula Method

Equation (3-20) is the basis of the formula method.

$$B = f_h \log (j_h I_h / g) \tag{3-20}$$

It states that process time (B) can be calculated if the following are known: (1) slope of the heating line (f_h), (2) retort temperature minus initial product temperature ($RT - T_{ih}$), (3) the factor identifying the thermal lag [$j_h = (RT - T_{pih})/RT = T_{ih})$], and (4) retort temperature minus cold point temperature at the end of the heating process (g). The first three parameters are either known from conditions of the process or can be obtained from a heat penetration curve. The fourth factor, g, is not easily obtained.

In developing the formula method, Ball [6] recognized that the g value obtained for a product and process was a function of several factors: (1) slope of the heating curve (f_h), (2) time of the process (expressed as the TDT of the microorganism at retort temperature), (3) z value, and (4) driving force for cooling (retort temperature minus cooling water temperature).

Ball related the g value to the dimensionless ratio f_h/F^z_{RT} for various z values and various driving forces for cooling. Since most processes are given in terms of the F^z_{250} value it was necessary to calculate the F^z_{RT} by knowing the TDT at 250°F. This was done as follows. Ball chose to call "U" the TDT at the retort temperature. The relationship between U and F^z_{250} can be easily comprehended from Figure 3-9. There are two TDT curves in Figure 3-9: one is the real TDT curve for the microorganism being considered and the other is a reference or phantom TDT curve. The reference TDT curve has the same slope as the real TDT curve but passes through 1 min at 250° F. In other words, the reference TDT curve represents the thermal resistance of the microorganism if F^z_{250} = 1.0 min. F^z_{RT} on the reference TDT curve is called F_i. When the TDT at retort temperature is called the U value, then the equation for the real TDT curve is:

$$\log U = \log F^z_{250} - (1/z)(RT - 250) \tag{3-23}$$

The equation for the phantom TDT curve is:

$$\log F_i = \log 1 - (1/z)(RT - 250)$$

or

$$\log F_i = -(RT - 250)/z \tag{3-24}$$

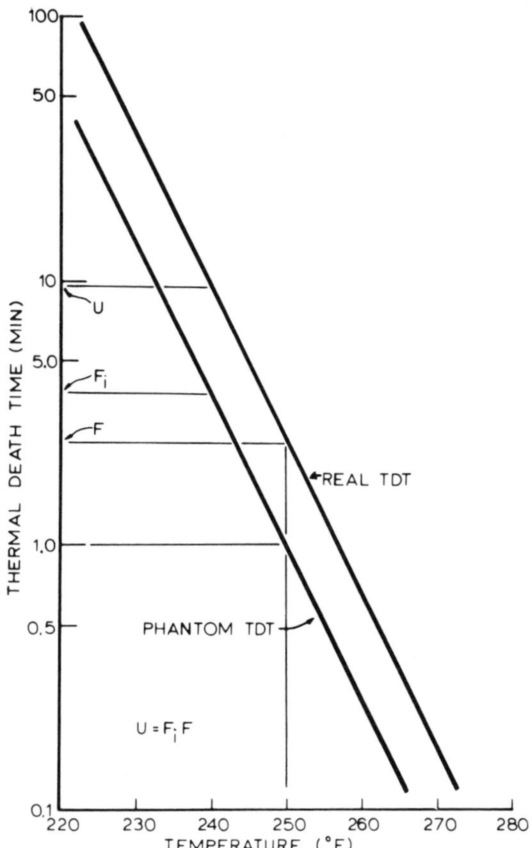

FIG. 3-9. TDT curve for defining formula method parameters.

Combining Eqs. (3-23) and (3-24) yields:

$$\log U = \log F_{250}^{z} + \log F_i \quad \text{or} \quad U = F_i F_{250}^{z} \tag{3-25}$$

Tables giving F_i as a function of z and retort temperature are available in Stumbo [58]. Thus U can be readily calculated if F_{250}^{z} is known.

The fourth factor that the g value depends on is the thermal driving force for cooling. Since the temperature difference between retort temperature and cooling water temperature is RT - CT, the driving force can be expressed in terms of I_c and g. If I_c is defined (analogously to I_h) as the center temperature at the start of cooling (T_{cc}) minus the cooling water temperature (CT), then $I_c + g = RT - CT$. Stumbo and Longley [59] made a series of calculations to determine the influence of variations in $I_c + g$ on calculated lethality values of processes. It was found that on the average, a difference of 10° F in $I_c + g$ caused about a 1% difference in the calculated F_{250}^{z} value. Thus, tables published in Stumbo [58] are for one value of $I_c + g$. Stumbo chose $I_c + g = 180°F$ as the basis of his tables since that would be fairly representative of values actually experienced. If the center temperature reached 240°-250°F at the end of heating and the cooling water temperature was 60°-70°F, then $170°F < I_c + g < 190°F$. If a correction is desired, the actual F^z_{250} value can be obtained by subtracting 1% from the theoretical F^z_{250} for every 10° the $I_c + g$ value is above 180°F, and by adding 1% for every 10° the $I_c + g$ value is below 180°F.

Stumbo and Longley [59] have suggested the incorporation of another parameter in establishing the $f_h/U:g$ charts, and this parameter relates to the cooling lag factor (j_c). Lag in temperature response of the product center can have a significant effect on overall process lethality since it occurs at temperatures where the lethal rate is very large. Consequently Stumbo and Longley [59] presented their f_h/U tables with j_c as an additional parameter. The j_c value is defined as

$$j_c = \frac{CT - T_{pic}}{CT - T_{ic}} \qquad (3\text{-}26)$$

and is determined from the heat penetration curve. Table 3-9 gives the $f_h/U:g$ correlations for various j_c values at $z = 18°F$. Extensive tables are given in Stumbo [58].

For a better grasp of the formula method an example will be given. A particular product is to be processed under the following conditions:

$T_{ih} = 140°F$

$RT = 240°F$

$z = 18°F$

$F^z_{250} = 7.0$ min

From a heat penetration test the following data were obtained:

$f_h = f_c = 25$ min

$j_h = 2.00$

$j_c = 1.40$

Calculate the process time.

Since process time is desired, Eq. (3-20) must be solved.

$$B = f_h \log (j_h I_h/g)$$

where $I_h = (RT - T_{ih}) = 240° - 140° = 100°F$.

To find g, calculate $U = F^z_{250} F_i$. From F_i tables in Stumbo [58, p. 210] at retort temperature of 240°F, and $z = 18°F$, $F_i = 3.594$. Therefore $U = (7)(3.594) = 25.1$ min.

From Table 3-9 for $f_h/U = 1.0$ and $j_c = 1.40$, $g = 0.608°F$ and $B = (25) \log [(2.00)(100)/0.608] = 25 \log 328.9 = 25(2.52)$. Thus, $B = 62.9$ min.

This is the process time required to achieve a lethality equivalent to $F^z_{250} = 7.0$ min. If the process time had been given, it would have been possible to solve for the F^z_{250} value by working the problem in reverse.

The calculated process time is applicable to all processes wherein the product at the start of the process is immediately put in contact with steam at the retort temperature. This is the case when continuous retorts are used. However, in batch retorts there may be a time lag in getting the retort to process temperature. In these circumstances the calculated process time is not the same as the operator's time. In this instance, the retort operator usually times the process from the time the retort reaches retort temperature. To correct for the lag, Ball [6] has determined that approximately 40% of the lag time contributes to the heating process and thus $0.4\,\ell$ (where ℓ = lag time) actually contributes to process time. Therefore, the process time (P_t) in a batch retort should be:

$$P_t = B - 0.4\ell \qquad (3\text{-}27)$$

The formula method utilizing Eq. (3-20) is applicable only to food products having linear heating curves on a semilogarithmic plot (simple heating curve). A method for calculating process times for products having broken heating curves is given in Stumbo [58] and is based on the same principles.

Equation (3-20) also gives the process time based on the assumption that every point receives an integrated lethal effect at least as great as that received by the cold point. This assumption is certainly true in products which heat by convection. For conduction heating products, however, this assumption is not necessarily true. Consider, for example, product with a temperature profile such as that shown in Figure 3-10. The center temperature may be considerably lower than retort temperature at the end of heating time. At the start of cooling the outer portion starts to cool but the center still sees a thermal driving force for influx of heat. Thus, an actual temperature inversion may be experienced in the cross-section of the container. Each point in the cross-section receives a different thermal process. Teixeira et al. [63] have shown that the location of the "least lethal point" in a conduction heating product is a function of container geometry and processing conditions. However, in thermal process calculations, the center temperature still can be regarded as the cold point since the probability of survival of a microorganisms does not vary appreciably within a small zone surrounding the center point. Nevertheless, innoculated pack experiments should be used to validate use of the calculated process.

When it is desirable to know average lethality for the entire container of a conduction heating food (e.g., determining average retention of nutrient or quality factors or probability of microbial survival), then the temperature profile must be considered. Several investigators have considered this problem and have developed suitable procedures [21, 25, 27, 31, 37, 58, 64]. The calculations are complex and generally require computer facilities.

The preceding discussion has centered on calculation of thermal processes based on microorganisms. The same procedures can be applied for calculating process times for destruction of enzymes and nutrients, provided the appropriate data are known. For nutrient destruction calculations, the process time is generally known, and it is desired to calculate the percent nutrient remaining. The formula method can be used by substituting D^z_{250} for F^z_{250}. The value obtained is the maximum retention since the process is based on the point receiving the least

TABLE 3-9

f_h/U : g Relationships when z = 18[a]

f_h/U	Values of g when j of cooling curve is: 0.40	0.60	0.80	1.00	1.20	1.40	1.60	1.80	2.00
0.20	4.09-05[b]	4.42-05	4.76-05	5.09-05	5.43-05	5.76-05	6.10-05	6.44-05	6.77-05
0.30	2.01-03	2.14-03	2.27-03	2.40-03	2.53-03	2.66-03	2.79-03	2.93-03	3.06-03
0.40	1.33-02	1.43-02	1.52-02	1.62-02	1.71-02	1.80-02	1.90-02	1.99-02	2.09-02
0.50	4.11-02	4.42-02	4.74-02	5.06-02	5.38-02	5.70-02	6.02-02	6.34-02	6.65-02
0.60	8.70-02	9.43-02	1.02-01	1.09-01	1.16-01	1.23-01	1.31-01	1.38-01	1.45-01
0.70	0.150	0.163	0.176	0.189	0.202	0.215	0.228	0.241	0.255
0.80	0.226	0.246	0.267	0.287	0.308	0.328	0.349	0.369	0.390
0.90	0.313	0.342	0.371	0.400	0.429	0.458	0.487	0.516	0.545
1.00	0.408	0.447	0.485	0.523	0.561	0.600	0.638	0.676	0.715
2.00	1.53	1.66	1.80	1.93	2.07	2.21	2.34	2.48	2.61
3.00	2.63	2.84	3.05	3.26	3.47	3.68	3.89	4.10	4.31
4.00	3.61	3.87	4.14	4.41	4.68	4.94	5.21	5.48	5.75
5.00	4.44	4.76	5.08	5.40	5.71	6.03	6.35	6.67	6.99
6.00	5.15	5.52	5.88	6.25	6.61	6.98	7.34	7.71	8.07
7.00	5.77	6.18	6.59	7.00	7.41	7.82	8.23	8.64	9.05
8.00	6.29	6.75	7.20	7.66	8.11	8.56	9.02	9.47	9.93
9.00	6.76	7.26	7.75	8.25	8.74	9.24	9.74	10.23	10.73
10.00	7.17	7.71	8.24	8.78	9.32	9.86	10.39	10.93	11.47
15.00	8.73	9.44	10.16	10.88	11.59	12.31	13.02	13.74	14.45
20.00	9.83	10.69	11.55	12.40	13.26	14.11	14.97	15.82	16.68
25.00	10.7	11.7	12.7	13.6	14.6	15.6	16.5	17.5	18.4
30.00	11.5	12.5	13.6	14.6	15.7	16.8	17.8	18.9	19.9
35.00	12.1	13.3	14.4	15.5	16.7	17.8	18.9	20.0	21.2
40.00	12.8	13.9	15.1	16.3	17.5	18.7	19.9	21.1	22.3
45.00	13.3	14.6	15.8	17.0	18.3	19.5	20.8	22.0	23.2

50.00	13.8	15.1	16.4	17.7	19.0	20.3	21.6	22.8	24.1
60.00	14.8	16.1	17.5	18.9	20.2	21.6	22.9	24.3	25.7
70.00	15.6	17.0	18.4	19.9	21.3	22.7	24.1	25.6	27.0
80.00	16.3	17.8	19.3	20.8	22.2	23.7	25.2	26.7	28.1
90.00	17.0	18.5	20.1	21.6	23.1	24.6	26.1	27.6	29.1
100.00	17.6	19.2	20.8	22.3	23.9	25.4	27.0	28.5	30.1
150.00	20.1	21.8	23.5	25.2	26.8	28.5	30.2	31.9	33.6
200.00	21.7	23.5	25.3	27.1	28.9	30.7	32.5	34.3	36.2
250.00	22.9	24.8	26.7	28.6	30.5	32.4	34.3	36.2	38.1
300.00	23.8	25.8	27.8	29.8	31.8	33.7	35.7	27.7	39.7
350.00	24.5	26.6	28.6	30.7	32.8	34.9	37.0	39.0	41.1
400.00	25.1	27.2	29.4	31.5	33.7	35.9	38.0	40.3	42.3
450.00	25.6	27.8	30.0	32.3	34.5	36.7	38.9	41.2	43.4
500.00	26.0	28.3	30.6	32.9	35.2	37.5	39.8	42.1	44.4
600.00	26.8	29.2	31.6	34.0	36.4	38.8	41.2	43.6	46.0
700.00	27.5	30.0	32.5	35.0	37.5	39.9	42.4	44.9	47.4
800.00	28.1	30.7	33.3	35.8	38.4	40.9	43.5	46.0	48.6
900.00	28.7	31.3	34.0	36.6	39.2	41.8	44.4	47.0	49.7
999.99	29.3	31.9	34.6	37.3	39.9	42.6	45.3	47.9	50.6

[a] Reprinted from Ref. [58], p. 260, by courtesy of Academic Press.

[b] 4.09-05 means 4.09×10^{-5}.

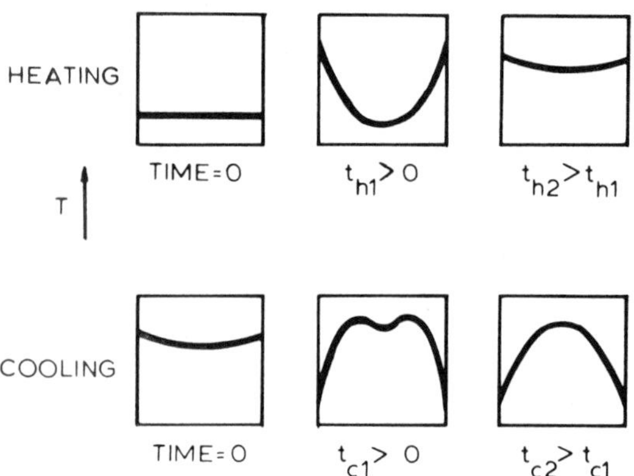

FIG. 3-10. Temperature profiles in conduction heating foods.

thermal process. For other methods which have been developed for estimating average nutrient retention see Stumbo [58], Jen et al. [31], and Hayakawa [25].

In discussing the dependence on temperature of the reaction rate constant for thermal destruction, it was pointed out that the Arrhenius equation was applicable. However, both the improved general method and the formula method are based on the empirical TDT curve, and little attempt has been made to develop a system based on the Arrhenius equation. Recently, however, Lenz and Lund [37] have presented a method using the Arrhenius equation which is particularly applicable for conduction heating foods.

VI. RECENT DEVELOPMENTS AND PROCESSES

The objective of extending the storage life of foods through thermal processing is to make available a varied diet that provides adequate nutrients. Thus the thermal process ultimately used should meet two criteria: (1) it must accomplish the process objectives from a microbial or enzymatic standpoint and (2) it should yield a maximum retention of nutrients. Although a detailed analysis of the optimization of thermal processes for nutrient retention is beyond the scope of this discussion, some general observations on optimization of blanching, pasteurization, and commercial sterilization processes can be made. For this purpose, the relative heat resistances of nutrients, enzymes, and microorganisms given in Table 3-7 are used along with a consideration of the objectives of the thermal process.

A. Optimization of Thermal Processes for Nutrient Retention

In blanching, where enzyme inactivation is desired, it is difficult to predict an optimum thermal process since the z values for thermally resistant enzymes, nutrients, and quality factors are approximately the same. Thus, blanching at higher temperature for a shorter time may not have an advantage over blanching at lower temperatures and longer times. However, in the blanching operation thermal degradation of nutrients and quality factors may not be the primary mechanism of loss. For example, blanching in hot water can result in a considerable loss of nutrients due to leaching. Similarly, losses due to oxidation can result during blanching in hot air. Considering leaching and oxidative losses, HTST would result in greater nutrient retention since contact time, not temperature, is the more important factor.

For pasteurization and commercial sterilization, optimization of the thermal process is important since microbial inactivation has a significantly different temperature dependence than destruction of nutrients.

For foods or food fluids which are pasteurized, high-temperature and short-time processes results in maximum nutrient retention. This occurs because a given increase in process temperature accelerates microbial destruction more than nutrient degradation.

For commercial sterilization, optimization of a thermal process is not so straightforward. The mode of heating within the product becomes an important factor. For those products which heat by convection, the high-temperature and short-time process results in optimum nutrient retention. Again, a comparison of z values indicates that the rate of destruction of nutrients is less temperature dependent than is the rate of destruction of microbial spores. Therefore, the high-temperature and short-time process favors nutrient retention. However, as the temperature is raised and the process time is shortened, a point is eventually reached where the process destroys microbial spores but does not destroy heat-resistant (HR) enzymes. This occurs because of a difference in reaction rate constants for the destruction of HR enzymes as compared to spores. At low temperature, the destruction rate for HR enzymes is greater than it is for spores, but as the temperature is raised the destruction rate for spores increases faster than that for HR enzymes. Therefore, some high temperature X exists at which the destruction rates for the HR enzymes and spores are equal. Above temperature X the destruction rate of spores is greater than that for HR enzymes, and a thermal process based on the destruction of spores does not totally inactivate HR enzymes. Therefore, spoilage of the product due to enzyme activity may occur. In these instances, the process must be based on inactivation of HR enzymes rather than on inactivation of spores. The temperature at which the destruction rate for HR enzymes equals that for spores is generally in the range of $270^\circ - 290^\circ$F. For products that are processed above these temperatures, it is difficult to accurately calculate conditions that result in inactivation of HR enzymes and spores while causing minimum damage to nutrients and other quality attributes.

For all products which heat by convection, one important assumption is made concerning the thermal process; i.e., pieces in the brine receive the same lethal treatment as the brine. This assumption is valid when the food particulates are sterile in the center and they have large surface heat-transfer coefficients. However, food particulates do exist that are not sterile in the center (e.g., foods that contain meat balls, stuffed noodles, etc.) and which have surface heat-transfer coeffients that are small (thus, surface temperatures are significantly lower than the brine temperature). In these instances, it cannot be assumed that the lethal treatment in the particulate equals or exceeds that in the brine [30]. Under these conditions and when most of the nutrients are located in particulates, the problem of designing the thermal process for optimization is basically the same as that for conduction heating foods.

In conduction heating foods, as pointed out earlier, each point in the cross-section of the container receives a different thermal process than every other point, and the thermal process is based on the slowest heating point. At the conclusion of heating and the start of cooling, a complexity arises since the slowest heating point (center) can continue to increase in temperature for a significant period. In this situation, the overall lethality for the process is the integrated effect of the heat treatment at every point in the container. Methods developed for calculating lethality of a thermal process in conduction heating foods have been applied to optimize thermal processes for nutrient retention. All of these methods are time consuming and apply only to conduction heating foods. For present purposes, the paper of Teixeira et al. [64] is most useful. These authors presented (Fig. 3-11) a relationship between various time-temperature treatments providing equal microbial lethality and retention of nutrients with different z values. It can be seen that retention of a nutrient with a small z value is favored by a low-temperature and long-time process; whereas retention of a nutrient with a large z value is favored by a high-temperature and short-time process. The curve for $z = 45°F$, $D = 188$ min, represents thiamine destruction in green bean puree. Note that the optimum process in green bean puree is 90 min at 248° F, very close to that used commercially. More significantly, note that as the temperature of the process is raised, thiamine retention decreases rapidly. In conclusion, determining thermal processes that provide optimum nutrient retention in conduction heating foods is not simple and requires considerable computational effort.

It should be pointed out that these methods for conduction heating foods cannot be applied to those retorting and heating conditions where the product is intimately mixed, such as in the "Orbitort" and in scraped-surface heat exchangers. Under these circumstances, the product usually receives sufficient mixing so that every point in the food receives nearly the same lethal heat treatment. Consequently, a high-temperature and short-time process favors nutrient retention.

Table 3-10 summarizes the methods for optimizing blanching, pasteurization, and commercial sterilization processes for nutrient retention.

B. Process Equipment

1. Blanching

Most blanching operations are accomplished by putting the product in contact with either hot water or steam for an appropriate time. The time is dependent on the objectives of the process; that is, whether enzymes are to be inactivated or whether partial cooking is desired.

Lee [36] presented an excellent evaluation of hot water and steam blanching from a nutrient viewpoint. In general steam blanching results in greater retention of water-soluble nutrients since less leaching occurs.

Steam blanching can be carried out as a batch process; however, in most modern canneries, continuous systems are used. In steam blanching, generally the material is layered on a belt which moves the product through a steam tunnel. The product is heated by steam jets from below and above. With leafy vegetables, such as spinach, care must be taken not to overload the belt since upon heating these products tend to wilt and mat. Heat transfer through this mat is very slow and underblanching could occur.

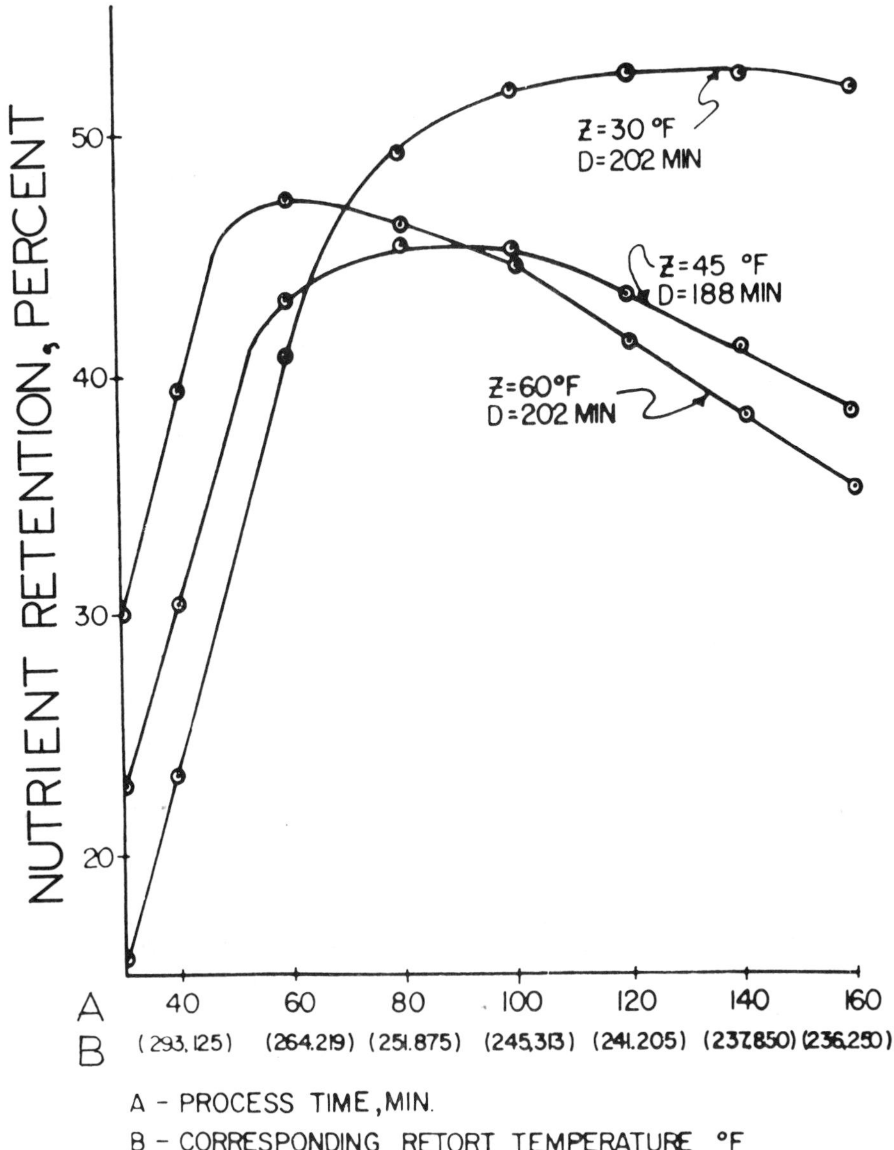

FIG. 3-11. Multi-nutrient optimization. Percent nutrient retention vs process time with corresponding retort temperature. Reprinted from Ref. [63], p. 850, by permission of Institute of Food Technologists.

TABLE 3-10

Optimization of Thermal Processes for Nutrient Retention

Process	Method of optimization
Blanching	Base on considerations other than thermal losses (e.g., leaching losses, oxidative degradation, damage to product)
Pasteurization	HTST, if heat-resistant enzymes are not present
Commercial sterilization	Convection heating foods and aseptic processing: HTST until heat-resistant enzymes become important
	Conduction heating foods: not HTST, necessarily; difficult but not impossible calculation

The thermal screw also has been used to steam blanch products. Here the product is conveyed in a trough by a closely fitting helical screw. Steam injected at regular intervals is used to heat the product. Similar designs use hot water as the transfer medium, and this reduces abrasion and damage to such sensitive products as mushrooms.

A unique steam blanching system has been developed for reducing the effluent and heating inequities that can occur in deep-bed steam blanching [35]. The system is called IQB (individual quick blanch) and consists of two steps: (1) a heating step wherein the product mass-average temperature is raised and (2) an adiabatic holding step wherein thermal gradients within the product are allowed to disappear. Effluent reduction can be further accomplished by predrying the product by 6% (w/w) prior to steam heating. This allows the steam condensate to remain on the piece and not to run off and leach solids from the product.

Water blanching also can be conducted on a batch basis by simply dipping a batch of product in hot water (usually 200°-210°F) for the required time. Continuous hot water blanchers can be screw types, drum types, or pipe types. The drum type consists of a perforated drum fitted internally with a helical screw. The residence time is regulated by the rpm of the screw and agitation is provided by action of the screw. The water can be heated by direct injection of steam into the blancher, or the water can circulate outside the blancher and be steam heated. Water blanchers of this type generally have large waste streams since fresh water must be continually added to dilute the solids content of the blanch water. Concentrations in excess of 2-3% solids in the blanching water can contribute to off flavors in the final product.

The pipe blancher can be used with solid food particles that can be pumped in water. Water and product are combined using 1-1.25 gal water per pound of product and pumped through 4-6 in. diameter pipe at a rate of approximately 3 fps. At this velocity most vegetables remain fluidized and yet agitation is not so vigorous as to damage the product. At the discharge end, the product is separated from the water at a dewatering reel and the water is recycled. Some makeup water is required so the solids content of the water does not get too high.

Hot gas blanching using combustion gas as the heat-transfer medium has been successfully applied to spinach and other vegetables [50]. The main advantage of the method is that there is practically no effluent since there is no steam condensation. In fact the product generally loses considerable water in the process and the hot gas stream must be humidified. Since partial dehydration can be accomplished in hot gas blanching the method is particularly well suited for products that are subsequently to be dried. The method is not well suited for such products as corn since the high temperatures employed can cause surface browning.

Microwave blanching also has been investigated but no commercial units are in operation [14]. Microwaves or microwaves in combination with hot water or steam give satisfactory blanching but equipment cost is prohibitive.

2. Pasteurization

Since pasteurization is normally accomplished at temperatures less than 212° F, solid food products can be pasteurized in the same type of equipment as that used for blanching. For acid food products or meat products a water bath is the simplest pasteurization equipment. Crates of the packaged food product are placed in steel tanks for heating with hot water. At the end of the process cold water is added for cooling. A continuous water bath pasteurizer is used in the pickle industry and in fruit processes. The equipment consists of a long tank through which the product moves on a belt.

Continuous water spray equipment is used for pasteurizing bottled beer and fruit juices. In this unit, the product is conveyed on a belt through several temperature zones where water is sprayed onto the containers. The zones are: first preheat, second preheat, pasteurization, precool, and final cool. Water economy is accomplished by using the water from the first preheat section to precool and then using this water again for preheating.

When glass containers are processed, care must be taken to avoid thermal shock. This can be done best by keeping the driving force for heat transfer around 70°F. For cooling, the driving force should be around 50° F.

Steam pasteurizers are also used but are more amenable to metal containers than glass because of the danger of thermal shock. Tunnels can be designed with various temperature zones by controlling the steam-air mixture in each zone. Cooling is accomplished by water sprays or by immersing the containers in water.

When metal and glass containers are cooled, it is desirable to remove them from the cooling zone at temperatures near 100° F. At these temperatures the surface water evaporates, avoiding corrosion problems during storage.

For fruit, fruit juices, and tomatoes, the continuous, agitating, atmospheric cooker is widely used. Operation of the unit is similar to the continuous, agitating, pressure cooker wherein the cans are screw conveyed through the unit. The unit operates at atmospheric pressure with either steam, hot water, or a combination of steam and hot water as the heating medium. Processing rates of up to 250-300 cans per minute can be achieved.

Pasteurization of unpackaged liquids is most commonly accomplished in indirect heat exchangers. In these units the product is passed along one side of a wall while the heating medium (steam or hot water) is passed along the other. A plate-type indirect heat exchanger is most commonly used for pasteurizing liquids. In this type of heat exchanger the product is separated from the heating medium by a thin metal plate. The plate is usually indented to promote turbulent flow, thereby resulting in large heat-transfer coefficients. Film heat-transfer coefficients on the order of 1000 to 2000 Btu hr^{-1} ft^{-2} $^{\circ}F^{-1}$ are not uncommon with relatively non-viscous fluids, such as milk. Plate heat exchangers are used extensively for pasteurization of milk and beer. Added economy in heating is achieved by a process called regenerative heating. In this process, the hot pasteurized fluid is used to partially heat the cold incoming product.

Direct injection of steam has been used to pasteurize milk. However, its use is limited because culinary steam is required and localized overheating can occur at the point of injection. Localized overheating can result in undesirable precipitation of such inverse solubility salts as $Ca_3(PO_4)_2$ or denaturation of protein.

3. Commercial Sterilization

There are two basic methods used to obtain commercial sterility in foods: (1) heating the food after it has been placed in the container and (2) heating and cooling the food and then packaging it aseptically. The first process is the conventional canning method and is, in principle, the same method that was used by Appert, the Frenchman who invented canning (conventional canning thus is sometimes referred to as "Appertizing"). The second method is generally referred to as aseptic canning. Although the principles of aseptic canning have been known for many years, the technology is rather recent.

Brody [10] has presented an excellent review of rigid containers and flexible packages used for thermal processing.

a. Conventional Thermal Processing. The conventional canning operation consists of (1) preparing the food (cleaning, cutting, grading, blanching, etc.), (2) filling the container, (3) sealing the container, and (4) subjecting the container to a heat-cool process sufficient for commercial sterility. There are several methods of accomplishing the thermal process and all are governed by principles of heat transfer and sterilization that have been discussed above. Since it is desirable to increase production rates while causing minimal damage to product quality, most recent developments have dealt with improved rates of heating.

The still retort is the oldest type of equipment used for thermal processing. It is currently used in large canning plants for processing large containers of conduction heating foods and for glass packed products, and in small canning plants for all types of products.

Basically, the method consists of loading crates of containers into an appropriately designed iron vessel (retort), closing the vessel, and heating the containers with steam. The temperature is regulated by a controller and the length of the process is determined by the rate of heat transfer into the containers. Cooling is accomplished by shutting off the steam and adding cold water to the retort. For large metal containers and glass jars with press-on lids, overriding air pressure must be used at the start of cooling because the pressure difference between the inside and outside of the container can otherwise cause, respectively, buckled cans and dislodged lids.

Introduction of steam into the retort should be done with care since it is necessary to displace all of the air in the retort. Presence of air during thermal processing can result in under processing since steam-air mixtures result in lower heat-transfer rates than steam or water alone.

Still retorts are either "horizontal" or "vertical" and are batch operated. A vertical retort is shown in Figure 3-12. Design and operation must meet approved standards and each canning plant must have personnel present at all times who have received specific training in retort operation. Procedures have been established for reprocessing food that has been underprocessed.

Improvements in batch-type retort systems have centered on the mechanics of handling the containers. In a conventional system, containers are loaded into crates for easy handling. To circumvent the container handling system, a "crateless" retort has been developed. The retort is filled with water which acts as a cushion for cans which are added by means of a conveyor. Steam is used to remove the water after the retort has been closed. After they are processed, the cans are deposited onto a shaking screen and are then conveyed to an unscrambler. The primary advantage of this system is the reduction in labor costs.

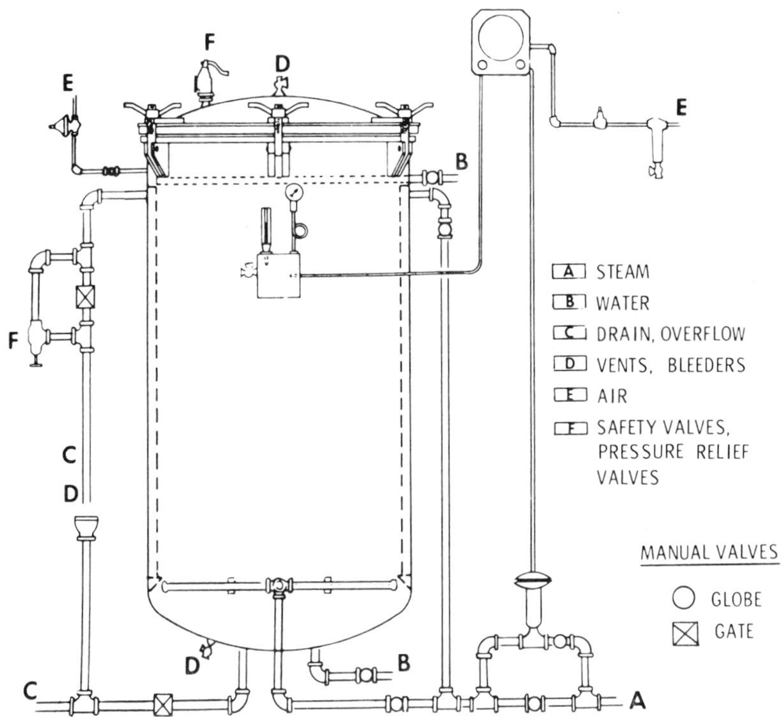

FIG. 3-12. Vertical retort.

Continuous retorts offer two distinct advantages over batch-type retorts: (1) the production rate is greater and the labor cost is lower, and (2) the rate of heat transfer is greater in products which transmit heat energy primarily by convection. The second advantage occurs because the product is agitated as it moves through the retort.

There are currently at least four types of continuous retort systems that use steam as the heat-transfer medium. Trade names or identifying terms for the processes are: (1) Sterilmatic, (2) Orbitort, (3) Hydrostatic, and (4) Hydrolock.

In the Sterilmatic system cans are admitted into the retort through a mechanical pressure lock and are conveyed through the retort by means of a horizontal, helical spiral. The cans remain at the periphery of the helix and each can rotates about its own axis for a portion of each rotation. By operating the reel at 5 rpm each can rotates about 100 rpm around its own axis as it moves through the retort. This rotation provides the agitation and improved heat transfer mentioned above.

The Orbitort is similar to the Sterilmatic except it is used with more viscous products such as cream-style corn. Once the retort is loaded, each can is locked into position. That is, each can is prevented from rotating about its own axis. At a critical reel speed of about 35 rpm, the headspace bubble in the can follows an elliptical path through the center of the container significantly increasing the rate of heat transfer. This operation is particularly suited to retorting No. 10 (603 x 700) cans of cream-style corn since Sterilmatic or batch cookers result in long process times and relatively poor product quality.

The Hydrostatic cooker employs a unique method of confining high-pressure steam. Steam is confined and controlled by adjustable water legs at the entrance and exit of the retort (Fig. 3-13). For each 14.7 psi, a water column approximately 34 ft high is required. In operation, a conveyor chain elevates the containers in air in the entrance leg and then lowers them through a U-shaped water leg. Water temperature at the entrance of the leg is close to 180° F and at the exit is 225° - 245°F. As the containers move down the leg, the temperature gradually increases, making this system particularly attractive for containers subject to thermal shock. After the containers pass the bottom of the "U," they soon enter the steam chamber where they reside for the appropriate length of time. The can must travel at least two passes in the steam chamber (one up and one down) and then exit through another U-shaped water leg. To aid heat transfer, equipment manufacturers have provided means of agitating the containers as they travel through the cooker.

The features of a hydrostatic cooker also are available in a unique system that overcomes the need for tall water legs [18]. The system is called a "helical pump" and is based on the principle of additive pressure in the legs of a rotating coil (Fig. 3-14). Although this system is not currently in commercial use, it offers interesting possibilities for processing rigid containers and pouches.

The Hydrolock system provides continuous agitation of containers by rolling them along tracks (Fig. 3-15). The containers enter and leave through the same pressure lock, which maintains pressure partly by a water seal and partly by a mechanical seal. The versatility of the Hydrolock system makes it suitable for either rigid containers or flexible pouches.

Direct flame sterilization of food in rigid containers has been successfully achieved with cans of small sizes. In this system the cans are conveyed on a live conveyor (each can rotates as it is conveyed) across a bank of gas burners operating at 2000° F. Since there is no overpressure, the containers must be able to

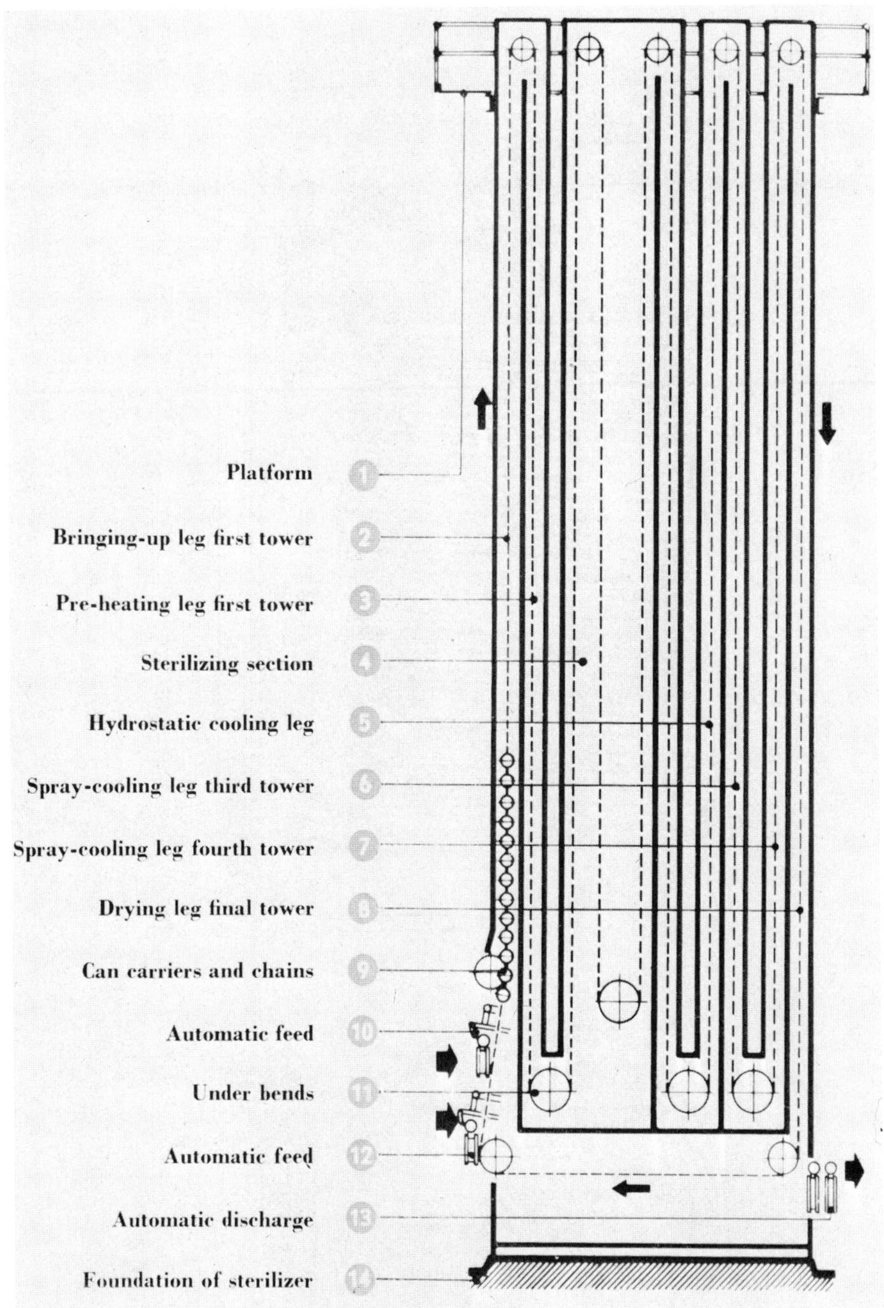

FIG. 3-13. Hydrostatic cooker. Reprinted by permission of Chemetron Corporation.

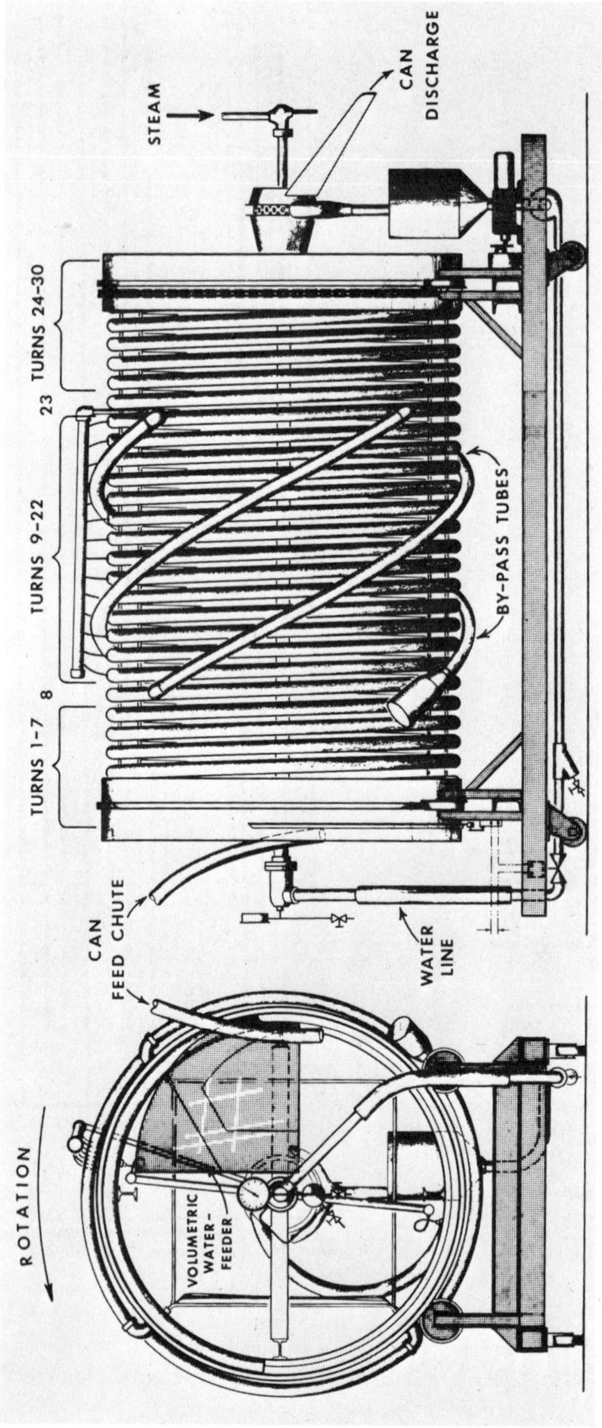

FIG. 3-14. Detailed view of helical pump sterilizer. Reprinted from Ref. [54], p. 1252, by permission of Institute of Food Technologists.

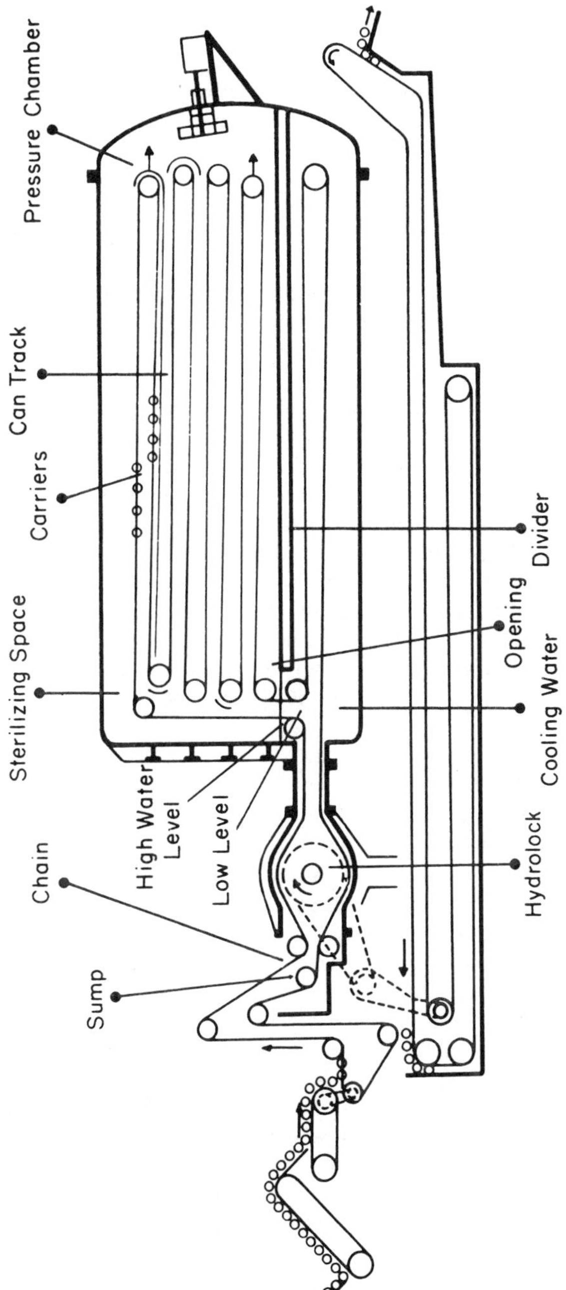

FIG. 3-15. Hydrolock sterilizer. Reprinted from Ref. [34], p. 75, by permission of Food Engineering.

withstand internal steam pressures of 10-25 psig. Since cans larger than 211 or 300 diameter cannot withstand this amount of internal pressure, applications of this system are somewhat limited. With mushrooms, extremely high heat-transfer rates are attained, resulting in decreased process times and improved product quality.

Although it has not been used commercially, a fluidized bed retort has been developed [39]. Sand or ceramic pellets heated by a gas flame and mixed with rapidly moving air is used to fluidize cans. Damage to the seals and surface of the container are unsolved problems.

b. Aseptic Processing. Sterilizing product prior to packaging involves the same principles as discussed with respect to pasteurization prior to packaging. The only differences between aseptic commercial sterilization and pasteurization are the severity of the thermal process and the temperatures used. Since very high temperatures are employed (270°-350°F) in aseptic commercial sterilization, these processes are generally referred to as UHT (ultrahigh temperature) processes. Processing at these temperatures usually requires that the process be based on enzyme inactivation rather than microbial destruction since some enzymes exhibit greater heat resistance in this temperature range than do microorganisms.

Heat-exchange equipment used for aseptic processing include scraped surface heat exchangers, plate heat exchangers, tubular heat exchangers, and equipment involving direct steam injection [3]. Systems have been developed for liquids and for liquids containing small particles, but success has not yet been achieved with fluids that contain large particles (peas or corn in brine). The primary problems associated with particles are development of a pumping system which does not damage the particles and assuring a failsafe residence time. Since UHT processing is on the order of seconds, the residence time must be precisely controlled to avoid underprocessing.

Direct steam injection has been used to sterilize milk for canning [12]. The process is called Uperization and has found use in those countries where refrigeration is not readily available. The condensate is flashed off in the cooling step so there is no permanent dilution of the product.

Recent developments in aseptic processing generally have centered on a post-sterilization packaging operation. One system, Flash 18, is used for filling solid foods which have been directly sterilized in steam. The operation is carried out in a room pressurized to 18 psig with air. Since the boiling point of water at 18 psig is 255°F, product with temperatures of 255°F can be handled without boiling. After the product is heated at 255°F with steam, the product is filled into nonsterile cans under flowing steam, sealed, and held for a time sufficient to achieve sterilization. The cans are then cooled and transferred out of the pressurized chamber. The system has been applied successfully to sterilization of meat and fish salads; however, the equipment is expensive and it is difficult to maintain a staff willing to work under the unusual pressure conditions.

A discussion of continuous aseptic packing of sterile products is beyond the scope of this chapter. Brody [10] has presented an excellent review of various current systems.

SYMBOLS

a	initial number of microorganisms
b	number of microorganisms after thermal processing
B	thermal process time (min)
c	concentration (number of organisms per unit volume or moles or grams per unit volume)
CT	cooling water temperature (°F)
D	decimal reduction time (min)
E_a	Arrhenius activation energy (cal/mole or calories per organism)
f	time (min) for the heat penetration curve to traverse one log cycle (min)
F	thermal death time (min)
F_o	thermal death time at 250°F for organisms with z = 18° (min)
g	retort temperature-product cold point temperature at the end of the heating cycle (°F)
I	retort temperature-initial product temperature = $(RT - T_i)$ (°F)
j_c	lag factor for cooling = $(CT - T_{pic})/(CT - T_{ic})$
j_h	lag factor for heating = $(RT - T_{pih})/(RT - T_{ih})$
k	first-order reaction rate constant (min^{-1} or hr^{-1})
l	lag time for heating the retort (min)
P_t	processor's time (min)
PT	product temperature (°F)
Q_{10}	rate of reaction at (T + 10)°C/rate of reaction at T°C
R	gas law constant (cal/K mole
RT	retort temperature (°F)
s	Arrhenius frequency factor (sec^{-1})
t	time (min or hr)
T	temperature (°C, °F, K, or °R)
T_c	center or cold point temperature = PT (°F)
T_i	initial product temperature (°F)
T_{pi}	pseudo-initial product temperature (°F)
T_s	surface temperature = RT (°F)
TDT	thermal death time (min)
U	thermal death time at retort temperature (min)
z	°F temperature change required to change the TDT by a factor of 10 (°F)

Subscripts

c	cooling center or cold point
h	heating
T	temperature (°F)

Superscripts

z	see above

REFERENCES

1. H. W. Adams and E. S. Yawger, Food Technol., 15, 314 (1961).

2. E. B. Anderson and L. J. Meanwell, J. Dairy Res., 3, 182 (1936).

3. Anonymous, Technical Digest Cb-201, 2nd ed., Cherry Burrell Corp., Chicago, Ill., 1969.

4. F. E. Atkinson and C. C. Strachan, Can. Dept. Agr. Exptl. Farms Service. Dominion Exptl. Sta. Progr. Repts., 63 (1951).

5. C. O. Ball, Univ. Calif. Publ. Public Health, 1, 230 (1928).

6. C. O. Ball, Bull. Natl. Res. Council, 7(1), No. 37, 7 (1923).

7. C. O. Ball and F. C. W. Olson, Sterilization in Food Technology, McGraw-Hill, New York (1957).

8. G. H. Bendix, D. G. Heberlein, L. R. Ptak, and L. E. Clifcorn, J. Food Sci., 16, 494 (1951).

9. W. D. Bigelow, J. Infect. Dis., 29, 528 (1921).

10. A. L. Brody, Crit. Rev. Food Technol., 2, 187 (1971).

11. H. Burton, J. Dairy Res., 30, 217 (1963).

12. H. Burton and A. G. Perkins, J. Dairy Res., 37, 309 (1970).

13. J. Davidek, J. Velisek, and G. Janicek, J. Food Sci., 37, 789 (1972).

14. W. C. Dietrich, C. C. Huxsoll, J. R. Wagner, and D. G. Guadagni, Food Technol., 47, 87 (1970).

15. P. R. Elliker and W. C. Frazier, J. Bacteriol., 36, 83 (1938).

16. J. B. Enright, W. W. Sadler, and R. C. Thomas, J. Milk Food Technol., 19, 313 (1956).

17. J. R. Esty and K. F. Meyer, J. Infect. Dis., 31, 650 (1922).

18. D. F. Farkas, M. E. Lazar, and W. C. Rockwell, Food Technol., 23, 180 (1969).

19. E. Feliciotti and W. B. Esselen, Food Technol., 11, 77 (1957).

20. E. R. Garrett, J. Amer. Pharm. Assoc., 45, 171 (1956).

21. T. G. Gillespy, in Recent Advances in Food Sciences (J. Hawthorn and J. M. Leitch, eds.), Vol. II, Butterworths, London (1962), pp. 93-105.

22. H. J. Gold and K. G. Weckel, Food Technol., 13, 281 (1959).

23. S. M. Gupte, A. M. El-bisi, and F. J. Francis, J. Food Sci., 29, 379 (1964).

24. R. S. Harris and H. von Loesecke (eds.), Nutritional Evaluation of Food Processing, AVI Publ. Co., New York, 1960.

25. I.-K. Hayakawa, Can. Inst. Food Technol. J., 2, 165 (1969).

26. J. Herrmann, Ernahrungsforschung, 15, 279 (1970).

27. E. W. Hicks, Food Technol., 6, 175 (1952).

28. L. R. Hackler, J. P. van Buren, K. H. Steinkraus, I. El Rawi, and D. B. Hand, J. Food Sci., 30, 723 (1965).

29. M. Ingram, in The Bacterial Spore (G. W. Gould and A. Hurst, eds.), Academic Press, New York, 1969, Chapter 15, pp. 549-610.

30. R. A. Jacobs, L. L. Kempe, and N. A. Milone, J. Food Sci., 38, 168 (1973).

31. Y. Jen, J. E. Manson, C. R. Stumbo, and J. W. Zahradnik, J. Food Sci., 36, 692 (1971).

32. L. B. Jensen, Microbiology of Meats, 2nd ed., Garrard Press, Champaign, Ill., 1945.

33. L. I. Katzin, L. A. Sandholzer, and M. E. Strong, J. Bacteriol., 45, 265 (1943).

34. F. K. Lawler, Food Eng., 39(7), 73 (1967).

35. M. E. Lazar, D. B. Lund, and W. C. Dietrich, Food Technol., 25, 684 (1971).

36. F. A. Lee, Advan. Food Res., 8, 63 (1958).

37. M. K. Lenz and D. B. Lund, unpublished work, 1974.

38. J. J. Licciardello, C. A. Ribich, J. T. R. Nickerson, and S. A. Goldblith, Appl. Microbiol., 15, 344 (1967).

39. A. Lopez, A Complete Course in Canning, The Canning Trade, Baltimore, Md., 1969.

40. G. Mackinney and M. A. Joslyn, J. Amer. Chem. Soc., 63, 2530 (1941).

41. T. Mansfield, unpublished work, 1974.

42. E. A. Mulley, C. R. Stumbo, and W. M. Hunting, Paper presented at the 34th Annual Meeting of the Institute of Food Technologists, New Orleans, 1974.

43. National Canners Association, Laboratory Manual for Food Canners and Processors, Vol. I. Microbiology and Processing, AVI Publ. Co., Westport, Conn., 1968.

44. F. C. W. Olson and H. P. Stevens, Food Res., 4, 1 (1939).

45. G. T. Peterson, Methods of Producing Vacuum in Cans, Bull. No. 18, Continental Can Company, 1947.

46. I. J. Pflug and W. B. Esselen, in Fundamentals in Food Processing Operations, (J. L. Heid and M. A. Joslyn, ed.), AVI Publ. Co., Westport, Conn., 1967, (Chapter 11, pp. 258-327.

47. I. J. Pflug and C. F. Schmidt, in Disinfection, Sterilization and Preservation, (C. A. Lawrence and S. S. Block, eds.), Lea and Febiger, Philadelphia, 1968, Chapter 6, pp. 63-105.

48. J. D. Ponting, D. W. Sanshuck, and J. E. Brekke, J. Food Sci., 25, 471 (1960).

49. O. Rahn, Bacteriol. Review, 9, 1 (1945).

50. J. W. Ralls, H. J. Maagdenberg, N. L. Yacoub, D. Hommick, M. Zinnecker, and W. A. Mercer, J. Food Sci., 38, 192 (1973).

51. T. V. Ramakrishnan and F. J. Francis, J. Food Sci., 38, 25 (1973).

52. R. B. Read, Jr. and J. G. Bradshaw, J. Dairy Sci., 49, 202 (1966).

53. R. A. Roberts and A. D. Hitchers, in The Bacterial Spore (G. W. Gould and A. Hurst, eds.), Academic Press, New York, 1969, Chapter 16, pp. 611-670.

54. W. C. Rockwell and D. F. Farkas, Food Technol., 25, 1250 (1971).

55. A. D. Russell, in Inhibition and Destruction of the Microbial Cell (W. B. Hugo, ed.), Academic Press, New York, 1971, Chapter 11, pp. 451-612.

56. O. T. Schultz and F. C. W. Olson, Food Res., 5, 399 (1940).

57. C. N. Stark and P. Stark, J. Bact., 18, 333 (1929).

58. C. R. Stumbo, Thermobacteriology in Food Processing, 2nd ed., Academic Press, New York, 1973.

59. C. R. Stumbo and R. E. Longley, Food Technol., 20, 109 (1966).

60. H. Sugiyama, J. Bact., 62, 81 (1951).

61. H. Taira, H. Taira, and Y. Sukurai, Jap. J. Nutr. Food, 18, 359 (1966).

62. S. S. Tanchev and N. Joncheva, Unters. Forsch, 153, 37 (1973).

63. A. A. Teixeira, J. R. Dixon, J. W. Zahradnik, and G. E. Zinsmeister, Food Technol., 23, 352 (1969).

64. A. A. Teixeira, J. R. Dixon, J. W. Zahradnik, and G. E. Zinsmeister, Food Technol., 23, 845 (1969).

65. G. E. Timbers, Ph.D. Thesis, Rutgers, The State University, New Brunswick, New Jersey, 1971.

66. C. T. Townsend, J. R. Esty, and J. C. Baselt, Food Res., 3, 323 (1938).

67. J. H. von Elbe, I.-Y. Maing, and C. H. Amundson, J. Food Sci., 39, 334 (1974).

68. J. R. Whitaker, Principles of Enzymology for the Food Sciences, Marcel Dekker, New York, 1972.

69. M. P. Williams and P. E. Nelson, J. Food Sci., 39, 457 (1974).

GENERAL REFERENCES

A. E. Bender, Nutritional effects of food processing. J. Food Technol., 1, 261 (1966).

A. L. Brody, Updates aseptic processing. Part I, Food Eng., 45, 94; Part II, Food Eng., 45, 108 (1973).

H. Cheftel and G. Thomas. Principles and Methods for Establishing Thermal Processes for Canned Foods, Bull. No. 14, Israel Program for Scientific Translations, IPST Cat. No. 1385, USDA-NSF, Office of Technical Services, U.S. Department of Commerce, Washington, D.C., 1963.

R. R. Ernst, Sterilization by Heat, in Disinfection, Sterilization and Preservation (C. A. Lawrence and S. S. Block, eds.), Lea and Febiger, Philadelphia, 1968, Chapter 43, pp. 703-740.

S. A. Goldblith, Thermal processing of foods: A review. World Rev. Nutr. Diet., 13, 165 (1971).

S. A. Goldblith, M. A. Joslyn, and J. T. R. Nickerson (eds.), Introduction to Thermal Processing of Foods, AVI Publ. Co., Westport, Conn., 1961.

F. H. Johnson, H. Eyring, and M. J. Polissar, Kinetic Basis of Molecular Biology, John Wiley and Sons, New York, 1954.

K. Lang, Influence of cooking on foodstuffs. World Rev. Nutr. Diet., 12, 266 (1970).

D. B. Lund, T. P. Labuza, G. E. Livingston, C. Y. Ang, C. M. Chang, P. A. Lachance, A. S. Ranadive, and J. Matas, Symposium: Effects of processing, storage and handling on nutrient retention in foods, Food Technol., 27(1), 16 (1973).

J. H. Perry (ed.), Chemical Engineer's Handbook, 3rd ed., McGraw-Hill, New York, 1963.

H. A. Schroeder, Losses of vitamins and trace minerals resulting from processing and preservation of foods. Amer. J. Clin. Nutr., 24, 562 (1971).

F. H. Slade, Food Processing Plant, Vol. I, CRC Press, New York, 1967.

Chapter 4

RADIATION PRESERVATION OF FOODS

Marcus Karel

CONTENTS

I. INTRODUCTION

Research directed toward the use of radiation for the preservation of foods began in 1945, and several institutions have invested a great deal of effort in this potentially important method of food preservation. Massachusetts Institute of Technology; the United States Army, Quartermaster Corps; the Federal Research Center for Food Preservation in Karlsruhe, Germany; and Michigan State University are a few of the important centers of past work [15,19].

It has been estimated that the research and development efforts expended on this method of preservation have exceeded 60 million dollars [9]. In spite of this massive effort, radiation is still not a widely used preservation method. The Food and Drug Administration in the United States, prompted in part by the 1958 Food and Drug Law which classifies radiation treatment a food additive, has demanded an animal testing program exceeding any previous requirements for establishing safety of a process. These tests are currently in progress. On the basis of preliminary data the wholesomeness of irradiated foods is likely to be established in the next few years, and commercial preservation of food by ionizing radiation is likely to follow. For this reason it appears important to include in the present book a discussion of principles of use of ionizing radiations for sterilization, pasteurization, and deinfestation of foods.

II. PROPERTIES OF IONIZING RADIATIONS

A. Major Types of Ionizing Radiations

Several radiation types have the characteristic ability to ionize individual atoms or molecules, thereby producing an electron and a positively charged ion:

$$M \rightsquigarrow M^{+} + e^{-} \qquad (4\text{-}1)$$

where M is an atom or molecule, and the symbol $\rightsquigarrow$ indicates the action of radiation. Among the important radiations of either historical or practical interest in food preservation are electromagnetic waves, including x-rays and γ-rays; electrons, including β-rays and cathode rays; and α-rays. Some of the pertinent properties of these radiations are listed in Table 4-1. Of the various ionizing radiations only electromagnetic waves and electrons appear suitable for food processing. Neutrons, deuterons, and α particles cause too much damage to the food and also have a significant potential for inducing radioactivity.

B. Electromagnetic Radiations

Light, x-rays, and γ-rays are all part of the electromagnetic radiation spectrum, which also includes infrared radiation, uv radiation, and radio waves as shown in Figure 4-1. Electromagnetic radiations possess a dual nature. According to the quantum theory, they may be considered as waves with a characteristic frequency υ, a wavelength λ, and a velocity which is independent of frequency. In vacuum, the velocity is equal to 3×10^{10} cm/sec. They are also, and with equal justification, considered as packets of energy, or photons. Energy, wavelength, and frequency are related by Eqs. (4-2) and (4-3)

$$\upsilon = c/\lambda \qquad (4\text{-}2)$$

TABLE 4-1

Nature and Properties of Some Ionizing Radiations

Radiation	Nature of rays	Typical energy levels (eV)	Penetration[a]
x-Rays	Photons produced in x-ray machines	$10^4 - 10^7$	20-150
γ-Rays	Photons from radioactive isotopes	$10^5 - 2 \times 10^6$	30-100
β-Rays	Electrons from isotopes	$10^4 - 10^6$	0.1-0.5
Cathode rays	Electrons from accelerators	$10^4 - 10^7$	0.1-5
Protons	Fast protons from accelerators	MeV range	0.003
Neutrons	Neutrons from fission or isotopes	MeV range	~10
α Particles	Helium nuclei from isotopes	$10^6 - 10^7$	~0.005

[a] Depth of absorber of unit density (water) required to reduce radiation intensity by a factor of 100.

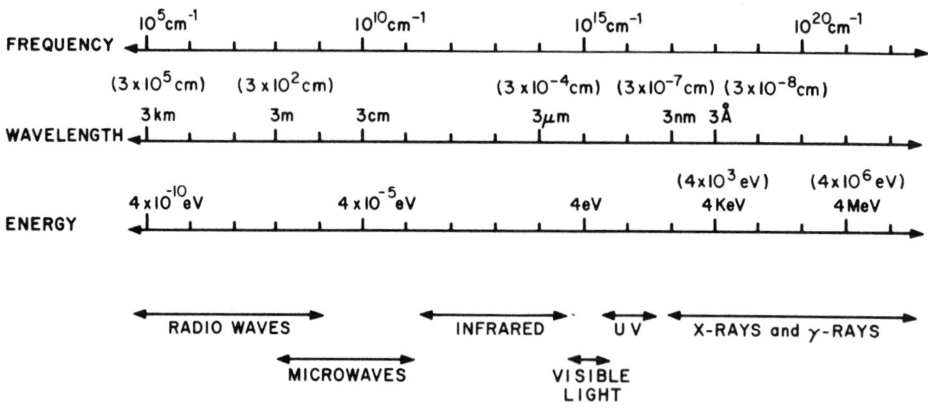

FIG. 4-1. The electromagnetic spectrum.

where

c is velocity of light in vacuum (cm/sec)
υ is frequency (sec^{-1})
λ is wavelength (cm)

$$E = h\upsilon \qquad (4\text{-}3)$$

where

E is energy of photon (ergs)
h is Planck's constant (6.626 x 10^{-27} erg sec)

Energy of radiation is often measured in electron volts (eV). One electron volt is the energy acquired by an electron falling through a potential of 1 V (1/eV = 1.602 x 10^{-12} erg; 1/keV = 10^3 eV; 1/MeV = 10^6 eV). The total effect of electromagnetic radiations upon a material depends on the energy of each photon and on the number of photons impinging on the material. As is evident from Figure 4-1 the lower the wavelength, the higher the energy per photon.

The types of interactions between matter and various electromagnetic radiations depend on the energy level of the photons. x-Rays and relatively low-energy γ-rays (electromagnetic radiations with energies of 5 x 10^4 eV to 3 x 10^6 eV) dissipate their energies in passing through matter by ionization through the so-called "photoelectric effect" and the "Compton effect." Figure 4-2 shows how these effects contribute to total ionization by γ-rays as a function of the γ-ray energy. At high γ-ray energies, energy dissipation through "pair production" is possible (the γ-ray energy is converted into formation of two particles, a negative electron and a positive positron), but this effect is of little importance at γ-ray energies applicable to food processing.

The photoelectric effect consists of absorption of the photon energy by an atom, with the resulting ejection of an orbital electron. The photon disappears. The energy of the photon is expended partly in overcoming the binding forces holding

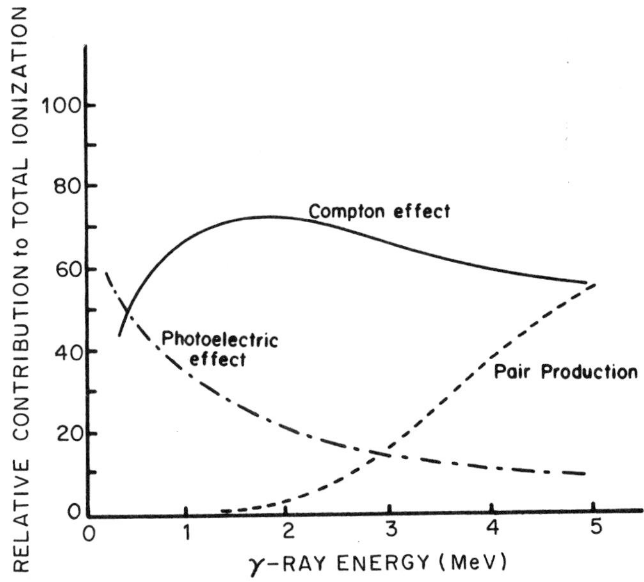

FIG. 4-2. Interaction of γ-rays with matter.

the electron and is partly converted into kinetic energy of the ejected electron:

$$h\upsilon = \phi + 1/2\ mv^2 \tag{4-4}$$

where

ϕ is binding energy of the electron (ergs)
m is mass of the electron (g)
v is velocity of the electron (cm/sec)

In the Compton effect the photon behaves as a particle undergoing collision. As shown in Figure 4-3 a photon undergoes a collision with an orbital electron. The electron is ejected with a kinetic energy derived from the photon and the photon then continues with a reduced energy:

$$h\upsilon = \phi + 1/2\ mv^2 + h\upsilon' \tag{4-5}$$

where υ' is the frequency of the scattered photon. In spite of the existence of the above three distinct modes of interaction, the absorption of x-rays and of γ-rays by matter can be represented by a single exponential equation:

$$I = I_o e^{-\mu X} \tag{4-6}$$

where

I is the intensity of radiation at distance X within the absorber
I_o is the intensity at the surface of the absorber

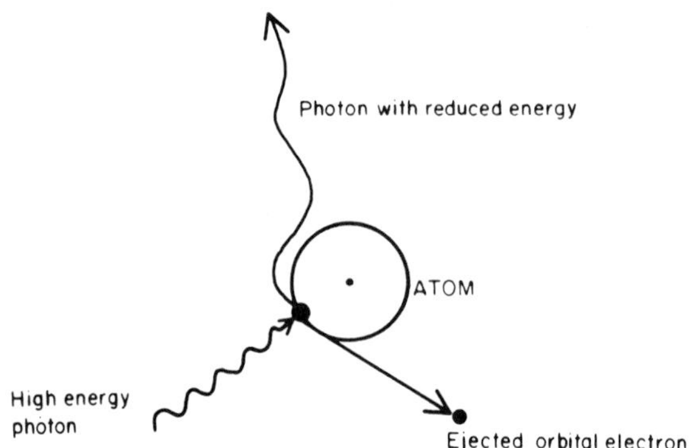

FIG. 4-3. Schematic representation of the Compton effect.

μ is the coefficient of absorption, which depends on the frequency of radiation and on the nature of the absorber

The coefficient of absorption depends on the density of the absorber and is, in fact, almost directly proportional to density for most substances. Coefficients of absorption for several materials are listed in Table 4-2.

The exponential nature of Eq. 4-6 results in an intensity attenuation pattern shown in Figure 4-4. It is evident that because of the logarithmic ordinate scale the intensity never reaches zero and that each additional equal thickness of absorber reduces the intensity by the same factor. A useful value is the "half-thickness," or the thickness of the absorber which reduces the intensity of radiation by a factor of two. The "half-thickness" is equal to $0.693/\mu$.

C. High-energy Electrons

Electrons emitted from a hot cathode can be accelerated to a very high velocity by one of several devices, some of which are discussed in Section V.B.2. High-velocity electrons can produce ionization effects similar to those produced by γ-rays. Electrons are negatively charged particles with a mass at rest of 9.1×10^{-28} g or about 0.00055 mass units. I refer here to mass at rest because at velocities approaching the speed of light, the electrons, as well as any other material object, undergo an increase in mass, along with contraction of length and slowing down in time. The relation between electron energy and velocity is shown in Table 4-3. In addition to acceleration in man-made devices, high-energy electrons are also produced by disintegration of some radioactive materials. As an example, ^{24}Na emits an electron and is converted to ^{24}Mg. Electrons produced by radioactive substances are called β-rays in contrast to cathode rays, which are electron beams produced by man-made machines. There is a difference between these two radiations: cathode rays can be produced with a uniform energy, but β-rays are produced with a spectrum of energies.

Electrons, whether from radioactive nuclei or accelerators, lose energy and are slowed down as they interact with matter; as more and more of the electrons in a beam are slowed down to negligible velocity, the number of fast electrons

TABLE 4-2

Absorption Coefficient (μ) for Several Materials at Different Photon Energies

	Absorption coefficient (cm)		
	0.5 MeV	1.0 MeV	4.0 MeV
Air (STP)	0.00008	0.00005	0.00004
Water	0.09	0.067	0.033
Aluminum	0.23	0.16	0.082
Iron	0.63	0.44	0.27
Lead	1.7	0.77	0.48

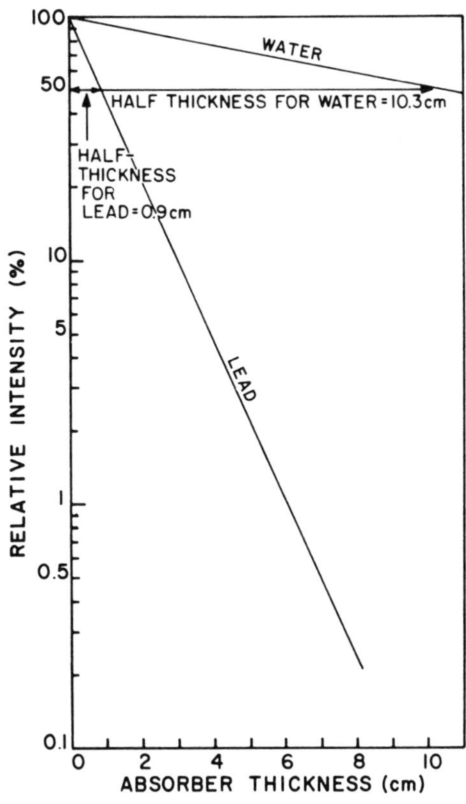

FIG. 4-4. Absorption of γ-rays by water and by lead.

TABLE 4-3

Relation Between Electron Velocity and Energy

Electron velocity		Electron energy
cm/sec x 10^{10}	% of speed of light	(eV)
0.417	13.9	5 x 10^3
1.1	36.6	5 x 10^4
2.6	86.7	5 x 10^5
2.82	94.06	10^6
2.985	99.57	5 x 10^6
2.994	99.869	10^7

remaining decreases with depth in an absorber. If a monoenergetic beam of electrons enters the absorber, the relative number of fast electrons remaining initially falls linearly with depth. However, not all of the electrons at a given depth have the same energy. As a consequence the relative ionization (or energy dissipation) at any point does not vary linearly with distance. Figure 4-5 shows how the relative number and relative ionization of electrons vary with depth in an absorber. In the case of β-rays, which have a distribution of energies, the intensity of ionization varies with depth in a manner similar to that of x-rays and γ-rays.

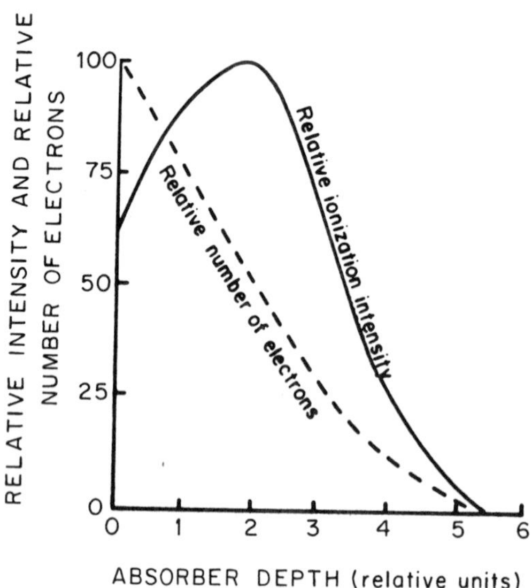

FIG. 4-5. Absorption of a monoenergetic electron beam by matter.

The maximum range of an electron beam is related to its energy by the Feather equation, Eq. (4-7), which is valid for energies greater than 0.8 MeV.

$$R_{max} = \frac{0.542E - 0.133}{\rho} \qquad (4\text{-}7)$$

where

R_{max} is maximum range (cm)
E is electron energy (MeV)
ρ is absorber density (g/cm^3)

D. Dosimetry of Ionizing Radiations

It should be evident from the preceding discussion of interaction of radiations with matter that the amount of energy absorbed by matter and the amount of ionization produced depend on a number of variables, including number of particles passing through an absorber, energy of the particles, and the nature of the absorber. Measurement of the radiation dose received by matter is therefore a complex matter. The unit first suggested for this purpose is the roentgen (R), which is defined as that quantity of radiation which produces one electrostatic unit per cubic centimeter of air under standard conditions. The amount of energy delivered by x-rays or γ-rays in delivering a dose of 1 R to a gram of dry air is 83.8 ergs. This dose led to the development of the unit called "roentgen equivalent physical," which was originally defined as the amount of energy absorbed in biological tissue when it was exposed to 1 R. This was originally to be 83 ergs/g as in air but measurements soon established that 1 R of γ- or x-rays produced 93 ergs/g in water or wet tissues and as much as 150 ergs/g in bone. It became clear therefore that a different basis for definition was needed and the International Commission on Radiological Units proposed term "rad." A rad is that quantity of radiation which results in absorption of 100 ergs/g at the point of interest.

A different kind of unit basic to radiation measurement is the curie. The curie (cu) is a measure of radioactivity of a substance. It is defined as 3.7×10^{10} disintegrations per second. Since it does not take into consideration the number and the energy of particles emitted in each disintegration, two radioactive materials with the same activity as measured in curies may produce very different radiation fluxes.

There are a number of methods available for detecting and measuring radiation. When the primary purpose is the detection and visualization of tracks of individual radioactive particles, the cloud chamber technique or darkening of photographic films constitute useful approaches. The cloud chamber has little relevance to the measurement of intensities useful in radiation processing of foods, and photographic film is useful only as a personnel monitoring device. (Photographic film and its response to radiation is, however, extremely useful in autoradiography and other techniques utilizing radioactive tracers in foods.)

An important series of measuring devices is based on the ionization chamber principle. By definition ionizing radiations produce ions of both signs [Eq. (4-1)]. By filling a chamber with a gas and by establishing in the chamber an electric field between two electrodes, it should be possible to collect these ions at the electrodes. Radiation can then be measured, since it results in a pulse of charged particles at the electrodes.

The size of the pulse for a given amount of radiation varies with the applied voltage. At low voltages the travel of the ions is slow and recombination occurs, thus reducing pulse size. At very high voltages multiplication of ions occurs, because the electrons produced by the primary ionization are accelerated by the applied voltage and thus acquire enough energy to produce many additional ionizations before they reach the electrodes. Different chamber designs and associated electronic systems are used, depending on whether each single particle is to be counted or the energy of ionization expended by many particles or photons is to be measured, and depending on radiation levels to be measured. Perhaps best known by laymen is the Geiger-Mueller counter. This counter, along with other types, uses the multiplication principle referred to above. A cylindrical chamber with a central thin wire (usually about 0.005 in. tungsten wire) collecting electrode is used. The field near the wire is very intense and electrons produced elsewhere in the chamber drift rapidly into this field. Each primary electron produces an avalanche of secondary electrons, and light quanta are also produced through a variety of interactions. The Geiger-Mueller counter operates in a voltage region where the pulse size does not vary with voltage, and it is a very sensitive and efficient counter of radiation. Because of these advantages it has been used for a wide variety of purposes, including prospecting for radioactive minerals and monitoring of safety in locations utilizing radiation sources for industrial or medical purposes.

For measurement of radiation absorbed by food during radiation processing the most useful technique involves use of chemical dosimeters. These in turn must be calibrated against some primary standard. Such a standard is often a calorimetric procedure. Measurement of heat is a much less sensitive technique than the collection and amplification of ions but it has the advantage of directly measuring the energy absorbed in the food or in water, which is after all usually the major food component.

The amount of heat produced by 10^6 rad absorbed by 1 g of water has been calculated to result in a temperature increase of 2.68°C. Davison et al. [4] have used the calorimetric technique to calibrate a ^{60}Co source used for research on food irradiation and also to calibrate chemical dosimeters. The chemical dosimetric systems are available in liquid or solid form. Among the important types are oxidation-reduction systems, such as certain metal or organic ions, and various transparent solids, including plastics and glass which darken on exposure to radiation. Some of the solid dosimeters develop color upon irradiation but this color fades later. The color may sometimes be stabilized by heat treatment. Characteristics of important dosimetric systems are summarized in Table 4-4.

III. CHEMICAL EFFECTS OF IONIZING RADIATIONS

Ionizing radiations are capable of initiating a vast array of chemical changes in gaseous, liquid, and solid systems. A complete, or even summary discussion of these changes is impossible to undertake within the scope of this chapter, but some features of the chemical effects of ionizing radiations in foods can be mentioned.

TABLE 4-4

Characteristics of Some Dosimeters

Dosimeter	Reaction	Applicable dose range	Precision (%)
Solution of ferrous sulfate	Oxidation	10^5-10^7	± 5
Solution of Methylene Blue	Reduction	10^4-10^7	± 10
Cobalt glass	Darkening	10^4-2×10^6	± 2
Polyvinyl chloride film	Darkening	10^6-10^7	± 15
Cellophane	Darkening	10^6-10^7	± 15
Radioluminescence of bibenzyl	Luminescence	10^5-10^7	± 10

Ionizing radiations split, or as it is called more technically "radiolyze," water. Since many biological systems as well as most foods are aqueous systems, this effect on water is of key importance. The passage of ionizing radiations through water results first of all in formation of the following intermediates:

Excited water	$(H_2O)^*$
Free radicals	$OH^\cdot$ and $H^\cdot$
Ionized water molecules	$(H_2O)^+$
Hydrated electron	e^-_{aq}

These species then react among themselves or with other components of the system. In pure water and in the presence of air they produce in particular the following:

Hydrogen gas	H_2
Hydrogen peroxide	H_2O_2
Water	H_2O
Hydronium ion	H_3O^+
Hydroxide ion	OH^-

The reactions that these intermediates can undergo with food components are too numerous to list. Every class of food constituents, including carbohydrates, proteins and other nitrogenous compounds, fats and oils, vitamins, enzymes, and pigments, can react with at least some of the intermediates to produce new intermediate compounds, many of which are highly reactive themselves. Oxidation reactions, free-radical reactions, and reduction reactions are particularly significant in this respect.

In addition to reactions mediated by water radiolysis, radiation has direct and significant effects on organic compounds, especially in nonaqueous systems.

In direct action on hydrocarbon chains, for instance, a number of primary events can occur. Hydrocarbon chains are of course present in lipids, as well as in many polymers of food packaging materials. The most important event is the abstraction of a hydrogen and the concurrent formation of a free radical:

$$-(CH_2)- \rightsquigarrow -(CH\cdot)- \quad +H\cdot \tag{4-8}$$

The hydrocarbon radicals can then undergo a very large number of reactions, among which those involving reactions with atmospheric or dissolved oxygen and crosslinking are the most important.

In polymers, as well as in some compounds of lower molecular weight, scission by the ionizing radiation is also possible. Different groups in polymers and different amino acid residues in proteins have different radiation sensitivities. Aromatic groups in polymers are the most stable, and groups containing halogens (Cl and F) are among the least stable. In food-related systems polysaccharides and, in particular, starch and pectin are very sensitive to radiation [1,3].

In proteins hydrogen abstraction usually results in free radicals on the α carbons or on the sulfur of the cysteyl residues. Among specific residues, histidine, sulfur-containing amino acids, phenylalanine, and tyrosine are the most susceptible to irradiation. Deamination, oxidation, polymerization, and decarboxylation have all been implicated in protein changes during irradiation. Reactions of proteins exposed to radiation have similarities to those caused by organic peroxides [13]. In both cases free radicals of proteins can either crosslink, recombine with hydrogen, or result in scission depending on various environmental factors, including the presence of water and oxygen (Fig. 4-6).

In lipids, including fats, many of the reactions are akin to "oxidative rancidity," that is, peroxidation by molecular oxygen, but various radiation-specific reactions also can be initiated.

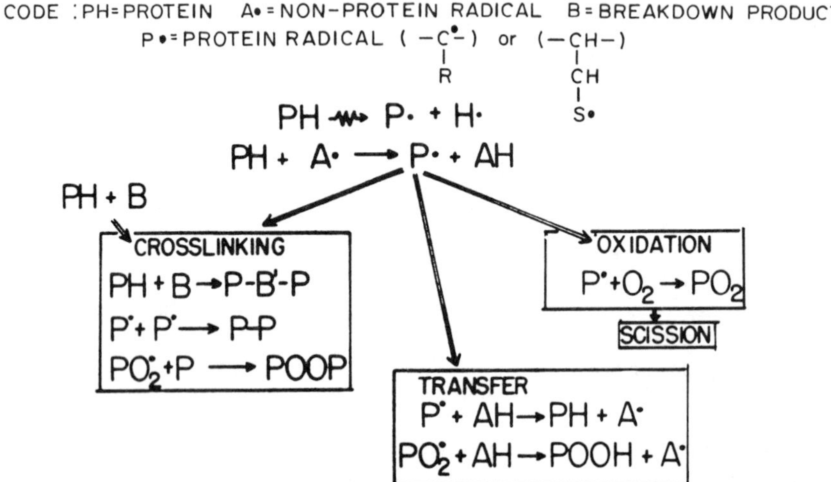

FIG. 4-6. Radiation-initiated and oxidation-initiated free-radical reactions of proteins.

Carbohydrates are very sensitive to radiation. Oxidative reactions predominate as well as some condensation reactions similar to nonenzymatic browning. It has been suggested that some products of irradiation of sucrose may have toxic effects on cells.

The effects of radiation-induced chemical reactions on the sensory quality, nutritive value, and wholesomeness of foods are discussed in Section VI of this chapter.

IV. EFFECTS OF RADIATION ON LIVING ORGANISMS

A. Direct and Indirect Effects

The potentially lethal effects of ionizing radiations were realized soon after the discovery of x-rays by Roentgen and of radioactivity by Becquerel in the late nineteenth century. The dose of radiation required to produce a lethal effect varies with the complexity of the organisms. The most complex organisms tend to be the most sensitive to radiation. A dose of several hundred rads is lethal to man but destruction of some microorganisms requires millions of rads. Figure 4-7 shows a schematic representation of the sensitivity ranges of various living organisms.

There are two basic potential mechanisms by which radiation affects living organisms: direct and indirect. In the direct, or target, mechanism it is assumed that there is a cellular or subcellular entity which when hit by a single high-energy particle produces lethality. This theory has applicability to microorganisms and other unicellular organisms, since the kinetics of their destruction are often in

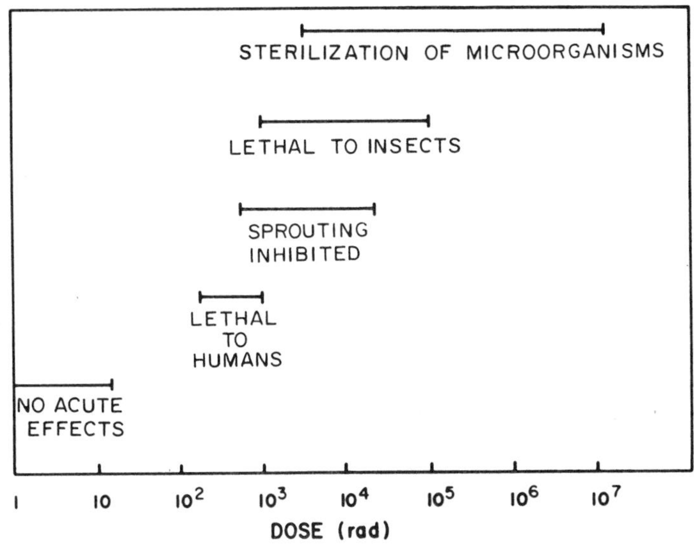

FIG. 4-7. Comparison of radiation sensitivity of different organisms.

conformity with those predicted by the direct action mechanism. This theory does not apply equally well to multicellular organisms, unless a single cell or group of cells is regarded as a composite target. Even in unicellular organisms, however there are indications that indirect effects may be occurring via free radicals produced elsewhere than in a primary target or via chemical reactions originating outside the cell. Radiation effects on cells are complex and some repair of damage as well as occurrence of nonlethal damage (some of which may lead to mutations) must be considered.

In spite of these complex effects it is possible, just as in the case of heat sterilization which also leads to complex effects on cells, to analyze radiation sterilization and radiation pasteurization in terms of relatively simple relationships.

B. Lethal Effects on Microorganisms

In the case of microorganisms exposed to a given lethal temperature the fraction surviving decreases logarithmically when plotted against time. "Death" of microorganisms when exposed to radiation also can be evaluated by plotting the logarithm of the fraction surviving against dose. ("Death" is in quotation marks because in the case of microorganism we are concerned with the ability of an individual cell to divide and reproduce into a colony of new cells. Thus, a cell which is still capable of some metabolic activities, but not of reproduction, would be counted as dead.)

In the case of thermal sterilization, however, the effect did not depend solely on the quantity of heat absorbed by the cell; it depended in a complex manner on the intensity factor—temperature, and on time. In this respect radiation sterilization is simpler. The intensity factor in this case is called "dose rate," or the amount of radiation absorbed by the cell per unit time. Dose rate has some effects, but in general it is possible to relate radiation effects to dose only:

$$n = n_0 e^{-D/D_0} \tag{4-9}$$

where

n is the number of live organisms after irradiation
n_0 is the initial number of organisms
D is the dose of radiation received (rad)
D_0 is a constant depending on type of organism and environmental factors

The constant D_0 (pronounced D zero) can be also thought of as that dose of radiation which causes death of 63% of the organisms. A plot of ln (n/n_0) is linear with dose, and it is obvious that for equal doses the population decreases by a given factor. Other constants can therefore be used to describe radiation sensitivity of microorganisms and, in particular, a value causing a reduction in number by a factor of 10, or D_{10} dose.

Figure 4-8 shows typical relations between dose and survival or microorganisms, and Table 4-5 lists some D_{10} values for a variety of microorganisms. An important aspect of radiation processing is that even under identical conditions of pH and composition of suspension media, and using the same cultures of

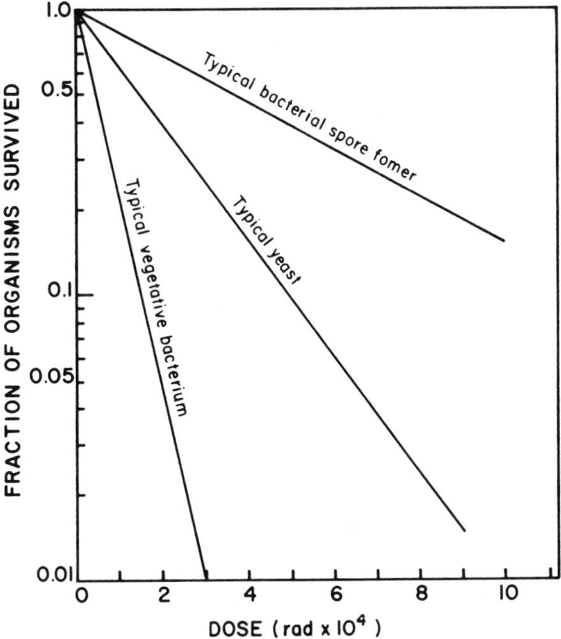

FIG. 4-8. Typical dose-response curves for different types of microorganisms.

microorganisms, the sensitivity to radiation of different organisms is not necessarily correlated with their sensitivities to heat. Table 4-6 compares relative sensitivities of different microorganisms to heating at 250°F and to radiation. In each case C. botulinum is used as the reference organism and is assigned a sensitivity of 1. The other organisms are assigned relative sensitivities based on the ratios of their D_{10} values to that of C. botulinum.

Environmental factors during irradiation, as well as physiological and genetic differences between strains and cultures of microorganisms, can affect radiation sensitivity [12]. The significance of these factors varies, as indicated below:

1. Type of radiation (whether electron, γ-rays, or x-rays) does not have a significant direct effect on dose response [8].
2. Dose rate (intensity of radiation) does not seem to affect sensitivity, except as it affects other environmental factors (for instance, temperature).
3. The presence of oxygen does have a significant effect, as do pH and the type of medium in which the microorganisms are irradiated. Table 4-7 summarizes some of these effects of composition.
4. Irradiation in the frozen state reduces radiosensitivity of some organisms. Table 4-8 compares radiation sensitivity in liquid and in frozen foods. The United States Army Natick Labs has developed an irradiation process in which food is sterilized in the frozen state to minimize chemical and flavor changes and has determined that a greater sterilizing dose is needed in the frozen state.

TABLE 4-5

Radiation Resistance of Some Microorganisms

Organism and type	Medium in which irradiated	D_{10}(rad)
Bacterial spores		
C. botulinum A36	Buffer	0.33×10^6
B53	Buffer	0.33×10^6
B53	Canned chicken	0.37×10^6
C. sporogenes 367	Water	0.22×10^6
Vegetative bacteria		
Salmonella typhimurium	Buffer	2.0×10^4
	Egg yolk	8.0×10^4
Escherichia coli	Buffer	9.0×10^3
Staphylococcus aureus	Buffer	2.0×10^4
Yeast		
Saccharomyces cerevisiae	Saline	5.0×10^4
Mold		
Aspergillus niger	Saline	4.7×10^4
Virus		
Organism responsible for foot and mouth disease	Frozen solution	1.3×10^6

TABLE 4-6

Radiation Sensitivity and Heat Sensitivity of Different Microorganisms Relative to C. botulinum (Type A)

Organism	Relative sensitivity[a]	
	To heat	To radiation
C. botulinum	1.0	1.0
B. subtilis	0.3	5.5
C. sporogenes 3679	0.16	1.5
B. coagulans	3.0	0.5
E. coli	$\sim 10^{13}$	36

[a] Radiation sensitivity compared on the basis of sensitivity to γ-rays; heat sensitivity based on sensitivity to heating at 121° C.

TABLE 4-7

Summary of Effects of Composition of Media on Radiation Resistance of Microorganisms

Factor	Organism	
	C. sporogenes	B. subtilis
Effect of chemical composition	More sensitive in saline than in pea puree	More sensitive in pea puree than in saline
Effect of oxygen	Least sensitive in absence of O_2	Least sensitive in absence of O_2
Effect of pH	Least sensitive at neutral pH	No consistent effect

TABLE 4-8

Radiation Sensitivity of Some Organisms in Frozen and Unfrozen States

Organism	Medium	D_{10} dose (rad)	
		Frozen	Unfrozen
E. coli	Pea puree	57,000	57,000
E. coli	Water	50,000	20,000
E. coli	1M glycerol	27,000	30,000
S. aureus	Water	53,000	10,000
B. thermoacidurans	Saline	150,000	150,000

5. Temperature has an effect on sterilizing requirements. This effect forms the basis for potential combination processes utilizing both heat and radiation. Table 4-9 demonstrates the effects of preirradiation on heat sensitivity, and Table 4-10 illustrates the effect of temperature of irradiation on resistance to irradiation.

Table 4-10 shows a fairly large and complicated effect of temperature on radiation sensitivity of C. botulinum in buffer but a lesser effect of temperature in frozen versus unfrozen food. In general, preirradiation sensitizes microorganisms to heat but the effect of temperature on radiation sensitivity is not large, except in the following two situations.

a. Microorganisms are less sensitive in frozen media.
b. Very high temperatures add their own lethal effects to those of irradiation.

TABLE 4-9

Effect of Preirradiation with γ-Rays on Subsequent Sensitivity of Microorganisms to Heat

Organism	Radiation dose (rad)	Relative number of organisms surviving a given heat treatment
C. sporogenes	0	100
	279,000	40
	930,000	0.0005
B. cereus	0	100
	279,000	12.5
	560,000	0.001
C. botulinum	0	100
	600,000	1

TABLE 4-10

Effect of Irradiation Temperature on Radiation Sensitivity of C. botulinum Irradiated with 900,000 rad of γ-Rays [a]

Irradiation temperature (°C)	Relative number of organisms surviving	
	In beef dinner	In buffered solution
-196	18	10
-100	18	3.2
- 50	18	0.05
+ 20	3	0.1
+ 60	1	3
+ 90	1	< 0.001

[a] After Grecz et al. [10].

C. Effects of Radiation on Insects

As mentioned previously insects are more sensitive to radiation than any of the microorganisms. Studies on radiation-caused sterilization of adult insects and on direct lethal effects of radiation on larvae, pupae, adults, and eggs of insects infesting stored food products have been carried out by the author as part of radiation researches led by Proctor et al. [18]. We found that immediate lethality required fairly substantial doses of radiation but reproduction could be prevented by doses of only a few thousand rad. Figure 4-9 shows the survival of some adult insects after exposure to γ-rays, and Table 4-11 gives the mean survival times of

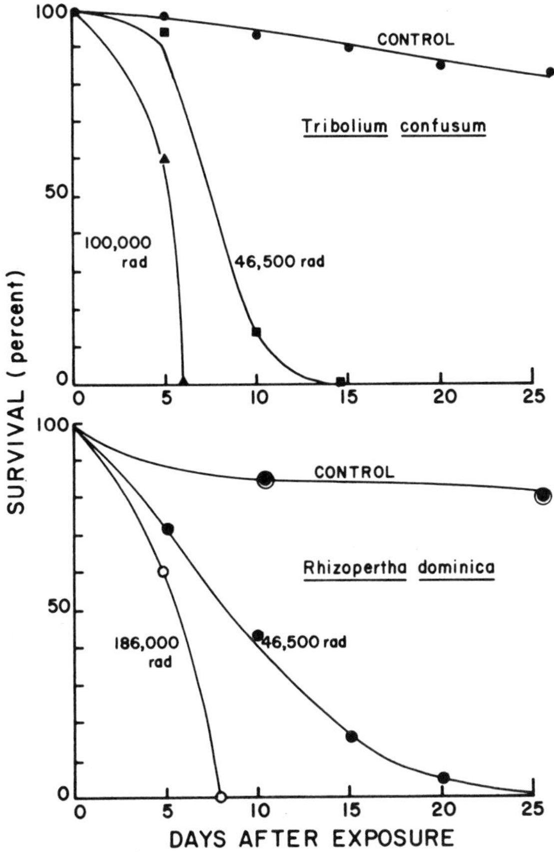

FIG. 4-9. Survival of irradiated adults of two stored food insects.

insects exposed to γ-rays or to 2.5 MeV electrons from a Van de Graaf electrostatic generator. None of the adult groups which received any of the radiation doses shown in Table 4-11 produced viable eggs. Control (nonirradiated) adults produced several eggs per adult, and over 60% of these eggs developed into larvae. Eggs from control adults could be prevented from hatching by relatively low doses of radiation as shown in Table 4-12.

V. TECHNOLOGICAL ASPECTS OF RADIATION PROCESSING

A. Applications of Radiation in Food and Related Industries

Ionizing radiations may have applications in several areas of food processing and related activities. Among the potential applications are:

1. Sterilization of foods in hermetically sealed containers. This process has been given the name "radappertizing" and is directed to production of foods

TABLE 4-11

Effects of Radiation on Some Insects

Insect	Stage	Type of radiation	Dose (rad)	Mean survival time (days)
Flour beetle (<u>Tribolium confusum</u>)	Adult	Control	—	Over 70
		γ-Ray	23,250	7
		γ-Ray	200,800	1.7
		Cathode rays	242,000	2.4
	Larva	Control	—	Over 30
		γ-Rays	23,250	6.5
		γ-Rays	186,000	1.2
		Cathode rays	204,000	1.5
Grain beetle (<u>Oryzaephilus surinamensis</u>)	Adult	Control	—	Over 70
		γ-Rays	158,000	1.1
		Cathode rays	167,400	1.1
	Larva	γ-Rays	158,000	1.5
		Cathode rays	158,000	1.5

TABLE 4-12

Effect of Cathode Rays on Development of Eggs of _T. confusum_ (Confused Flour Beetle)

Dose (rad)	Status following irradiation			
	Day 5		Day 25	
	Unhatched eggs[a]	Live larvae	Unhatched eggs[a]	Live larvae
Control	79	21	36	60
23,250	98	2	98	0
46,500	100	0	100	0

[a] Initial count 100 eggs in each group.

stable at room temperature for an indefinite period of time. It may be preferable to canning for those foods which are adversely affected by heating but can tolerate sterilizing doses of radiation. At present this process is under extensive investigation by the United States Army Natick Labs. The doses required for this application are between 2 and 6 million rad.

2. Extension of shelf life of various foods to be distributed and stored at refrigerated temperatures. There are a number of food items which are distributed at refrigerated temperatures. These include fresh fish, poultry, meats, fresh vegetables and fruits, milk, eggs, and cheese. For many of these the shelf life can be extended by reducing the microbial populations and/or eliminating potentially pathogenic microorganisms. This is of course done by heat pasteurization in the case of milk, and by bacteriostatic chemicals in some products. Irradiation can achieve the same aims in many cases. For example, irradiation with doses of several hundred thousand rad could allow a much longer shelf life at refrigerator temperatures for fish and other seafoods and could improve the safety of poultry by eliminating potentially dangerous organisms. In the case of fresh fruits and vegetables, irradiation can destroy surface molds and yeasts and thus make the commodities more stable.
3. Treatment of water, sewage, and food-processing wastes [6]. A number of applications are possible in which the food itself is not treated by radiation but some materials involved in the food-processing plants or in the waste disposal system are so treated. The doses required, of course, must vary with the intended end result, which is either sterilization or pasteurization. An interesting application in this respect is the treatment of cooling water for fish on board ships. By maintaining the microbial count in this water at a low level, without heating the water, it should be possible to retard deterioration of fish prior to the time of their delivery for subsequent processing at the harbor.
4. Deinfestation of grain or dried fruits and of other commodities subject to insect attack. As we have seen in Section IV. C, insects are destroyed readily by relatively low doses of radiation. Doses of 25,000 rad, for

instance, would in several days kill adults and larvae of the major insects infesting grain and cereals and, more importantly, would immediately prevent further reproduction. Provided that steps are taken to prevent recontamination, deinfestation can be an extremely useful step in preserving grain supplies and in preventing insect infestation of other products.

5. Inhibition of sprouting. Onions, potatoes, and carrots are commonly stored extensively at relatively high temperatures for long periods of time. Sprouting is a major cause of loss in storage. This can be prevented by chemical treatments, but an alternative method is provided by ionizing radiation. Only about 15,000 rad is required and the chemical effects of these low doses are negligible.
6. Other food applications. Several other applications to food processing have been suggested. One of these is irradiation of dehydrated vegetables. Irradiation causes scission of polymer chains contributing to the structure of these materials, and this scission is beneficial in accelerating rehydration of otherwise poorly rehydrating vegetables. Another interesting application is the possibility of accelerated aging of Scotch whisky. The flavor and appearance of aged whisky are due to oxidative changes which occur slowly under normal conditions but may be greatly accelerated by irradiation.

Two general aspects of the technology of irradiation are worthy of additional discussion, namely: irradiation sources and effects of radiation on eating quality of foods. Two examples of processes which have been developed for specific purposes shall be examined in detail, namely (a) sterilization of army rations, and (b) treatment of potatoes to inhibit sprouting.

B. Irradiation Sources

1. Cobalt-60

Radiation sources for food processing are either radioisotopes or machine sources. The only radioisotope presently considered for commercial processing is cobalt-60. This radioisotope is produced by irradiation of cobalt-59 in a nuclear reactor. The usual procedure is to machine cobalt metal into shapes suitable for the ultimate source design, encapsulating it in a material that does not become radioactive during neutron irradiation, and then exposing it to the intense radiation in the reactor. A second encapsulation follows the irradiation, and this must be performed by remote control since exposure of man to the now radioactive cobalt results in certain death. The United States Atomic Energy Commission has developed sophisticated remote control methods for production, inspection, shipment, and installation of cobalt-60 [14].

This radioisotope has a half-life of 5.27 yr and emits two γ-rays per disintegration with energies of 1.17 and 1.33 MeV.

The sources can be produced in various shapes and can have different intensities depending on how long they have been irradiated. A presently widely used form is the plaque source, consisting of flat strips which may contain more than 10,000 cu per strip. A number of these strips can be assembled by remote control operation into sources of megacurie activity. There are many variants of the cobalt-60 irradiators, depending on the desired medical or laboratory research or industrial application. Two of the large-scale cobalt-60 food irradiators have been installed in Massachusetts: one at the Gloucester Technological Laboratory of the Bureau of Commercial Fisheries, and the other at the United States Army Natick Laboratories. The source in Gloucester has an intensity of 250,000 cu and the one at Natick an activity of 3,600,000 cu. The Natick installation contains sources built up from nickel-plated "wafers" 0.745 in. in diameter and 0.040 in. thick. The wafers are combined into sources of 15 wafers each, doubly encapsulated in stainless steel. The source is immersed in a 25 ft deep storage pool of water when not in use. The water shields operators who inspect or manipulate the submerged sources from the ionizing radiation. The sources are raised out of the water by remotely controlled equipment and a conveyor carries packaged food between radiation walls at a rate designed to deliver the desired dose.

The original Natick source had an activity of 1,300,000 cu and was designed for a throughput of 3000 lb/hr at a sterilizing dose of 5×10^6 rad. The present source has a much larger capacity, as mentioned above. The Gloucester source has a capacity to process 6 tons/hr at the pasteurizing dose of 250,000 rad. Figure 4-10 shows the Natick source in its storage position under water and Figure 4-11 shows the source in the operational position.

Irradiation of a product from two sides with cobalt-60 γ-rays has the advantage of providing a fairly uniform dose throughout the irradiated food containers at an energy level which precludes formation of induced radioactivity.

FIG. 4-10. The United States Army Natick Labs cobalt-60 source in submerged position (courtesy United States Army Natick Labs).

FIG. 4-11. The United States Army Natick Labs cobalt-60 source being raised by remote control to the position for irradiation of conveyed food containers (courtesy United States Army Natick Labs).

2. Machine Sources of Radiation

A number of electron accelerators have been proposed for radiation processing of foods [16]. Table 4-13 lists the characteristics of some of these accelerators. Several types have been used for research on food and commercial nonfood applications, such as processing of drugs, plastics, and paints.

The most advanced machine for food irradiation is the linear electron accelerator at the United States Army Natick Laboratory. Figure 4-12 shows the principle of the electrostatic accelerator and the linear accelerator.

One of the disadvantages of electrons is their lower penetration as compared to γ-rays. This can be overcome in part by using high-energy electrons, provided the energy is below the level capable of inducing radioactivity. It is presently considered that electrons with energies below 10 MeV fulfill this condition. By irradiation from two sides it is also possible to maximize energy utilization by having a fairly uniform intensity, compared to irradiation from one side only

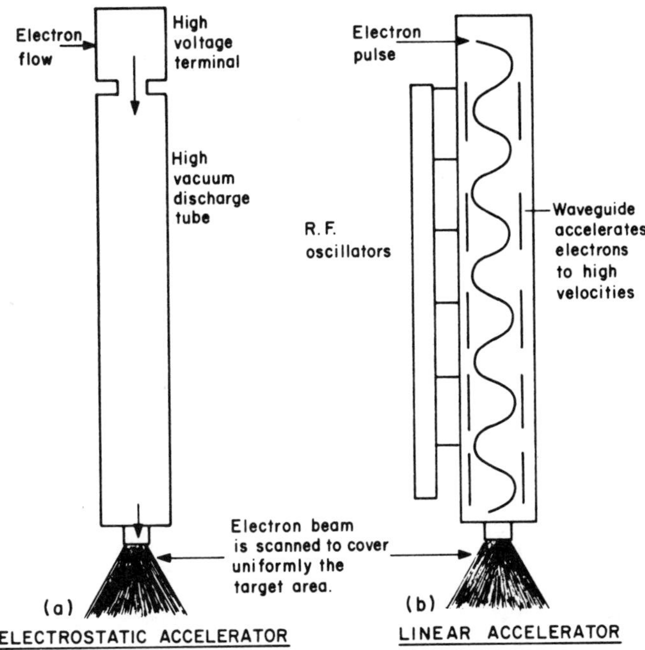

FIG. 4-12. Schematic representation of electron acceleration in the electrostatic accelerator and the linear accelerator. (a) A high static voltage accelerates electrons down a discharge tube. (b) An electron pulse is accelerated as it travels down the beam tube. The electrons are trapped in a travelling rf wave and accelerated to high speeds.

(Fig. 4-13). Lateral variations in intensity are minimized by scanning the electron beam over the package. Figure 4-14 shows how a scanned electron beam provides uniform dose delivery across a package area.

In the present arrangement for irradiation with electrons at Natick, packages are irradiated from one side only. The product is brought to the electron beam by a conveyor system which takes a tortuous path among shielding walls, assuring that no radiation can leak out of the processing room (Fig. 4-15).

C. Effects of Irradiation on Individual Foods

All food-preservation operations have an element of compromise between maximizing the efficiency of the process and maintenance of optimum product quality. In the case of irradiation there are undesirable side effects which can cause impaired flavor, texture, and appearance. As in the case of other processes the side effects can be minimized, or in some cases eliminated, by changes in process conditions and the seriousness of the problem various with different foods. Table 4-14 lists some major organoleptic problems of key classes of foods and dose ranges at which they appear when no modifications in the process are taken.

TABLE 4-13

Characteristics of Some Electron Generators Proposed for Use in Food Processing

Name	Typical energy level (MeV)	Typical power output (kW)	Type of beam	Principle of operation
Van de Graff electrostatic generator	3	3	Continuous	A moving belt continuously brings charge to an insulated high-voltage terminal
Resonant transformer	3	20	Fluctuating	High voltage obtained by a step-up transformer
Dynamitron	3	15	Intermittent pulse of microsecond duration	Impulse generator with a large dc potential built up by a cascaded rectifier stack
Linear accelerator (several types)	10 or more	30	Pulsed beams	Electrons accelerated to high energy by traveling with an rf wave

TABLE 4-14

Quality Defects Produced by Side Effect of Ionizing Radiation and Methods of Prevention

Food	Defects	Dose at which the side effects become important (rad)	Method(s) of prevention
Beef	Off flavor	$\sim 2 \times 10^6$	Irradiation in frozen state
Pork	Off flavor	$\sim 3 \times 10^6$	Irradiation in frozen state
Milk	Off flavor	less than 10^5	Distillation of volatile off flavor
Poultry	Off flavor, pink discoloration	$\sim 3 \times 10^6$	Irradiation in frozen state
Fish	Off flavor	$\sim 10^6$	Irradiation in frozen state
Flour	Poor baking performance	$\sim 0.5 \times 10^6$	Dose reduction
Vegetables	Softening, discoloration	10^6 to 2×10^6	Dose reduction
Fruits	Softening	$< 10^6$	Dose reduction

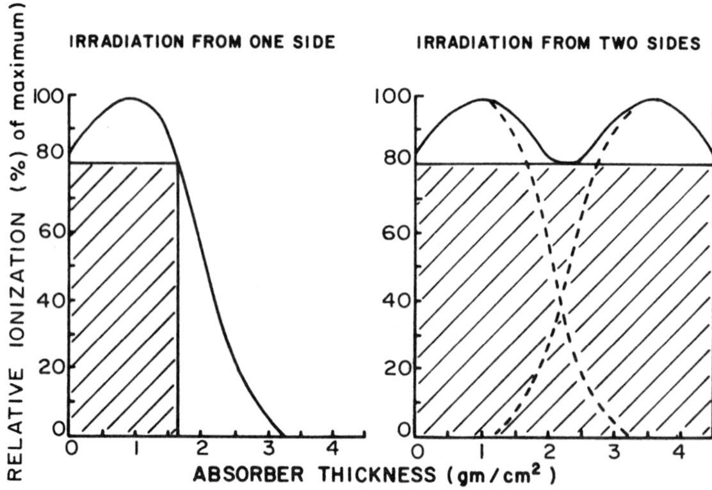

FIG. 4-13. Comparison of relative ionization within a food due to an electron beam applied from one or from two sides of a slablike food package.

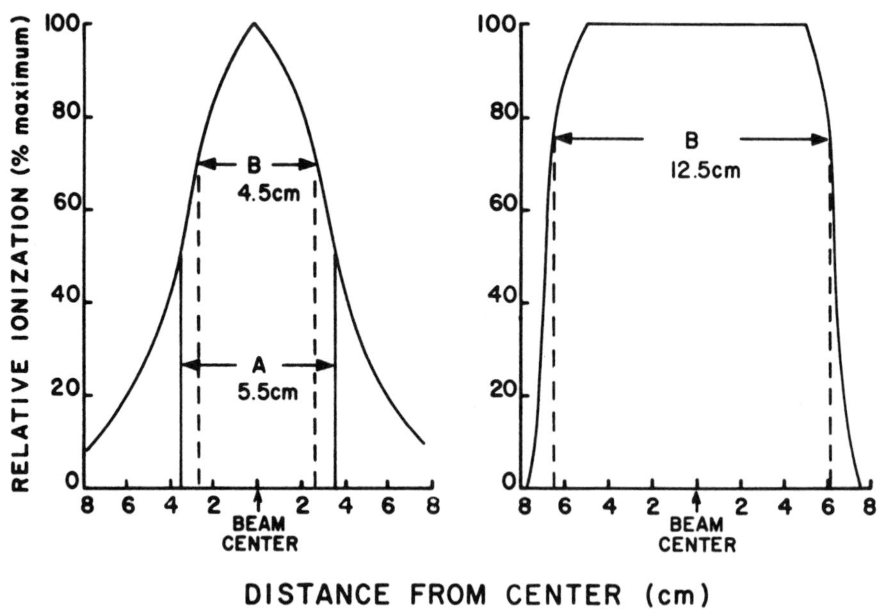

FIG. 4-14. Schematic representation of the effect of scanning of an electron beam on the lateral variation of beam intensity. The figure on the right represents irradiation with scanning. In the figure on the left, "A" represents the width receiving at least 50% of maximum dose, "B" the width receiving at least 70%. In the figure on the right, "B" represents the width receiving at least 75%.

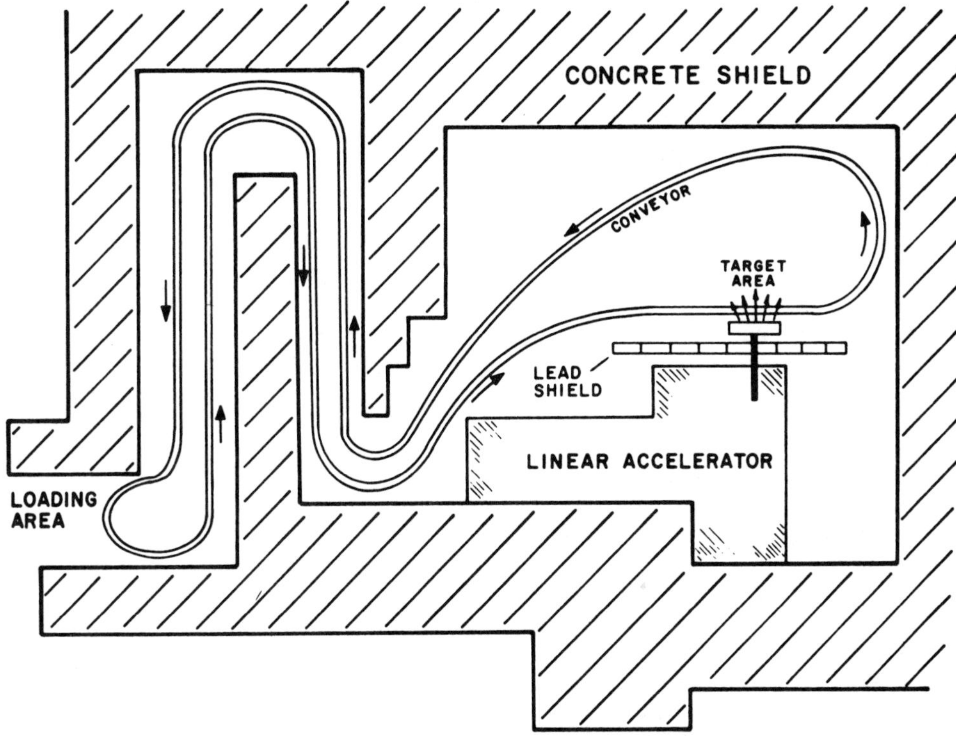

FIG. 4-15. Schematic representation of the path taken by foods irradiated in the United States Army Natick Linear Accelerator Facility (Based on blueprints provided by United States Army Natick Labs).

In general, off flavors are the major organoleptic problem in meats, fish, and poultry. The chemistry of off-flavor production is only partially understood but oxidation of fat as well as breakdown of proteins may be involved. Beef is considered to be most sensitive and pork and poultry the least sensitive. Pigment changes also occur, with red pigments of raw meat turning brown and cooked meat and poultry pinkish; some softening of texture also occurs.

In the case of fresh fruits and vegetables, textural changes are important as well as off flavor, and undesirable textural side effects occur at doses of 10^5 rad or more.

Some foods, notably dairy products, are particularly sensitive to radiation-induced off flavors. Milk, for instance, acquires an unpleasant off flavor at doses as low as 10,000-50,000 rad and some types of chocolate acquire a detectable off flavor at doses below 5000 rad [11].

The undesirable side effects may be greatly minimized or eliminated entirely by changes in process conditions. Two major approaches are possible, each with several variations. The first is to reduce the required radiation dose. This may be achieved by using radiation in combination with heat treatment or by using radiation together with chemical preservatives. The second approach is to use the full sterilizing dose but to reduce the severity of side effects of irradiation in the frozen state and in an inert atmosphere or to add free-radical scavengers, such as ascorbic acid. Irradiation in the frozen state increases the dose required for sterilization somewhat but this is outweighed entirely by the almost complete elimination of intolerable off flavors. Table 4-15 demonstrates that even in the case of the very sensitive beef, an acceptable sterile product can be produced by irradiation in the frozen state.

D. Sterilization of Army Rations and Other Shelf-stable Foods

United States Army scientists developed sterilization technology for several foods and conducted extensive wholesomeness studies which in 1965 led the United States Army Surgeon General to conclude that foods irradiated with up to 5.6×10^6 rads of cobalt-60 γ-rays or with electrons up to 10^7 eV have been found wholesome. In 1968, however, the Food and Drug Administration, in reviewing the data presented by the Army in support of its petition for approval of irradiation-sterilized ham, concluded that these data were insufficient to prove wholesomeness [9]. The FDA revoked its prior approval for sterilized bacon and the Army withdrew the petition for approval of ham. An unprecedented program of safety testing has since been initiated. This is still underway and is to continue for several years.

In the meantime, however, the technology of the process has been refined and, in fact, due to the very large quantities required for feeding of the experimental animals, the operations have been conducted on a semiworks scale. It is possible therefore to outline the features of radiation sterilization processes as they are likely to develop once FDA approval is granted [5].

1. Products

Among the products tested are bacon, beef, chicken, ham, pork, shrimp, fish cakes, and pork sausages. Typical dose requirements and conditions of irradiation are shown in Table 4-16. These requirements are based on reducing a potential population of C. botulinum spores to 10^{-12} of the original population. Other products are being actively considered and significant variations in composition may arise out of the process. As an example, it may be possible to produce cured meat products with a much reduced nitrite level. This accomplishment may have significance because of the current concern about potential dangers of nitrite in food.

2. Preheating

Irradiation is not a particularly effective means of destroying enzymes in foods, whereas heat is. Radiation-sterilized foods are therefore subjected to a mild preheating treatment which is adequate to inactivate undesirable enzymes but does not affect adversely the eating quality of food. The severity of the treatment varies from food to food. In the case of beef 2 - 5 min at 70° C is adequate.

TABLE 4-15

Effect of Irradiation Temperature on Quality of Irradiated Beef[a]
(Dose = 4.5 x 10^6 to 5.6 x 10^6 rad)

Temperature of irradiation (° C)	Intensity of defects[b]		
	Flavor	Color	Texture
10	4.1	3.9	2.9
-80	2.1	2.3	2.0
-180	1.5	2.3	1.6
Control	1.0	1.2	1.4

[a] After Wierbicki et al. [20].

[b] 1 = no defect; 9 = extreme defect.

TABLE 4-16

Sterilizing Dose for Several Foods

Food	Irradiation temperature	Dose (rad)
Bacon	15° C	2.3 x 10^6
Beef	-30° C	4.7 x 10^6
Beef	-80° C	5.7 x 10^6
Ham	15° C	2.9 x 10^6
Ham	-30° C	3.7 x 10^6
Pork	15° C	4.6 x 10^6
Pork	-30° C	5.1 x 10^6

3. Packaging

Packaging in evacuated cans or in evacuated laminated pouches has been evaluated by the United States Army Natick Laboratory. On the basis of extensive extraction tests five polymeric materials were established as safe for doses up to 6 x 10^6 rad and approved for use by the FDA. These included polyethylene, polyethylene terephthalate, Nylon -6, polyvinyl chloride-polyvinyl acetate, parchment paper, and metal cans with suitable lacquers. The packaging for the process envisioned by the Army is complicated by two factors:

a. Irradiation produces gas from the usual food components. In particular, hydrogen is a major component of the atmosphere produced by radiation. This gas has no adverse effects provided the containers are evacuated prior to irradiation. If the pressure is high before irradiation, however, then the added pressure due to evolved gases may cause container damage.
b. The present procedure requires irradiation at very low temperatures, with subsequent rewarming of the containers to room temperature. This procedure imposes the additional requirement that containers be resistant to the imposed temperature changes.

The present packaging concepts involve both metal cans and laminated pouches made of 0.001 in. Nylon-6, 0.00035 in. aluminum foil, 0.0005 in. Mylar, and 0.002 in. medium-density polyethylene. Extensive tests have established the adequacy of these containers but many other container forms are suitable and are likely to appear should commercial production materialize [17].

E. Inhibition of Sprouting of Potatoes, Onions, and Carrots

The sterilization process as discussed in the preceding section represents the most difficult application of radiation to food. Inhibition of sprouting presents the other extreme, a process with the fewest problems. The reasons for its relative simplicity are primarily two.

1. The required dose is very low and therefore the safety considerations both with respect to operation and to product wholesomeness are easily met.
2. Inhibition of sprouting can be achieved by a single application of radiation and does not require subsequent prevention of contamination or special packaging and storage precautions.

The rationale of the process is as follows. Immediately after harvest, potatoes are in a dormant state and, under ideal environmental conditions, can be stored for many months without sprouting or mold growth. Under commercial conditions the ideal environment of about 7°C and 95% relative humidity is difficult to maintain. For this reason potatoes must be treated to avoid sprouting, and at the present time this is achieved mainly by chemical treatment with maleic hydrazide or other inhibitors. An alternative is the irradiation treatment.

Extensive research on potatoes has shown that sprouting can be inhibited by application of 8,000-10,000 rad. Irradiation does produce some damage to the capability of potato tissue to heal quickly after bruises and other injuries and does produce some temporary changes in normal respiration of the potato tuber. These defects, however, do not prohibit utilization of the treatment.

Wholesomeness of potatoes and other foods irradiated with the low doses required for inhibition of sprouting is considered acceptable and a number of countries, including Canada, Denmark, France, Israel, Japan, the Netherlands, the Phillipines, Spain, the United States, and The USSR, have granted clearances for this process.

Steps toward commercial production have been taken in a number of countries and varied plant designs are possible, ranging from a portable irradiator to a permanent installation located at the site of a massive storage facility. The portable irradiator is essentially a truck and a trailer containing a shielded source of cobalt-60. The truck and trailer unit can be brought to a location at which the potatoes are to be treated and a conveyor system set up to allow the continuous conveying of potatoes through the irradiation chamber located in the trailer. Radiation safety personnel and technical experts are part of the mobile system and can service widespread agricultural locations. A mobile irradiator was tested in Canada in the early 1960's, and one of the lessons learned was the need to handle the potatoes carefully during the radiation treatment to avoid bruising. One proposed solution was the irradiation of potatoes in pallet boxes. The boxes are about 1 m^3 in volume, and because of the dose reduction toward the center of the box the process must be designed to deliver 15,000 rad at the periphery. Permission for this dose level was granted in Canada in 1963. Demonstration of the feasibility of the mobile irradiator may lead to further improved concepts in the future.

Other large-scale trials of sprout inhibition by irradiation have been carried out in many locations around the world. All have demonstrated the technical feasibility of the process. The economic feasibility, however, depends on a variety of nontechnical factors and may lead to some failures as well as successes. The quantities of processed potatoes, onions, and carrots produced to date have not been large and additional experience is required. In Israel, in Hungary, and in some other countries, test marketing of several hundred tons of treated vegetables was considered successful. In Canada a large-scale plant was built in Montreal in the 1960's designed for irradiation of about 15 million pounds of potatoes per year. However this venture met with failure largely because of some serious economic and technical miscalculations [2]. The second generation of experimental potato irradiation plants is expected to provide information to correct these miscalculations. New plants include (1) a 150,000 cu plant near Moscow, USSR, built in 1964 and having a capacity of several thousand tons per year, and (2) a plant with a capacity of 10,000 tons/month in Hokkaido, Japan.

In any case, the irradiation of potatoes, carrots, and onions to achieve sprout inhibition appears a process ready for use when economics dictate its suitability.

VI. WHOLESOMENESS OF IRRADIATED FOODS

As indicated previously, commercialization of food irradiation hinges on establishing rigorous proof of the wholesomeness of resulting products. Radiation is the first preservation process to be subjected to the stringent requirements of the 1958 Food and Drug legislation. Five aspects of wholesomeness must be considered, the first two of which are unique to irradiation; the other three apply equally well to processing by heat. These five aspects are:

1. Can the process induce radioactivity?
2. Can a safety program prevent exposure of personnel to radioactivity during processing?
3. Do the changes in surviving microflora create a potential hazard?
4. Does the process result in nutrient losses of public health significance?
5. Does the energy input by radiation result in chemical reactions producing toxic or carcinogenic compounds?

The first four questions can be answered by relatively short-term studies involving physical, chemical, and microbiological tests. The fifth requires long-term and very expensive animal testing.

A. Induced Radioactivity

The induction of radioactivity in food depends on (a) the type of radiation and its energy, (b) the dose applied to the food, and (c) the abundance of specific elements in the food. In addition the half-life of the induced radioisotopes is important since if it is very short these isotopes disappear before reaching the consumer. Very extensive studies have shown the cobalt-60 γ-rays do not induce radioactivity in foods and that below an energy level of 10 MeV, electrons do not produce measurable induced radioactivity even at very high doses.

Electrons with high energies (over 20 MeV) do produce measurable radioactivity, especially isotopes of Na, P, S, and Fe. The amounts produced, however, are much lower than the levels considered maximal for safety by the National Bureau of Standards.

However, since any added amount of radioactivity, no matter how small, is considered undesirable by most health authorities, future approvals of radiation processing are expected to limit electron energy levels to about 10 MeV.

B. Radiological Safety of Operating Personnel

This is an important consideration but one which has been satisfactorily resolved in countless installations using radiation, including hospitals using x-rays and radioisotope therapy, industrial installations processing radioisotopes, industrial installations using radiation for crosslinking of polymers and sterilization of medical supplies, and a variety of research laboratories. A safety system in food industries is expected to include the following factors [7]: (a) adequate interlock systems automatically preventing irradiation when personnel enter the radiation chamber, (b) adequate shielding to prevent radiation leakage, (c) continuous monitoring of radiation at all pertinent points, and (d) training of personnel in radiation safety procedures.

C. Danger of Changing Microflora

This problem is of potential consequence when partial destruction of a microbial population (pasteurization) rather than sterilization is carried out. It is applicable to heat pasteurization and chemical sterilization as well as to irradiation. Simply stated, the concern is based on the following reasoning. In nonpreserved foods there is a certain correspondence between growth of smell- or discoloration-producing bacteria and the growth of bacteria capable of producing disease. This natural correspondence may be disturbed by the preservation treatment so that an innocuous appearing food may become dangerous. While there are some obvious fallacies in this argument, notably the notion that nonprocessed foods have a built in warning system, radiation processing because it is a new process must, like Caesar's wife, be above suspicion. Accordingly, extensive microbiological studies were conducted on radiation-pasteurized refrigerated fish, since this was a product for which the potential for dangerous changes in microflora was considered most acute. The tests showed that the fears were not justified [9].

D. Nutrient Losses

Radiation processing does cause some nutrient losses. These losses are most prevalent in dilute systems and depend strongly on conditions of irradiation. They are similar to losses occurring in heat processing, but specific nutrients vary in their relative sensitivity to heat and to radiation. Conditions of irradiation may be adjusted to greatly minimize these losses. In particular, radiation in the frozen state greatly reduces losses of thiamine (Table 4-17) and exclusion of oxygen protects fat-soluble, radiation-sensitive vitamins and some oxidation-susceptible amino acids.

In general it may be concluded that the effects of ionizing energy on nutrients are of the same magnitude as those of other processes. However, such dietary measures as menu planning and nutrient labeling should take into consideration losses in processing.

E. Toxicity and Carcinogenicity

Present regulations consider radiation a food additive. Under the Food and Drug Law, it must therefore be proved that irradiated foods when fed in greatly exaggerated amounts to several species of animals produce no carcinogenic effects and, when fed in amounts even remotely likely to be consumed, produce no toxic effect. Note that with respect to potential for cancer production the law requires an extreme exaggeration with no effect whatsoever being observed. No

TABLE 4-17

Thiamine Retention in Processed Ham and Beef[a]

Food	Treatment	Dose (rad)	Irradiation temperature (° C)	Thiamine retention (%)
Beef	Irradiation	3×10^6	-80	90
	Irradiation	3×10^6	5	10
	Irradiation	6×10^6	-80	75
Ham	Irradiation	4×10^6	-80	69
	Irradiation	4×10^6	Room temperature	4
	Irradiation	5×10^6	-80	88
	Thermal sterilization	Commercial canned 1.5 lb ham	—	33

[a] After Wierbicki et al. [20].

other physical food preservation process has received this severe a test. Nevertheless these tests on irradiated, sterilized beef are now in progress. When concluded they are very likely to decide the fate of radiation processing as a commercial process.

All of the data obtained to date indicate that there are no acute toxic effects or obvious dangers of chronic toxicity or carcinogenicity. However a rigorous proof of safety is yet to come.

VIII. PROSPECTS FOR THE FUTURE

Some applications of radiation processing of foods are likely to be adopted commercially. In fact many are already approved by legal authorities in several countries (Table 4-18). In the United States, numerous research conferences, military volunteers, and most recently NASA astronauts have been fed irradiated meals. I have consumed a number of such meals over the past 25 yr. If and when the processes become commercial, however depends on a variety of complicated technical, economic, and legal factors [9, 13].

TABLE 4-18

Some Processes Utilizing Ionizing Radiation which Have Been Approved by Regulatory Agencies in Different Countries

Process	Dose required (rad)	Countries in which approved
Inhibition of sprouting in potatoes and/or onions	5000-15,000	Netherlands, USSR, Canada, Israel, USA, Japan, Hungary, Spain, Denmark, Thailland
Deinfestation of grain and dried fruit	10^4-10^5	USSR, Canada, USA
Pasteurization of fresh fruits and vegetables	10^5-0.5×10^6	(Experimental batches for testing only) Netherlands, USSR
Sterilization of food for hospitalized patients requiring sterile food	2.5×10^6 and up	Netherlands, United Kingdom

SYMBOLS

c	velocity of light in vacuum (3×10^{10} cm/sec)
e^-	electron
h	Planck's constant (6.626×10^{-27} erg sec)

m	mass of electron
n	number of microorganisms
n_0	initial number of microorganisms
v	velocity of electron
D	dose of radiation
D_0	dose of radiation producing 63% lethality
D_{10}	dose of radiation producing 90% lethality
E	energy of radiation (eV)
I	intensity of radiation
I_0	intensity of radiation at surface of absorber
M	an atom or molecule
M^+	an ion
R_{max}	maximum range of electrons (cm)
X	thickness of absorber
λ	wavelength (cm)
μ	absorption coefficient
υ, υ'	frequency (cm^{-1})
ρ	density (g/cm^3)
ϕ	binding energy

REFERENCES

1. American Chemical Society, Radiation Preservation of Foods, Advances in Chemistry Series, 65, American Chemical Society, Washington, D. C., 1967.

2. Anonymous, Background Notes. Food Irradiation Information, Number 2, International Project in the Field of Food Irradiation, 75 Karlsruhe, Germany, 1973, p. 43.

3. F. A. Bovey, The Effects of Ionizing Radiation on Natural and Synthetic High Polymers. Interscience, New York, 1958.

4. S. Davison, S. A. Goldblith, B. E. Proctor, M. Karel, B. Kan, and C. J. Bates, Nucleonics, 11(7), 22 (1953).

5. J. Deitch, J. W. Osburn, Jr., and H. W. Ketchum, Cost-Benefits Analysis. Potential Radiation Sterilized Military Subsistance Items, U. S. Department of Commerce, Washington, D. C. (1972).

6. C. G. Dunn, Sewage Ind. Wastes, 25(11), 1277 (1953).

7. S. A. Goldblith, in Food Processing Operations (Joslyn and Heid, eds.), Vol. 1, AVI Publ. Co., Westport, Conn., 1963, p. 401.

8. S. A. Goldblith, B. E. Proctor, S. Davison, B. Kan, C. J. Bates, E. M. Oberle, M. Karel, and D. A. Lang, Food Res., 18, 659 (1953).

9. S. A. Goldblith, J. Food Technol., 5, 195 (1970).

10. N. Grecz, J. Upadhyay, T. C. Tang, and C. A. Lin, in International Atomic Energy Agency Symposium on Microbiological Problems in Food Preservation by Irradiation, International Atomic Energy Agency, Vienna, Austria, 1967, p. 99.

11. T. Grünewald, Z. Lebensm. Unters. Forsch., 151, 13 (1973).

12. B. Kan, S. A. Goldblith, and B. E. Proctor, Food Res., 22, 509 (1957).

13. M. Karel, Proceedings of the Third International Congress of Food Science and Technology, Washington, D. C., August 1970, Institute of Food Technology, Chicago, Ill., 19 , pp. 551-560.

14. G. S. Murray, Nucleonics, 20(12), 50 (1962).

15. National Academy of Sciences-National Research Council, Radiation Preservation of Foods, NAS-NRC, Washington, D. C., 1965.

16. J. W. Olander, Mod. Plastics, 38(10), 105 (1961).

17. B. E. Proctor and M. Karel, Mod. Packag., 28(6), 137 (1955).

18. B. E. Proctor, E. E. Lockhart, S. A. Goldblith, A. V. Grundy, G. E. Tripp, M. Karel, and R. C. Brogle, Food Technol., 8, 530 (1954).

19. United States Army Quartermaster Corps, Radiation Preservation of Foods, U. S. Government Printing Office, Washington, D. C., 1957.

20. E. Wierbicki, A. Anellis, J. J. Killoran, E. L. Johnson, M. H. Thomas, and E. S. Josephson, High Dose Radiation Processing of Meat, Poultry and Seafood Products, paper presented at the Third International Congress of Food Science and Technology, Washington D. C., 1970, report available from the U. S. Army Natick Labs, Natick, Mass.

PRESERVATION THROUGH TEMPERATURE REDUCTION

Chapter 5

PRESERVATION OF FOOD BY STORAGE AT CHILLING TEMPERATURES

Owen Fennema

CONTENTS

I. INTRODUCTION

Storage of foods at temperatures above freezing and below 15° (59° F) is known as refrigerated or chilling storage. Chilling storage is widely used because it generally results in effective short-term preservation by retarding the following occurrences:

1. Growth of microorganisms
2. Postharvest metabolic activities of intact plant tissues and postslaughter metabolic activities of animal tissues

3. Deteriorative chemical reactions, including enzyme-catalyzed oxidative browning or oxidation of lipids and chemical changes associated with color degradation, autolysis of fish, and loss of nutritive value of foods in general
4. Moisture loss

Chilling is also used for such nonpreservative purposes as crystallization; aging of beef, wine, and cheese; and to facilitate such operations as pitting of cherries and peaches, cutting of meat, and slicing of bread.

II. CONSIDERATIONS RELATING TO STORAGE OF FOOD AT CHILLING TEMPERATURES

A. Normal Behavior of Food Stored at Chilling Temperatures

1. Plant Tissues (Fruits, Vegetables, Seeds, Nuts)

The most important characteristic of harvested, intact plant tissues is the persistence of aerobic respiration throughout the storage life of the product. Aerobic respiration involves metabolism of carbohydrates and organic acids in the presence of atmospheric oxygen, with the ultimate production of carbon dioxide, water, heat, and small amounts of organic volatiles and other substances [27]. For maximum storage life of plant tissues at chilling temperatures, it is desirable that (1) aerobic respiration be allowed to continue at a slow rate so that the maintenance processes associated with life continue to function and the natural protective coatings which hinder invasion of microorganisms remain intact, and (2) the temperature be suitably low so that major deteriorative reactions are slowed as much as possible. When the keeping qualities of various fruits and vegetables are compared, an association (not necessarily a cause and effect relationship) is often found between the rate of respiration and the rate of quality loss; i.e., commodities which normally respire rapidly often have short storage lives [37]. This relationship is evident from the data in Table 5-1, where most of the rapidly respiring (expressed in terms of heat evolution) commodities on the left half of the table are highly perishable. Two other points about the data in Table 5-1 are worthy of comment: (1) at a constant temperature, species (and even varieties) exhibit large differences in rates of respiration, and (2) the amount of heat generated is sufficient to influence refrigeration requirements. Thus any calculation of refrigeration requirements for intact plant tissues must include heat of respiration, sensible heat of the product, and heat leakage into the storage facility.

When temperature reduction is used to slow the rate of aerobic respiration, this usually delays senescence and decay of plant tissues (except for those susceptible to chilling injury) and enables some fruits to be ripened at controlled rates.

In addition to large differences in the rates at which various plant tissues respire, a given tissue frequently does not respire at a constant rate when held at a constant temperature. This irregularity in rate of respiration is most apparent in fruits which undergo a "climacteric" (a relatively abrupt but temporary rise in the rate or respiration). This behavior is shown in the upper curve of

TABLE 5-1

Rates of Respiration of Fruits and Vegetables Based on Evolution of Heat from 1 ton of Product during a 24 hr Period at 4.5° (40° F)[a]

Rapid (>10,000 Btu)	Moderately Rapid (5000-10,000 Btu)	Moderately Slow (2000-5,000 Btu)	Slow (<2000 Btu)
Asparagus	Beans, lima	Apples (yellow transparent)	Apples
Beans (green)	Brussels sprouts	Beets (topped)	Cabbage
Broccoli	Lettuce, leaf	Carrots (topped)	Cranberries
Corn, sweet	Raspberries	Cauliflower	Grapefruit
Okra	Spinach	Celery	Grapes
Peas (green)	Strawberries	Cherries (sour)	Lemons
		Lettuce, head	Onions, mature
		Potatoes, sweet	Oranges
		Turnips	Peaches
			Plums
			Potatoes, mature
			Tomatoes

[a] Hardenburg and Lutz [38].

Figure 5-1. The climacteric maximum frequently but not always occurs at optimum ripeness of the fruit and the sharp decline in respiration rate following the maximum is often associated with overmaturity, a stage referred to as senescence or the "state of growing old." Thus the climacteric maximum can be thought of as a point dividing growth and maturation from deterioration and death.

The exact nature (time after harvest, duration, extent of rise) of the respiration curve produced by fruits which exhibit a climacteric varies greatly depending on species, variety, maturity, and environmental conditions [6].

The climacteric phenomenon is not exhibited by vegetables and it does not occur in some fruits [22, 26]. Table 5-2 contains a listing of fruits for which a climacteric has been conclusively demonstrated and fruits for which a climacteric has never been observed or is doubtful. Nonclimacteric fruits generally exhibit a decline in respiratory activity during storage, as illustrated by the lower curve of Figure 5-1. The respiratory behavior of vegetables is similar to that of nonclimacteric fruits except that many vegetables undergo an abrupt decline in rate of respiration immediately following harvest, and then a more gradual decline in rate during storage [67]. Many members of the nonclimacteric group of fruits

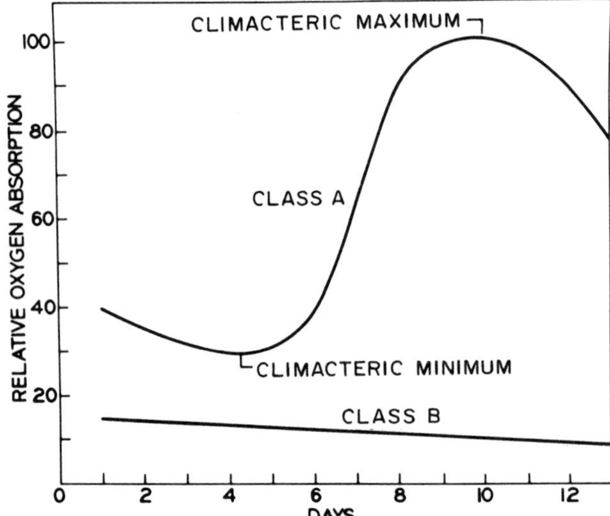

FIG. 5-1. Respiratory behavior of climacteric and nonclimacteric fruits. From Biale [6]; courtesy of Springer-Verlag. Class A represents climacteric fruits. Class B represents nonclimacteric fruits.

and vegetables respire more slowly than those of the climacteric group but there are exceptions to this behavior.

Fruits in general are harvested when fully mature and either ripe or unripe. Harvesting in an unripe state is often done with fruits that are capable of ripening during storage (some such as citrus fruits are not capable of ripening during storage), particularly with those fruits that are soft when ripe and with those that retain optimum ripeness for only a relatively short period (e.g., bananas, avocados, pears; [13]). Vegetables are frequently harvested when immature, tender and succulent.

2. Animal Tissues

Slaughter of an animal results in disruption of the natural protective coating and curtailment of most activities common to life. A major consequence is that the tissue is left with none of the life processes which formerly resisted deterioration. Lost is the ability to effectively resist invasion by microorganisms — an ability so effective in living animals that many interior tissues exist in a sterile or near sterile condition. Also lost is the efficient oxygen-carbon dioxide exchange system (breathing and blood circulation) which is necessary for normal aerobic respiration (glucose + O_2 →→ CO_2 + H_2O + energy). Following cessation of blood circulation, the internal supply of oxygen rapidly diminishes and aerobic respiration ceases. Anaerobic respiration (glycogen →→ glucose →→ lactic acid + energy) occurs in its place, causing the pH to fall from a physiological value somewhat above 7 to an ultimate value ranging from about 5.1 to 6.5. The exact ultimate pH depends on the type of muscle and how it has been handled before and

TABLE 5-2

Respiratory Characteristics of Fruits[a]

Climacteric (fully demonstrated)	Nonclimacteric (not demonstrated or doubtful)
Apple (Pyrus malus)	Cherry (Prunus avium)
Apricot (Prunus armeniaca)	Cucumber (Cucumis sativus)
Avocado (Persea gratissima)	Fig (Ficus carica)
Banana (Musa sapientum)	Grape (Vitis vinifera)
Cherimoya (Annona cherimola)	Grapefruit (Citrus paradisi)
Feijoa (Feijoa sellowiana)	Lemon (Citrus limon)
Mango (Mangifera indica)	Melon (Cucumis melo)
Orange (Citrus sinensis)	Pineapple (Ananas comosus)
Papaya (Carica Papaya)	Strawberry (Fagaria vesca americana)
Passion fruit (Passiflora edulis)	
Papaw (Asimina tribola)	
Peach (Prunus persica)	
Pear (Pyrus communis)	
Plum (Prunus americana)	
Sapote (Casimiroa edulis)	
Tomato (Lycopersicum esculentum)	

[a]Biale [6]; Trout et al. [90].

after slaughter [19]. All respiration ceases after a postmortem period of 1-36 hr (the time for completion depending on the same factors just mentioned for ultimate pH).

At some point during the short-term course of anaerobic respiration (anaerobic glycolysis) the ATP content of muscle decreases to a level such that rigor mortis ensues (the muscle changes from a soft, pliable condition to an inextensible, generally firm condition). The diminished level of ATP and the lowering of pH tend to destabilize muscle proteins, especially sarcoplasmic proteins, and the muscle suffers a decrease in water-holding capacity. A normal lowering of pH in properly chilled muscle is desirable, however, since this helps retard growth of microorganisms and favors retention of a pleasing color [42, 45]. To achieve good qualities (tenderness, water-holding capacity, color), carcasses should be promptly cooled following slaughter and the handling practices should encourage slow respiration so that the muscle is cool when the desired ultimate pH value is attained [42, 47].

Following onset of rigor, some muscles are stored for an additional period of time to allow for resolution of rigor and an associated improvement in tenderness and water-holding capacity [26, 45, 50].

3. Nontissue Foods (Milk, Shell Eggs, Manufactured Foods)

Since nontissue foods are physiologically inactive (the physiological activity of eggs, if it does in fact persist, is of no concern), the various complexities associated with the physiological activities of plant and animal tissues are not encountered. Although nontissue foods deteriorate at greatly different rates, effective short-term preservation can nearly always be accomplished by prompt cooling to near their respective freezing points. Highly perishable nontissue foods such as milk, must be handled under strict sanitary conditions and be promptly cooled.

B. Causes of Quality Loss in Food Stored at Chilling Temperatures

Loss of food quality during chilling storage can occur by one or more of the following means.

1. Microbiological Activity

Growth of microorganisms is a major cause of spoilage of most natural and manufactured foods. Animal tissues are especially susceptible to spoilage by microorganisms (especially bacteria) since (a) slaughtering destroys the muscle's ability to combat microorganisms, (b) dressing and cutting operations disrupt the protective coating and contaminate previously sterile or near sterile tissues with microorganisms from the intestines and exterior surfaces of the animal, and (c) muscle is an excellent substrate for growth of microorganisms. Plant tissues are moderately susceptible to spoilage by microorganisms, especially fungi.

2. Physiological and Other Chemical Activities

Depending on the product, quality loss resulting from physiological or other chemical activities can be slight, moderate, or severe. Fruits and vegetables, because of natural senescence, eventually become unacceptable under optimum conditions of chilling storage. Low nonfreezing storage temperatures can result in physiological disorders and damage to the quality of some fruits and vegetables (chilling injury; see Section II. C. 2). Unless proper precautions are observed, normal respiration can cause accumulation of carbon dioxide and depletion of oxygen to levels which instigate physiological disorders and quality defects in fruits and vegetables.

Autolysis is a major cause of damage in fish. Natural postslaughter metabolism of animal tissue, if improperly controlled, can result in excessive toughness, poor water-holding capacity, and poor color — the exact defects depending on the type of tissue and the conditions.

Chemical reactions, such as lipid oxidation, pigment degradation, protein denaturation, development of various off flavors, vitamin degradation, etc., can reduce the quality of both tissue and nontissue foods. Furthermore, chemicals which are added intentionally or unintentionally to the product or its environment (e.g., food additives, sulfur dioxide, ozone, ammonia, carbon dioxide, ethylene) can result in damage to some products. For example, ammonia is an unintentional additive (leaks in refrigeration equipment) the presence of which should be carefully avoided since it can cause severe damage to many food products [20, 37, 38, 94].

3. Physical Damage

Bruising is a major cause of damage to fruits. Dehydration, if excessive, can cause substantial damage to most food products (e.g., wilting or shrivelling of fruits and vegetables, enlargement of air sacs in eggs). Excision of muscle prerigor can result in excessive contraction, toughness, and poor water-holding capacity.

C. Temperature

1. Desirable Consequences of Chilling Temperatures

Control of temperature in the storage facility is important because it has a profound effect on growth of microorganisms, metabolic activities, other chemical reactions, and moisture loss. Growth of pathogenic microorganisms occurs rapidly in the range $10^\circ - 37^\circ$ ($50^\circ - 98^\circ$F) but only slowly in the range $3.3^\circ - 10^\circ$ ($38^\circ - 50^\circ$F). Below 3.3°(38°F) pathogenic microorganisms can no longer grow [24, 25]. Growth of mesophilic and thermophilic microorganisms is greatly retarded at chilling temperature. Psychrotrophic microorganisms, of course, grow well in the range $0^\circ - 15^\circ$ ($32^\circ - 59^\circ$F), but in most instances their growth is much slower in this range than at temperatures of $15^\circ - 45^\circ$ ($59^\circ - 113^\circ$F). Chilling temperatures, therefore, substantially retard spoilage caused by microorganisms.

Loss of quality resulting from physiological activity and other chemical reactions also can be effectively reduced by chilling storage. As the temperature is lowered within the range of chilling temperatures, decreases occur in the rates of respiration of those fruits and vegetables which are not subject to chilling injury. This occurrence, illustrated for representative products in Table 5-3, effectively retards the process of senescence and results in an extension of storage life.

Similarly, the ripening of fruits can be controlled by the storage temperature. Within the range of chilling temperatures, the rate of ripening usually declines as the temperature is reduced and ripening ceases at temperatures below about 4°

TABLE 5-3

Effect of Temperature on Rates of Respiration of Fruits and Vegetables[a]

Commodity	Btu per ton per 24 hr at:[b] 0° (32° F)	4.5° (40° F)	15.5° (60° F)
Apples	300-1500	600-2700	2300-7900
Peas, green	8200-8400	13,200-16,000	39,300-44,500
Strawberries	2700-3800	3600-6800	15,600-20,300

[a]Hardenburg and Lutz [38].

[b]Divide Btu by 220 to determine milligrams CO_2 evolved per hour per kilogram of product.

(39° F) (ripening ceases at higher temperatures for fruits subject to chilling injury). Below about 5° (41°F) failure to ripen is associated with the absence of a climacteric rise [29].

Several other examples will be cited to illustrate the great dependence of quality retention on temperature of storage. Figure 5-2 shows that prompt cooling of sweet corn is necessary for retention of its desirable sweetness. Data in Table 5-4 illustrate the general need for low handling and storage temperatures for retention of vitamin C in vegetables.

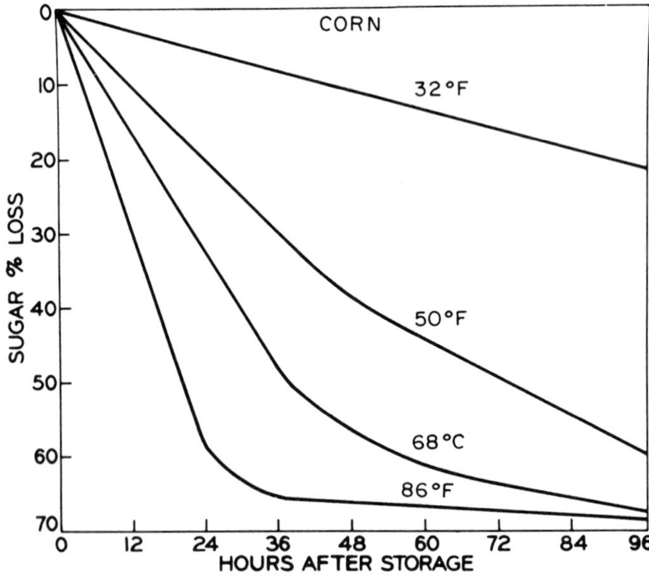

FIG. 5-2. Effect of temperature on sugar loss in sweet corn. Appleman and Arthur [2].

TABLE 5-4

Loss of Vitamin C in Fresh Vegetables During Storage[a]

Commodity	Temperature of storage (° C)	Total vitamin C Fresh (mg/100 g)	Loss in % after: 24 hr	48 hr
French beans	4	25.6	22	34
	12		26	40
	20		36	52
Peas	4	36	7	10
	12		18	29
	20		23	36
Spinach (spring)	4	39.8	20	32
	12		27	43
	20		34	54
Spinach (autumn)	4	72.3	27	33
	12		41	51
	20		56	79
Spinach (winter)	4	120.5	5	8
	12		5	8
	20		14	20

[a]Zacharias [98].

Surprisingly small changes in temperature sometimes result in marked changes in the rate of quality loss. For example, the storage life of cod and haddock is 22-29 days at -1° to -2° (28° - 30°F) as compared to 13 days at the temperature of melting ice [18]. Likewise, fresh poultry has a 40% longer (additional 12 days) storage life at -2° (28° F) than at 0° (32° F) [83], and Williams pears have a 12 week storage life at -1.5° (29° F) as compared to 10 weeks at -1° (30° F), 9 weeks at -0.5° (31°F), 7-1/2 weeks at 0° (32°F) and 6 weeks at 1° (34° F) [27]. Thus, careful control of handling and storage temperatures is necessary in order to achieve maximum storage life and minimal loss of nutrients.

2. Undesirable Consequences of Chilling Temperatures

There are some instances where damage can occur if cooling is too fast and/or if the temperature is reduced to near freezing. For example, peaches undergo damage to quality if cooled too rapidly. A serious defect referred to as "wooly texture" develops under this circumstance [27]. Cold shortening of muscle, chilling injury of plant tissues, and staling of bread are additional types of injury that can result from low chilling temperatures, and these defects are discussed below.

a. Cold Shortening of Animal Muscle. Animal muscle which is excised immediately after slaughter (not common) can undergo a detrimental occurrence known as "cold shortening" if it is promptly cooled prerigor to the range of 0° - 5° (32°-41° F). As the name implies, the excised muscle shortens greatly, resulting in poor water-holding capacity and excessive toughness (usually). Cold shortening has been observed with prerigor excised muscle from poultry [77], beef [48], and lamb [16]. It has been suggested that cold shortening results because low chilling temperatures impair the ability of the sarcoplasmic reticulum to regulate the distribution of calcium ions in muscle.

The possible adverse consequences resulting when freshly slaughtered animal carcasses are cooled very rapidly to 0°- 5° are less well documented than with excised muscle. It has been demonstrated, however, that intact longissimus dorsi muscles of lamb carcasses cooled rapidly to 0°-5° (32°- 41°F) are substantially tougher than similar muscles from carcasses which are cooled less rapidly [52]. Supporting evidence indicated that cold shortening was responsible for the toughening effect. Although the skeleton tends to restrict muscle shortening, Marsh and Leet [51] have indicated several means by which carcass muscles can undergo cold shortening.

1. Several muscles, such as the sternomandibularis (neck muscle), are relatively free to shorten because the neck has been severed.
2. The economically important longissimus dorsi muscle is comparatively free to shorten because its fibers in the natural state are firmly attached to the skeleton at one end only.
3. A single muscle attached at both ends may maintain a constant total length while certain portions contract and other portions stretch. This can easily result because of temperature differentials within a muscle.

Muscles in prerigor carcasses of animal species other than lamb probably are susceptible to cold shortening, especially if they contain large quantities of the sensitive red fibers.

b. Chilling Injury. A substantial number of fruits and vegetables, especially those of tropical or subtropical origin, develop physiological disorders when exposed to temperatures below their optimum storage temperatures but above their freezing points. This type of disorder is known as "chilling injury," "low-temperature injury," or "cold injury," and the mechanisms of damage are discussed in Volume 1 of this series. Care must be taken to store these products at temperatures above their respective critical temperatures (lowest temperature at which chilling injury does not occur) if more than a brief storage period (depends on the product) is anticipated.

The symptoms of chilling injury differ depending on the product. Common symptoms include browning (external or internal or both), pitting or other skin blemishes, excessive rotting, and failure to ripen (fruits). One of the first indications of chilling injury in apples stored at a constant temperature is an abnormal increase in the rate of respiration [28].

Factors which influence the onset of chilling injury are discussed below, and much of this material is derived from an excellent article by Fidler [28].

1. Temperature. Different species and varieties exhibit different critical temperatures for chilling injury, ranging from a high of 14.4° (58° F) for Lacatan bananas to a low of 1° (34° F) for certain varieties of apples. Critical temperatures for various susceptible fruits and vegetables are listed in Table 5-5. Many critical temperatures fall in the range of 8°-12° (46-54°F).
2. Time at subcritical temperature. A certain time must elapse at subcritical temperatures for symptoms of chilling injury to develop. Minimum exposure times needed to induce chilling injury range from 12 hr for bananas, to 9-12 days for tomatoes, to 2-3 months for citrus fruits [53]. The minimum time needed for symptoms of chilling injury to develop also depends on the subcritical temperature used [29]. Symptoms often develop more rapidly at high subcritical temperatures; however, given sufficient time, maximum injury occurs at the lowest nonfreezing temperature. This point was well illustrated by Van Der Plank and Davies [93]. They stored Marsh seedless grapefruit at temperatures ranging from -0.5° to 10° (31° - 50° F) and examined them at various storage times. Fruit stored at 10° was not damaged during the storage period used. After 25 days, fruit stored at 7.5° exhibited pitting of the surface, whereas fruit at -0.5° exhibited no damage. After 33 days fruit stored at 5° exhibited the most severe injury, whereas after 52 days fruit stored at -0.5° was most severely injured.
3. Metabolic state of the tissue. Susceptibility of fruits and vegetables to chilling injury frequently depends on maturity (some apples and bananas are most susceptible at the respiratory climacteric), composition (mineral and acid content of apples), and the growth environment (a given variety of apples can exhibit significantly different critical storage temperatures depending on the climatic conditions that prevailed during growth).
4. Composition of the atmosphere. Alterations in the carbon dioxide and oxygen contents of the storage atmosphere can influence the incidence of chilling injury, but reported results are inconsistent and generalizations are not possible.
5. Rate of water loss. Rates of water loss from fruits and vegetables can influence susceptibility to chilling injury, but results vary depending on the product.

c. Staling of Bread. Bread represents a notable exception to the rule that all quality attributes of nontissue foods are retained for longer periods at chilling temperatures than at room temperature. It is evident from Figure 5-3 that decreasing temperature in the range of 32° - 0° (90° - 32°F) accelerates firming of bread [63].

D. Relative Humidity

Control of relative humidity in the storage facility is essential if maximum storage life of foods is desired. Relative humidities which are greater than optimum (1) favor growth of microorganisms and result in excessive food spoilage or (2) result in abnormal splitting of some fruits (e.g., apples and plums) [96].

TABLE 5-5

Some Fruits and Vegetables which are Susceptible to Chilling Injury[a]

Commodity	Approximate lowest safe temperature °F	°C	Nature of injury when stored above freezing and below critical temperature[b]
Apples—certain varieties	36-38	2-3	Internal browning, brown core, soggy breakdown, soft scald
Avocados	40-55	4-13	Grayish-brown discoloration of flesh
Bananas, green or ripe	53-56	12-13	Dull color when ripened
Beans (snap)	45	7	Pitting and russeting
Cranberries	36	2	Rubbery texture, red flesh
Cucumbers	45	7	Pitting, water-soaked spots, decay
Eggplants	45	7	Surface scald, alternaria rot
Grapefruit	50	10	Scald, pitting, watery breakdown
Lemons	58	14	Pitting, membranous staining, red blotch
Limes	45-48	7-9	Pitting
Mangos	50-55	10-13	Grayish scaldlike discoloration of skin, uneven ripening
Melons			
Honey Dew	45-50	7-10	Pitting, surface decay, failure to ripen
Casaba	45-50	7-10	Pitting, surface decay, failure to ripen
Crenshaw and Persian	45-50	7-10	Pitting, surface decay, failure to ripen
Watermelons	40	4	Pitting, objectionable flavor
Okra	45	7	Discoloration, water-soaked areas, pitting, decay
Olives, fresh	45	7	Internal browning
Oranges, California and Arizona	38	3	Pitting, brown stain

TABLE 5-5 continued

Commodity	Approximate lowest safe temperature °F	°C	Nature of injury when stored above freezing and below critical temperature[b]
Papayas	45	7	Pitting, failure to ripen, off flavor, decay
Peppers, sweet	45	7	Sheet pitting, alternaria rot on pods and calyxes
Pineapples	45-50	7-10	Dull green when ripened
Potatoes	38	3	Mahogany browning (Chippewa and Debago), sweetening
Pumpkins and hardshell squashes	50	10	Decay, especially alternaria rot
Sweet potatoes	55	13	Decay, pitting, internal discoloration
Tomatoes			
Ripe	45-50	7-10	Watersoaking and softening, decay
Mature green	55	13	Poor color when ripe; alternaria rot

[a]Lutz and Hardenburg [49].

[b]Often these symptoms are apparent only after removal to warm temperatures, as in marketing. In many instances, temperatures lower than stated can be tolerated for periods which are relatively short in comparison to the maximum storage life of the product being considered.

Relative humidities which are less than optimum result in wilting or shrivelling of fruits and vegetables, damage to the appearance of animal tissues, and unnecessary economic losses because of a reduction in product weight. Moisture losses of 3-6% will result in a marked loss in quality of many vegetables [37]. Between the extremes discussed above, relative humidity can influence the severity of physiological disorders, especially flesh breakdown in apples and skin pitting of citrus fruits [73, 96].

The relative humidity in the storage facility is determined primarily by the temperature difference between the air and the refrigerating coils [27, 81]. A small temperature difference (about 0.5°-1° (1°-2°F) during storage of a cool product) is needed in order to maintain a suitably high relative humidity, and this is possible only if the surface area of the refrigerating coil is sufficiently large and air passage across the coil is sufficiently great to provide adequate heat transfer [81].

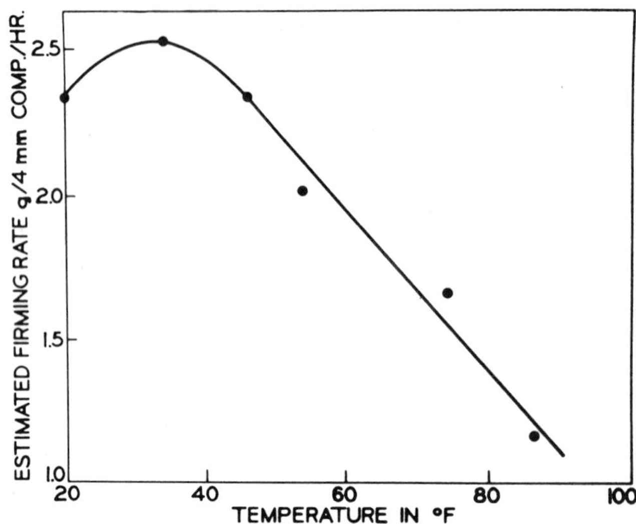

FIG. 5-3. Firming rate of commercial white bread as influenced by temperature. Pence and Standridge [63]; courtesy of the American Association of Cereal Chemists.

E. Air Circulation and Purification

Adequate circulation of air during chilling storage is necessary to (1) maintain a uniform temperature, (2) maintain a uniform composition of the atmosphere, (3) rapidly cool products which enter storage at temperatures warmer than desired, and (4) facilitate air purification when this treatment is desired.

Careful design of air flow patterns (size and location of air ducts, amount of product, and manner of stacking) is an absolute necessity if the above objectives are to be achieved [38, 62]. Moderate to rapid air circulation can result in some product dehydration, but this can be slight provided a proper relative humidity is maintained [81].

Purification of air is sometimes accomplished to remove undesirable odors and to remove volatiles emanating from fruit. In some instances, removal of volatiles can retard ripening, and some claim that it retards the onset of apple scald (stored apples sometimes develop this disorder, which is characterized by the surface formation of sharply defined brown patches that are often arranged in a ribbonlike pattern [15]). Passage of air through activated charcoal or simple ventilation are common methods of purification.

F. Adjuncts to Chilling Temperature

1. Light

As a general rule the storage facility should be dark except during periods of product entry or removal. Darkness or reduced light delay the sprouting of potatoes and onions and may, with certain products, help retard development of off flavors and discoloration.

Ultraviolet light can suppress growth of bacteria and molds on smooth surfaces and for this reason it is sometimes used during rapid aging of beef at temperatures well above freezing [91]. However, care should be used in applying ultraviolet light since it catalyzes oxidative reactions and thus can hasten the onset of off flavors and discoloration.

Exposure to ultraviolet light is of no advantage for fruits and vegetables and is decidedly detrimental to the quality of many high-fat products (cheese, lard, butter, cream [15]). Thus long-term application of ultraviolet light is undesirable for most food products.

2. Pasteurization

Mild heat treatments (pasteurization) can be used to reduce the population of microorganisms on fresh fruits (mainly those on or near the surface) and thereby retard decay during chilling storage. As compared to chemical treatments, pasteurization has the advantage of not leaving an undesirable residue.

Pasteurization is sometimes advantageous for such fruits as lemons, papayas, nectarines, peaches, and plums [15, 95]. The treatment involves immersing the fruit in water at 46°- 54° (115° - 130°F) for a period of 1-4 min.

3. Waxing or Oiling

Certain fruits and vegetables are waxed to either reduce dehydration (cucumbers, root crops) or to improve appearance. Waxing is common for rutabagas, cucumbers, mature green tomatoes, and cantaloupes and is not uncommon for peppers, turnips, honeydew melons, sweet potatoes, and certain other crops [37]. The wax consists of paraffin or a vegetable wax-paraffin combination.

The storage life of shell eggs can be effectively extended by dipping them in light mineral oil 12-24 hr after laying. This treatment retards both dehydration and loss of carbon dioxide. Maintaining the original carbon dioxide content of eggs is an important factor in preserving the quality of the egg white.

4. Oxygen and Carbon Dioxide Contents of the Atmosphere

Altering the oxygen and carbon dioxide contents of the storage atmosphere is done to decrease the rates at which intact fruits and vegetables respire aerobically, to retard growth of microorganisms, and sometimes to inhibit oxidative reactions. The first two goals deserve further elaboration.

a. Retarding Aerobic Respiration of Plant Tissues by Altering the Composition of the Atmosphere. For maximum storage life of plant tissue it is desirable that aerobic respiration be sustained at a slow rate. For some products this can be satisfactorily accomplished by reducing the oxygen content of the atmosphere, increasing the carbon dioxide content of the atmosphere, and reducing the temperature to a suitable level. Shown in Table 5-6 are the effects of various levels of carbon dioxide and oxygen on the rate of respiration (carbon dioxide evolution and oxygen

TABLE 5-6

Rates of Respiration of Apples at 3.3° (38° F), Relative to a Rate of 100 in Air[a]

$\% CO_2/\% O_2$	Rates, relative to air = 100	
	CO_2	O_2
Air	100	100
0/10	84	80
0/5	70	63
0/3-2	63	52
0/1.5	39	—
10/10	40	60
5/16	50	60
5/5	38	49
5/3	32	40
5/1.5	25	29

[a]Cox's Orange Pippin apples; other varieties exhibit similar behavior. From Fidler and North [31].

consumption) of apples stored at 3.3°. Apples provide the most valuable example since they are by far the most important crop that is stored in atmospheres of controlled composition. Reducing the oxygen concentration from the normal value of 21% to 10% or lower, decreases the rate of respiration, both in the presence and in the absence of carbon dioxide. Increasing the carbon dioxide concentration over the range of 0.03% (normal) to 10% also decreases the rate of respiration in the presence of a broad range of oxygen concentrations [31, 32]. At temperatures above 3.3° the results are qualitatively similar [32].

If the oxygen content of the storage atmosphere is reduced too drastically, normal aerobic respiration is suppressed and at least partially replaced by anaerobic respiration [68]. As a result, alcohol and other toxic end products of respiration accumulate in the plant tissues, the rate of respiration sometimes increases as compared to the rate obtained at oxygen levels just adequate to sustain aerobic respiration, physiological disorders eventually occur, and storage life generally is shortened [33, 68]. The threshold concentration for the onset of significant anaerobic respiration is dependent on temperature (oxygen needs are greater at higher temperatures), age of product (effect varies with different products), and the type of product (permeability of the tissue to oxygen, shape of the intact tissue, natural differences in respiration rate, etc.) [29, 32, 33, 68]. Examples of approximate oxygen threshold concentrations for anaerobic respiration are 1.8% oxygen for Cox's Orange Pippin apples at 3.5° [33]; 2.3% and 1.2% oxygen for asparagus at 20° and 10°, respectively; less than 1% oxygen for spinach at 20°; and 4% oxygen for carrots and peas at 20° [68].

Excessively high levels of carbon dioxide depress all metabolic activities (both aerobic and anaerobic respiration of plant tissues), either delay or eliminate the climacteric, and often result in rapid onset of quality defects [5, 43]. The first changes in quality are generally off flavors, and these are eventually followed by discoloration, breakdown of the tissues, and decay [10, 75].

b. Growth of Microorganisms as Influenced by the Carbon Dioxide Content of the Atmosphere. High levels of carbon dioxide (greater than 10% can significantly retard growth of microorganisms that exist on the surface of foods [9, 10, 12, 15, 17, 55, 57, 59, 75, 76, 84, 88, 89]. The minimum effective level of carbon dioxide depends on such factors as temperature (growth retardation can be achieved with less carbon dioxide at lower temperatures), exposure time, population of organisms, and nutritive quality of the food.

5. Other Chemicals

A variety of chemicals other than carbon dioxide and oxygen are used during the handling and storage of foods at chilling temperatures. These chemicals are applied in the form of vapors (or aerosols) or aqueous solutions, or by addition to packaging materials (box liners, wrappers, or shredded packing materials). Ideally these chemicals should act quickly, be convenient to use and economical, and should have no undesirable effects on wholesomeness, nutritive value, or quality. These chemicals, classed according to function, are listed below. Latest regulations should be consulted to determine legality of use and levels allowed.

a. Chemicals for Controlling Growth of Microorganisms and Insects. To retard or stop growth of microorganisms and insects, appropriate chemicals (e.g., chlorine, acetates, ozone, sulfur dioxide, methyl bromide, diphenyl) are commonly applied to fruits, occasionally to vegetables, and rarely to other foods [14, 15, 56, 76, 94, 95].

b. Chemicals for Controlling Ripening of Fruits [15, 22, 27, 94]. The ripening of bananas, pears, and some other fruits can be induced by adding small amounts of ethylene (e.g., 1 cu ft ethylene per 1000 cu ft of air is sufficient for bananas) to the atmosphere.

Ripening of mangoes can be retarded by 2,4,5-trichlorophenoxy acetic acid (2,4,5-T).

c. Chemicals for Controlling Physiological Disorders [14, 76]. Scald of apples can be retarded by applying diphenylamine (often applied to wrappers) or ethoxyquin prior to storage. Mineral oil added to the wrappers of apples also retards development of apple scald, presumably by absorbing volatile emanations.

d. Sprout Inhibitors. Phenyl carbamates (isopropylphenyl carbamate, 3-chloro-isopropyl-n-phenyl carbamate), maleic hydrazide, or vapors of nonyl alcohol are used to retard the formation of sprouts on such commodities as onions, potatoes, and carrots [15, 27, 38].

e. Antioxidants. Antioxidants, such as butylated hydroxyanisole (BHA), butylated hydroxytoluene (BHT), L-ascorbic acid, and many others, are frequently used to inhibit oxidation of susceptible commodities [15].

f. Deodorizers. Ozone is sometimes used to oxidize odorous volatile compounds and render them innocuous [38]. However, ozone must be used with care since it can promote undesirable oxidative reactions in susceptible products. Simple ventilation or absorption of odors on activated charcoal are safer procedures.

g. Color Modifiers. Ethylene gas, because it degrades chlorophyll, is sometimes used at a level of 10 ppm in air to degreen citrus fruit [22, 27, 94].

6. Ionizing Radiation

Substerilizing doses of ionizing radiation (see Chapter 4) can effectively extend the storage life of some foods which are stored at chilling temperatures. However, this technique currently is not permitted for most foods.

G. Summary of Product Characteristics which Influence Conditions Selected for Chilling Storage

Since behavioral characteristics of plant and animal tissues at chilling temperatures are discussed under several headings in the above discussion, it is appropriate to summarize the major points. This is done in Table 5-7.

III. APPLICATIONS AND PROCEDURES

A. Precooling

Precooling is a treatment administered prior to storage or long-distance shipping to promptly reduce the product temperature from that which exists at the time of slaughter or harvest to a chilling temperature that is approximately optimum. The purpose is to retard the rate of product deterioration and to reduce the refrigeration capacity needed in the transport vehicle or storage facility. Precooling is especially needed for products with short postharvest storage lives, such as most berries, asparagus, broccoli, sweet corn, and spinach, and for products where growth of microorganisms must be promptly retarded to avoid toxicological problems and quality defects (animal tissues). The profound effect that a reduction in temperature has on the storage life of perishable food products was discussed in Section II.

Precooling is less important for those fruits and vegetables that are only semiperishable and for those that require curing or ripening. This is especially true if the ambient temperature is relatively low at the time of harvest [4].

When precooling is recommended, it generally should be accomplished as rapidly as possible. Prompt cooling is nearly always advantageous, and the ultimate temperature ideally should be just above freezing for milk, eggs, those fruits and vegetables not subject to chilling injury, and those animal tissues not subject to cold shortening. The United States Department of Agriculture stipulates that poultry be chilled to less than 4.5° (40° F) in accordance with the following schedule [92]: (1) if 4 lb or less, less than 4 hr; (2) if 4-8 lb, less than 6 hr; and (3) if greater than 8 lb, less than 8 hr.

Products with small masses, such as peas, can be easily cooled in a few minutes, whereas cooling of large carcasses of beef sometimes involves 2-3 days.

Precooling can be accomplished with moving air, moving water (hydrocooling), ice or ice-slush, or by vacuumization [4, 72].

1. Precooling with Moving Air

This method is widely used because it is simple, economical, sanitary and relatively noncorrosive to equipment. Major disadvantages of precooling with air are the dangers of excessive dehydration and the possibility of freezing the product if air temperatures below 0° are utilized. Products which are precooled in air include mammalian meat, citrus fruits, grapes, cantaloupes, green beans, plums, nectarines, sweet cherries, and apricots.

2. Hydrocooling

Hydrocooling is a common method of precooling because of its simplicity, economy (least expensive method for rapid precooling), and rapidity. The product to be cooled is immersed in, flooded with, or sprayed with cool water (often near 0°) containing a mild disinfectant, such as chlorine or an approved phenol compound. Products such as celery, asparagus, peas, sweet corn, radishes, topped carrots, cantaloupes, peaches, and tart cherries are precooled by this method.

3. Precooling with Ice or Ice-Slush

Precooling with crushed ice or ice-slush is simple and effective if done properly. However, considerable labor is often involved when crushed ice is used. Products such as cabbage, cantaloupes, peaches, root crops (radishes, carrots), and fish are frequently cooled by direct contact with crushed ice.

Nearly all poultry and some fish are precooled by contact with an agitated mixture of ice and water (ice-slush). Precooling of poultry in ice-slush is preferred to precooling in air, because product weight is increased by the former method and decreased by the latter.

4. Vacuum Cooling [3, 4, 41]

Vacuum cooling is extremely effective for products possessing the following two properties: (1) a large surface to mass ratio (leafy vegetables), and (2) an ability to readily release internal water. This method involves exposing the product to a pressure of about 4-4.6 mm Hg until evaporative cooling yields the desired

TABLE 5-7

Product Characteristics which Influence the Selection of Conditions Used for Chilling Storage

Class of food	Presence of natural protective coating[a]	Physiological characteristics[a]				Major trend in quality following harvest, slaughter, or manufacture	Common means by which quality becomes unacceptable
		Normal type and duration of respiration	Climacteric	Heat generated	Possible injury caused by abnormal metabolism at chilling temp.		
Fruits	Yes	Aerobic for entire storage life of product	Yes for some	Moderate to substantial	Some are subject to chilling injury	Some improve temporarily (ripen) then deteriorate; others start to deteriorate immediately after harvest	**Microbiological. Physiological, and Pathological. Physical.**
Vegetables	Yes	Aerobic for entire storage life of product	No	Moderate to substantial	Some are subject to chilling injury	Deterioration	Microbiological, physiological, and pathological
Animal tissues	No	Anaerobic, for about 24 hr, then complete cessation	No	Slight	Some are subject to cold shortening	Deterioration, except for temporary improvement of those meats which are aged	Microbiological

Other products such as milk, eggs, and manufactured foods	No, except for shell of egg	None	None	None	None	Deterioration	Microbiological, chemical, physical

[a] Postharvest if plant tissue, postslaughter if animal tissue.

temperature. Cooling times for suitable products, such as lettuce, are less than 1 hr. The extent of cooling is proportional to the water evaporated (evaporation of 1 lb of water removes approximately 1060 Btu). Regardless of the product, the temperature is lowered about 5° for each 1% reduction in water content. Prewetting is beneficial for many products, especially those that have high initial temperatures and those with properties such that substantial amounts of the added water are retained on their surfaces until the vacuum is applied.

For suitable products, vacuum cooling is advantageous because it is rapid and economical as compared to direct icing, which is otherwise likely to be employed. The economy is derived from reductions in the cost of labor and packaging and from decreased product damage. The main disadvantage of vacuum cooling is the large capital investment required. Equipment needs are a large, strong chamber (some are large enough to accommodate an entire railroad car), a device for reducing the pressure (a mechanical vacuum pump or steam ejectors), and a device for condensing water vapors that emanate from the product (needed only when mechanical vacuum pumps are employed).

Large amounts of lettuce and moderate but increasing amounts of celery, cauliflower, green peas, and sweet corn are vacuum cooled.

B. Conditions Recommended for Storage of Food in Ordinary Air at Chilling Temperatures

Shown in Tables 5-8 and 5-9 are temperatures and relative humidities recommended for attainment of maximum storage lives of various perishable foods stored at chilling temperatures. When the optimum temperature of storage is substantially above the freezing point of the product, this indicates the product is susceptible to chilling injury. Air temperature should be maintained within ±1° (±2°F) of the recommended value [40]. Total variation in relative humidity generally should not exceed ±3 - ±5% [38].

In Table 5-10 products are grouped according to the approximate storage lives that can be expected, assuming optimum storage conditions prevail and the food is fresh and of good quality when it enters storage. For more detailed information on conditions leading to optimum storage lives of chilled foods, the reader is referred to the references cited at the bottom of Table 5-8 and to the Refrigeration Research Foundation [70].

Not fully apparent from Tables 5-8, 5-9, and 5-10 are the important influences that such factors as maturity, ripeness, date of harvest, variety, and origin can have on optimum storage temperatures and maximum storage lives for fruits and vegetables. For example, optimum storage temperature is influenced by the ripeness of bananas, lemons, pineapples, and tomatoes; the variety of apples and pears; the harvest date of potatoes; and the origin of tomatoes, grapefruit, oranges, cucumbers, and green beans. Similarly, maximum keeping quality is influenced by the maturity of carrots; the ripeness of lemons, tomatoes, and bananas; and the harvest date of potatoes.

During transit over relatively short distances, or during temporary storage in warehouses, it is usually not feasible to maintain optimum environmental conditions for each product because this necessitates an uneconomical number of separate storage and shipping units. A workable solution is arrived at by recalling that quality changes are dependent on time as well as temperature. Thus, during

TABLE 5-8

Chilling Temperatures Yielding Maximum Storage Lives of Various Foods[a]

Temperature[b]	Product
Just above freezing	Animal tissues — mammalian meats, poultry, fish
	Fruits — apples (some varieties), apricots, berries, lemons (yellow), nectarines, oranges (Florida), peaches, pears, plums
	Vegetables — asparagus, beets (topped), broccoli, brussels sprouts, cabbage (late), carrots (topped), cauliflower, celery, corn (sweet), peas (green), radishes, spinach
	Milk (fluid, pasteurized, grade A)
	Eggs (shell)
2° - 7° (34° - 45°F)	Fruits — apples (some varieties), melons, oranges (except Florida), pineapple (ripe)
	Vegetables — potatoes (Irish, early)
Above 7° (45°F)	Fruits — avocados, bananas, grapefruit, lemons (green), limes, mangoes, pineapples (mature, green), tomatoes
	Vegetables — beans (green), cucumbers, potatoes (Irish, late), potatoes (sweet)

[a]Hardenburg and Lutz [38]; International Institute of Refrigeration [40]; Patchen [62]. These references should be consulted for exact conditions.

[b]Atmosphere contains normal amounts of oxygen and carbon dioxide.

transit or temporary storage it is possible to deviate somewhat from optimum storage temperatures without incurring unacceptable changes in quality. A common course of action is to provide two storage environments, one at 0° (32°F) and 90% relative humidity and the other at 10° (50°F) and 85-90% relative humidity [7, 37]. The 0° facility accommodates eggs, milk, all animal tissues, and those fruits and vegetables not subject to chilling injury. The 10° facility accommodates all fruits and vegetables which are subject to chilling injury. Normal precautions, of course, must be observed to avoid storing incompatible products in the same room (e.g., highly odoriferous products, such as fish and onions, should not be stored in the presence of odor-absorbing products, such as butter).

TABLE 5-9

Relative Humidities Recommended for Various Foods Stored at Optimum Chilling Temperatures[a]

Relative humidity[b]	Product
Less than 85%	Butter, cheese, coconuts, dates, dried fruits, eggs (shell), garlic, hops, nuts, onions (dry)
85-90%	Animal tissues — poultry, mammalian meat except veal
	Fruits — bananas (yellow), citrus, melons, peaches, pineapples, plums, nectarines, tomatoes
	Vegetables — potatoes (sweet), potatoes (Irish, early)
90-95%	Animal tissues — veal, fish
	Fruits — apples (90%), bananas (green), berries, pears
	Vegetables — beans (green), corn (sweet), cucumbers, leafy vegetables, peas (green), potatoes (Irish, late), root crops (topped)

[a]Same references as Table 5-8. These references should be consulted for exact conditions.

[b]Atmosphere contains normal amounts of oxygen and carbon dioxide.

Exact recommendations of the International Institute of Refrigeration [39, 40], the United States Department of Agriculture (Redit, [69]), and the Refrigeration Research Foundation [70] should be followed when it is possible to tailor the environmental conditions during transit or temporary storage to the needs of individual products.

C. Conditions Recommended for Storage of Food in Chilled Atmospheres of Modified Composition

Lowering the level of atmospheric oxygen, raising the level of atmospheric carbon dioxide, or drastically reducing levels of both oxygen and carbon dioxide (nitrogen flushing), when used as supplements to low temperatures, can be significant aids to retaining marketable qualities of some products.

TABLE 5-10

Approximate Storage Lives of Various Foods Stored at Optimum Conditions of Relative Humidity and Chilling Temperature[a]

Maximum storage life[b]	Product
Less than 2 weeks	Animal tissues — fish, poultry, mammalian meat except beef
	Fruits — apricots, bananas, berries, cantaloupes, tomatoes (firm, ripe)
	Vegetables — asparagus, beans (green), broccoli, corn (sweet), cucumbers, peas (green), spinach
2-6 weeks	Animal tissues — beef
	Fruits — avocados, grapefruit, lemons (yellow), limes, nectarines, peaches, pineapples, plums, tomatoes (mature, green)
	Vegetables — lettuce, potatoes (early), radishes
1-4 months[c]	Celery, lemons (green), oranges
Greater than 3 months	Fruits — apples, pears
	Vegetables — beets (topped), cabbage, carrots (topped, mature), potatoes (Irish, late), potatoes (sweet)
	Eggs (shell)

[a]Same references as Table 5-8. These references should be consulted for more exact information.

[b]Atmosphere contains normal amounts of oxygen and carbon dioxide. It is assumed that foods are fresh and of good quality when placed in storage.

[c]Storage lives of individual products fall somewhere within the range stated.

1. Controlled-atmosphere Storage of Apples and Pears

Controlled-atmosphere (CA) storage involves the use of chilling temperatures combined with an atmosphere containing oxygen at an accurately controlled level substantially less than that of ordinary air, and carbon dioxide at an accurately controlled level substantially greater than that of ordinary air. Large quantities of apples and pears are stored under CA conditions and this attests to the effectiveness and practicality of this approach. CA storage is especially advantageous for apples (such as McIntosh) that must be stored several degrees above their freezing points in order to avoid the undesirable consequences of chilling injury. The desirable effect of CA storage on the keeping quality of apples is shown in Figure 5-4. However, it is important to recognize that the quality of apples in CA storage generally does not exceed the quality of apples in conventional chilling storage until approximately 90 days or more have elapsed [79].

Recommended conditions for CA storage of apples and pears are shown in Table 5-11. It is evident that recommended levels of oxygen and carbon dioxide depend not only on the species of fruit but also on the variety. Origin of the fruit is an additional factor influencing recommended levels of oxygen and carbon dioxide [40]. Unfortunately, the most effective composition of the atmosphere can be determined only by trial and error.

a. Features of CA Facilities for Fruit. A facility for CA storage of fruit should have the following characteristics [65, 78]:

1. Air tightness. Rooms must be specially constructed and tested to assure that air leakage is reduced to a low level. Excessive leakage can seriously interfere with maintenance of the desired levels of oxygen and carbon dioxide or can result in excessive operating costs if the desired atmosphere is being maintained by gases not originating from the product.
2. Concentrations of oxygen and carbon dioxide. These gases must be accurately maintained at recommended levels.
3. Temperature. Air temperature should be controlled to within $\pm 0.5^\circ$ ($\pm 1^\circ$F) of the recommended value, and in no instance should the product be allowed to fall below its freezing point.
4. Relative humidity. Relative humidity of the storage atmosphere should be about 90% for apples and 90-95% for pears.
5. Air circulation. The storage atmosphere should be circulated (1-1.5 cu ft/min/btu) and evenly distributed to provide rapid cooling of incoming fruit and to assure uniformity of both the ultimate product temperature and the atmospheric composition.
6. Removal of volatile emanations. Removal of volatiles is sometimes recommended if the fruit is in a preclimacteric stage [78, 97].

b. Methods for Establishing and Controlling the Concentration of Oxygen During CA Storage. Oxygen in the storage atmosphere can be reduced either by natural or by artificial means. The natural method relies on the ability of the stored plant tissue to consume oxygen. Excessively low levels of oxygen are avoided by introducing appropriate amounts of outside air. Advantages of the natural method are its simplicity and economy of operation. Disadvantages are (1) a long time is required to initially establish the recommended level of oxygen,

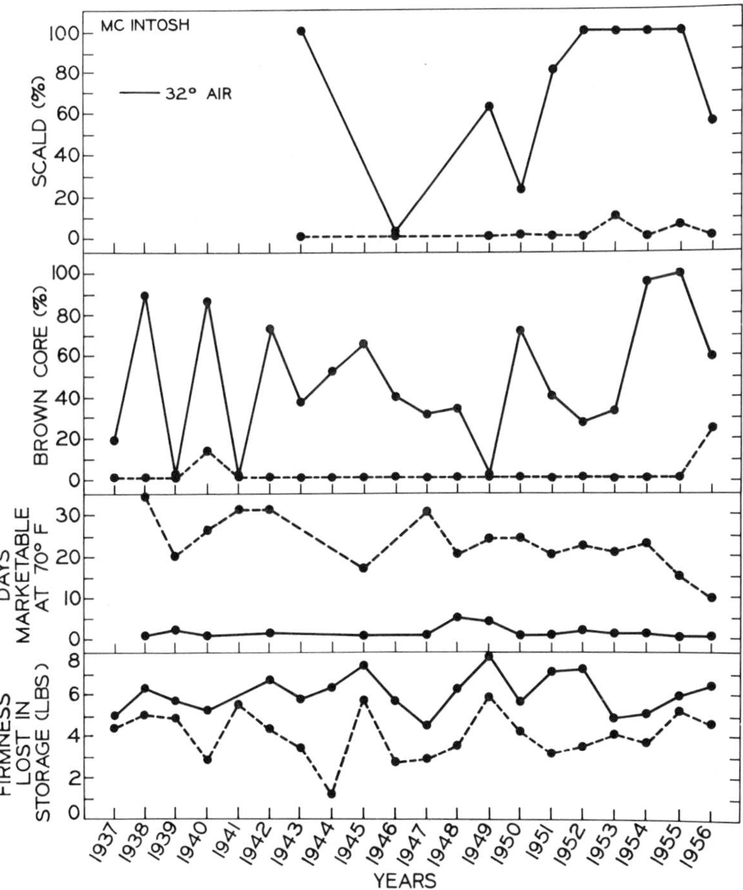

FIG. 5-4. Comparative keeping quality of McIntosh apples in CA storage [5% CO_2 and 3% O_2 at 3.3-4.4° (38-40°F)] and in conventional chilling storage [air at 0° (32° F)]. Measurements were made each year after about 6 months storage. From Smock (78).

especially if the fruit is quickly cooled (slows respiration); (2) the room must be exceptionally well constructed (otherwise too much oxygen leaks in); and (3) the room must be filled to capacity and should not be opened until the entire contents are removed [66, 87].

Artificial methods (also referred to as "externally generated atmospheres") for reducing the oxygen content of the storage atmosphere involve either flushing with pure nitrogen (liquid or compressed) or flushing with air from which most of the oxygen has been removed by catalytic combustion in the presence of added methane or propane [44, 66]. Advantages of the artificial method are (1) the desired level of oxygen can be established quickly (if liquid nitrogen is used, the rate of cooling also is increased); (2) the storage room need not be filled to capacity, and it can be partially emptied or refilled during the course of storage

TABLE 5-11

Conditions Recommended for CA Storage of Apples and Pears in the USA[a]

Variety	Carbon dioxide (%)	Oxygen (%)	Temperature °C (°F)
Apples			
Cortland	2 for 1 month, then 5	3	3.3 (38)
Golden Delicious	2	3	0 to -1 (30-32)
Delicious	2[b]	3	0 to -1 (30-32)
Jonathan	5[c]	3	0 (32)
McIntosh	2 for 1 month, then 5	3	3.3 (38)
Northern Spy	2-3	3	0 (32)
Rome Beauty	2-3	3	0 to -1 (30-32)
Stayman	2-3	3	0 to -1 (30-32)
Turley	2-3	3	0 (32)
Yellow Newtown	7	3	3.3-4.4 (38-40)
Pears			
Anjou	0.5-1	2-3	0 to -1 (30-32)
Bartlett	1-2[d]	2-3	0 to -1 (30-32)
Bosc	0.5-1[d]	2-3	0 to -1 (30-32)

[a]From Smock [80].

[b]Washington State recommends less than 1%.

[c]If stored with Delicious, the Delicious requirement must be met.

[d]Based on limited tests at Oregon State University and Cornell University.

(this is possible because of the rapidity with which the desired level of oxygen can be reestablished); and (3) proper levels of oxygen can be maintained even in rooms that are somewhat leaky (provided the artificial atmosphere is continuously added). This characteristic facilitates conversion of conventional storage units into units suitable for CA storage.

The main disadvantage of the artificial method is its high cost, especially if the storage rooms are moderately leaky.

Artificial methods of controlling the level of oxygen can be employed continuously (very expensive and usually not done), or as a supplement to the natural method (i.e., to speed establishment of initial conditions or to quickly reestablish desired conditions if the room develops excessive leaks or the room is opened for removal or addition of product).

c. Methods for Establishing and Controlling the Concentration of Carbon Dioxide During CA Storage. Intact plant tissues evolve substantial amounts of carbon dioxide. Accumulation of carbon dioxide can be beneficial up to a point, but excessive levels of carbon dioxide must be avoided or product damage occurs. The level of carbon dioxide can be controlled by simple ventilation or by methods which selectively remove carbon dioxide.

Simple ventilation removes excess carbon dioxide but simultaneously adds oxygen. In practice, of course, the extent of ventilation is restricted so that the carbon dioxide content of the storage atmosphere is much greater than that of normal air and the oxygen content is lower than that of normal air. The combined concentrations of carbon dioxide and oxygen always totals 21% by this method. Since this value is far from ideal, the results are less satisfactory than those obtained when optimum levels of oxygen and carbon dioxide are employed.

Methods for selective removal of carbon dioxide from air (generally referred to as "carbon dioxide scrubbers") can be classed as chemical or physical. Chemical methods involve an interaction between air and such chemicals as dry lime (calcium hydroxide), solutions of sodium hydroxide or ethanolamines, or suspensions of lime in water [30, 71, 75]. Bagged dry lime is popular because it is effective, easy to handle, and relatively noncorrosive [23].

Physical methods for controlling the concentration of carbon dioxide in air involve absorption in water and adsorption on molecular sieves. Since the water method is effective, simple, economical to operate, noncorrosive, and is in common use it is considered further here [60, 65, 82].

A schematic representation of a water absorption apparatus for removal of carbon dioxide is shown in Figure 5-5. Successful operation is based on two principles: (1) the solubility of carbon dioxide in water at constant temperature varies with the partial pressure of carbon dioxide in air; (2) the solubility of carbon dioxide in water at constant pressure varies inversely with temperature.

In the absorption section of the apparatus, water which has been equilibrated in normal air (0.03% carbon dioxide) is intimately mixed with air from the storage unit. This air, which contains a higher than normal amount of carbon dioxide (perhaps 5%) yields part of its carbon dioxide to the water and then returns to the storage unit.

In the desorption section, water that is rich in carbon dioxide comes into contact with normal air and the carbon dioxide previously acquired in the absorption section is now released from the water and discharged to the atmosphere. This water is then returned to the absorption section and the process is repeated. The trap located between the absorption and desorption sections prevents interchange of air between the two sections. Concentration of carbon dioxide in the storage unit is controlled primarily by the water temperature in the absorption unit and the duration of operation. A carbon dioxide concentration as low as 2% can be maintained by this method [64, 65].

Major disadvantages of the water absorption system for removal of carbon dioxide are (1) a room with a very low rate of leakage is required and (2) initial establishment of a low oxygen concentration requires somewhat longer with the water absorption system than with an alkaline absorption system. This situation arises because the water absorption system adds small amounts of oxygen during the process of removing carbon dioxide [64].

Carbon dioxide removed by any of the methods discussed above is replaced by air or an artificial atmosphere, thereby providing control of the oxygen concentration.

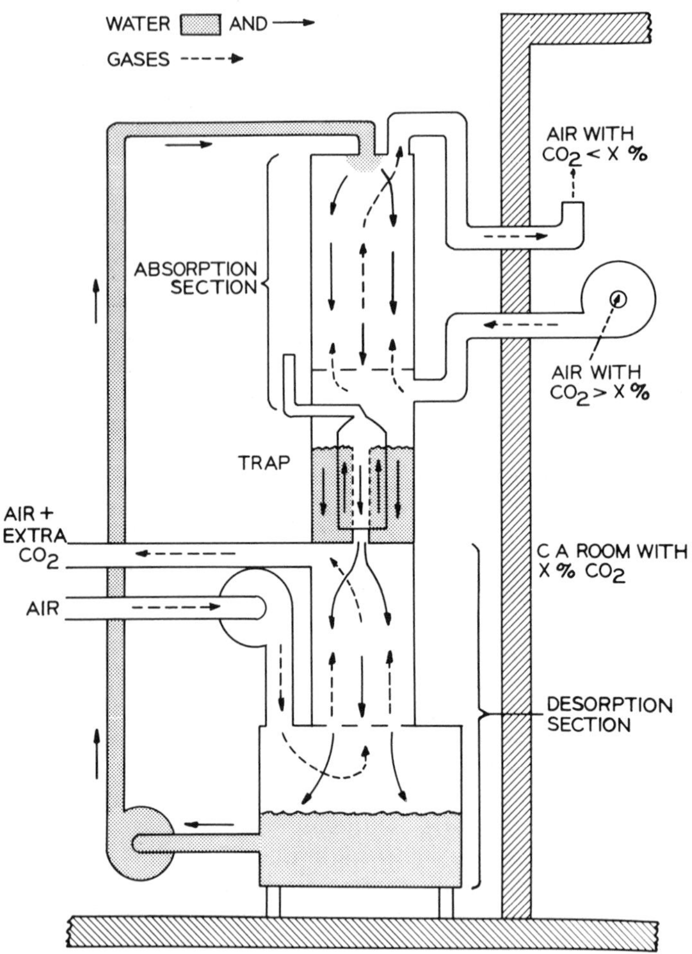

FIG. 5-5. Water absorption apparatus for removal of carbon dioxide. Redrawn from Pflug [64]; courtesy of American Society of Heating, Refrigerating, and Air-Conditioning Engineers.

Regardless of how the desired levels of oxygen and carbon dioxide are controlled, it is important to arrive at the desired storage conditions (temperature, relative humidity, levels of oxygen and carbon dioxide) as soon as possible following loading of the storage unit. The adjustment period should not exceed 10 days for apples and 5 days for pears, and ideally should be shorter [65, 49].

2. Other Applications of Modified Atmospheres During Storage of Foods at Chilling Temperatures

Some plant and animal tissues are transported at low, nonfreezing temperatures in the presence of atmospheres that have been modified with respect to oxygen, carbon dioxide, and nitrogen concentrations. Additional commercial applications are likely to develop as optimum conditions become better known, advantages become more clearly demonstrated, and improved handling techniques become available. The effects of various modified atmospheres on the quality of foods other than apples and pears are reviewed briefly below.

a. Mammalian Meat and Meat Products. The storage lives of meats and meat products often can be extended by the presence of greater than normal amounts of carbon dioxide in the atmosphere. Moran [54] and Moran et al. [55] have found that the storage life of chilled beef can be increased significantly by using an atmosphere containing 10-20% carbon dioxide rather than ordinary air. The desirable effect occurs primarily because the high level of carbon dioxide suppresses growth of microorganisms [89]. An atmosphere containing 10% carbon dioxide is used commercially during shipment of beef by sea from Australia and New Zealand to Great Britain. At a temperature of -1 to -1.5° (29°-30° F) this treatment increases the storage life of beef to approximately 70 days as compared to about 45 days in air [36, 75].

Oxidative rancidity and microbiological spoilage of fresh pork can be greatly reduced by using an atmosphere containing 50-100% carbon dioxide as compared to air [11, 12]. However, this treatment can result in some brown discoloration (formation of methemoglobin and metmyoglobin [8]). Levels of carbon dioxide ranging from 25-50% provide some protection against growth of microorganisms but do not retard oxidative rancidity.

The storage life of bacon at 5° (41° F) is approximately doubled by using an atmosphere of 100% carbon dioxide rather than air [12]. The storage life of frankfurters increases as the carbon dioxide content of the atmosphere is increased up to 50% [58].

For beef and pork stored for 6-8 days at either 3° or 7° (37° or 45° F), the following atmospheres provide better preservation than air [61]: (1) nitrogen (with oxygen less than 0.5%), (2) 70% carbon dioxide and 30% air, and (3) 90% nitrogen and 10% carbon dioxide. Similarly, red meats apparently deteriorate less rapidly when transported in an atmosphere of nearly pure nitrogen at an accurately controlled temperature (±0.5°) than they do when transported in air at the same but less well-controlled temperature [21].

b. Poultry. Microbiological spoilage of poultry occurs at a decreasing rate as the carbon dioxide content of the atmosphere is increased up to 25% [57]. However, discoloration and off flavors can occur at carbon dioxide levels above approximately 15% [57, 88].

Poultry apparently deteriorates less rapidly when stored in nearly pure nitrogen at an accurately controlled temperature (±0.5°) than it does in air at the same but less well-controlled temperature [21].

c. Fish. The keeping quality of chilled postrigor haddock is reportedly doubled by storage in an atmosphere containing 50-100% carbon dioxide as compared to air at the same temperature [84]. This effect can be explained largely on the basis that 20% carbon dioxide at 0° effectively retards growth of those bacteria responsible for spoilage of fresh fish [17].

d. Eggs. Quality attributes of chilled shell eggs are apparently retained better in an atmosphere containing 0.55% carbon dioxide than in ordinary air [74]. However, oiling of eggs is more practical and achieves similar benefits [15].

e. Fruits. Deterioration of some freshly harvested fruits can be significantly reduced by brief exposure (1-2 days) to atmospheres containing at least 20% carbon dioxide. This technique has resulted in decidedly favorable experimental results (less decay, and firmer and fresher than controls in air) with sweet cherries, plums, peaches, Bartlett pears, raspberries, dewberries, blackberries, figs, grapefruit, and oranges [9, 34]. However, off flavors can result if the carbon dioxide concentration is too high or the exposure time is too long. For many other fruits the carbon dioxide treatment described above offers no advantage or is injurious (e.g., green tomatoes fail to ripen following removal from the modified atmosphere).

During a period of 6-9 weeks at 0°, peaches and nectarines keep better in an atmosphere containing 1% oxygen and 5% carbon dioxide than they do in air [1].

By packaging of sweet cherries in sealed polyethylene bags (box liners), a beneficial atmosphere containing an elevated level of carbon dioxide and a reduced level of oxygen can develop and persist during refrigerated storage [35].

Some fruits, such as tomatoes, strawberries, grapes, and melons, apparently deteriorate less rapidly when transported in an atmosphere of nearly pure nitrogen at an accurately controlled temperature (±0.5°) than they do in air at the same but less well-controlled temperature [21].

f. Vegetables. Quality loss of some vegetables can be slowed by short-term (1-7 days) exposure to atmospheres containing 5-50% carbon dioxide. On an experimental basis this treatment has been shown to benefit carrots, cauliflowers, sweet corn, and asparagus [9, 10, 46]. However, many vegetables are either not benefited or are injured by the treatment described above [9, 86].

Lettuce and asparagus apparently deteriorate somewhat less rapidly when transported in an atmosphere of nearly pure nitrogen at an accurately controlled temperature (±0.5°) than they do in air at the same but less well-controlled temperature [21, 85].

D. Handling of Food Following Removal from Chilling Storage

When food products are moved from chilling storage to a warm environment, the product temperature frequently is below the dewpoint of the air, condensation occurs, and growth of microorganisms is encouraged. Occurrence of undesirable condensation can be predicted from the temperature and relative humidity of the warm air and the temperature of the product [40]. Condensation can be avoided by warming the product gradually, but this is often not practical. If condensation does occur, ventilation should be adequate to warm and dry the product or alternatively the product should be consumed without undue delay.

REFERENCES

1. R. E. Anderson, C. S. Parsons, and W. L. Smith, Jr., U. S. Dept. of Agr. Mktg. Res. Rept. No. 836, U. S. Government Printing Office, Washington, D. C., 1969.

2. C. O. Appleman and J. M. Arthur, J. Agr. Res., 17, 137 (1919).

3. W. R. Barger, Mktg. Res. Rept. No. 600, Department of Agriculture, U. S. Government Printing Office, Washington, D. C., 1963.

4. A. H. Bennett, in ASHRAE Guide and Data Book — Applications 1971, American Society of Heating, Refrigerating and Air-Conditioning Engineers, New York, 1971, Chapter 22.

5. J. B. Biale, Advan. Food Res., 10, 293 (1960).

6. J. B. Biale, in Encyclopedia of Plant Physiology (W. Ruhland, ed.), Vol. XII/2, Berlin, Gottingen, Heidelberg, Springer Verlag, 1960, p. 536.

7. R. K. Bogardus and J. M. Lutz, Agr. Mktg., 6(12), 8 (1961).

8. J. Brooks, Proc. Roy. Soc., Ser. B, 118, 560 (1935).

9. C. Brooks, C. O. Bratley, and L. P. McColloch, Technical Bull. No. 519, U. S. Department of Agriculture, U. S. Government Printing Office, Washington, D. C., 1936, 24 pp.

10. C. Brooks, E. V. Miller, C. O. Bratley, J. S. Cooley, P. V. Mook, and H. B. Johnson, Technical Bull. No. 318, U. S. Department of Agriculture, U. S. Government Printing Office, Washington, D. C., 1932, 60 pp.

11. E. H. Callow, J. Soc. Chem. Ind., 51, 116T (1932).

12. E. H. Callow, Food Investigation Board Report 1932, Great Britain Department of Sci. Ind. Research, 1933, pp. 112-116.

13. W. H. Chandler, Deciduous Orchards, 3rd ed., Lea and Febiger, Philadelphia, 1957.

14. L. L. Claypool, K. E. Nelson, T. W. Kendall, D. H. Dewey, and F. G. Mitchell, in ASHRAE Guide and Data Book — Applications 1971, American Society of Heating, Refrigerating and Air-Conditioning Engineers, New York, 1971, Chapter 27, pp. 363-370.

15. H. T. Cook, in ASHRAE Guide and Data Book — Applications 1971, American Society of Heating, Refrigerating and Air-Conditioning Engineers, New York, 1971, Chapter 38, pp. 471-476.

16. C. F. Cook and R. F. Langsworth, J. Food Sci., 31, 497 (1966).

17. F. P. Coyne, Proc. Roy. Soc., Ser. B, 113, 196 (1933).

18. J. A. Dassow and J. W. Slavin, in ASRAE Guide and Data Book — Applications 1971, American Society of Heating, Refrigerating and Air-Conditioning Engineers, New York, 1971, Chapter 25, pp. 323-376.

19. D. De Fremery and M. F. Pool, Food Res., 25, 73 (1960).

20. D. H. Dewey, Ice and Refrig., 123(Sept.), 19 (1952).

21. D. A. Dixon and K. Robe, Food Proc. Mktg., March, 138 (1966).

22. R. B. Duckworth, Fruits and Vegetables, Pergamon, London, 1966.

23. C. A. Eaves, in Agr. Ext. Bull. No. 422 (W. J. Lord, ed.), University of Massachusetts, Amherst, 1964, pp. 16-21.

24. R. P. Elliott, personal correspondence, 1963.

25. R. P. Elliott and H. D. Michener, Agr. Res. Service Publ. ARS-74-21, U. S. Department of Agriculture, U. S. Government Printing Office, Washington, D. C., 1960.

26. N. A. M. Eskin, H. M. Henderson, and R. J. Townsend, Biochemistry of Foods, Academic Press, New York, 1971, Chapter 1, pp. 1-29.

27. J. C. Fidler, Rec. Advan. Food Sci., 1, 269 (1962).

28. J. C. Fidler, in Low Temperature Biology of Foodstuffs (J. Hawthorn and E. J. Rolfe, eds.), Pergamon Press, London, 1968, pp. 271-283.

29. J. C. Fidler, in Current Trends in Cryobiology (A. U. Smith, ed.), Plenum Press, New York, 1970, Chapter 2, pp. 43-60.

30. J. C. Fidler and G. Mann, in Storage of Fruits and Vegetables; Cold Stores, Vol. 1966-1, International Institute of Refrigeration, Paris, 1966, pp. 41-45.

31. J. C. Fidler and C. J. North, in Storage of Fruit and Vegetables; Cold Stores, Vol. 1966-1, International Institute of Refrigeration, Paris, 1966, pp. 93-100.

32. J. C. Fidler and C. J. North, J. Hort. Sci., 42, 189 (1967).

33. J. C. Fidler and C. J. North, J. Hort. Sci., 46, 213 (1971).

34. F. Gerhardt and A. L. Ryall, Technical Bull. No. 631, U. S. Department of Agriculture, U. S. Government Printing Office, Washington, D. C., 1939, 20 pp.

35. F. Gerhardt, H. Schomer, and T. R. Wright, Agr. Mktg. Ser. Publ. AMS-121, U. S. Department of Agriculture, U. S. Government Printing Office, Washington, D. C., 1956, 8 pp.

36. K. C. Hales, Advan. Food Res., 12, 147 (1963).

37. R. E. Hardenburg and T. W. Kendall, in ASHRAE Guide and Data Book — Applications 1971, American Society of Heating, Refrigerating and Air-Conditioning Engineers, New York, 1971, Chapter 29, pp. 381-394.

38. R. E. Hardenburg and J. M. Lutz, in ASHRAE Guide and Data Book — Applications 1971, American Society of Heating, Refrigerating and Air-Conditioning Engineers, New York, 1971, Chapter 37, pp. 459-470.

39. International Institute of Refrigeration, Recommended Conditions for Land Transport of Perishable Foodstuffs, 2nd ed., the Institute, Paris, 1963.

40. International Institute of Refrigeration, Recommended Conditions for Cold Storage of Perishable Produce, 2nd ed., the Institute, Paris, 1967.

41. F. M. Isenberg and J. Hartman, Agr. Ext. Bull. No. 1012, Cornell University, Ithaca, New York, 1958.

42. A. W. Khan and R. Nakamura, J. Food Sci., 35, 266 (1970).

43. F. Kidd, Proc. Roy. Soc., Ser. B, 89, 136 (1917).

44. J. W. Lannert, Agr. Eng., 45, 318 (1964).

45. R. A. Lawrie, Meat Science, Pergamon Press, London, 1966.

46. W. J. Lipton, U. S. Dept. of Agr. Mktg. Res. Report No. 817, U. S. Government Printing Office, Washington, D. C., 1968.

47. D. Lister, in The Physiology and Biochemistry of Muscle as a Food (E. J. Briskey, R. G. Cassens, and B. B. Marsh, eds.), Vol. 2, University of Wisconsin Press, Madison, 1970, pp. 705-734.

48. R. H. Locker and C. J. Hagyard, J. Sci. Food Agr., 14, 787 (1963).

49. J. M. Lutz and R. E. Hardenburg, Agr. Handbook No. 66, U. S. Department of Agriculture, U. S. Government Printing Office, Washington, D. C., 1968.

50. W. W. Marion and H. M. Goodman, Food Technol., 21, 307 (1967).

51. B. B. Marsh and N. G. Leet, J. Food Sci., 31, 450 (1966).

52. B. B. Marsh, P. R. Woodhams, and N. G. Leet, J. Food Sci., 33, 12 (1968).

53. L. P. McColloch, in U. S. Dept. Agr. Yearbook, U. S. Department of Agriculture, U. S. Government Printing Office, Washington, D. C., 1953, pp. 826-830.

54. T. Moran, Food Ind., 9, 705 (1937).

55. T. Moran, E. C. Smith, and R. G. Tomkins, J. Soc. Chem. Ind., 51, 114T (1932).

56. H. D. Nelson, Mktg. Res. Report No. 886, U. S. Department of Agriculture, U. S. Government Printing Office, Washington, D. C., 1970.

57. W. S. Ogilvy and J. C. Ayres, Food Technol., 5, 97 (1951).

58. W. S. Ogilvy and J. C. Ayres, Food Technol., 5, 300 (1951).

59. W. S. Ogilvy and J. C. Ayres, Food Res., 18, 121 (1953).

60. R. G. Palmer, Amer. Fruit Grower, 79, 20 (1959).

61. W. Partmann, H. K. Frank, and J. Gutschmidt, Kältetechnik, 23, 113 (1971) (in German).

62. G. O. Patchen, Mktg. Res. Rept. No. 924, U. S. Department of Agriculture, U. S. Government Printing Office, Washington, D. C., 1971.

63. J. W. Pence and N. N. Standridge, Cereal Chem., 32, 519 (1955).

64. I. J. Pflug, ASHRAE J., 3(12), 65 (1961).

65. I. J. Pflug, in Proceedings of the First International Congress on Food Science and Technology (J. M. Leitch, ed.), Vol. 4, Gordon and Breach, New York, 1965, pp. 461-472.

66. I. J. Pflug and D. Gurevitz, in Storage of Fruit and Vegetables; Cold Stores, Vol. 1966-1, International Institute of Refrigeration, Paris, 1966, pp. 171-178.

67. H. Platenius, Plant Physiol., 17, 179 (1942).

68. H. Platenius, Plant Physiol., 18, 671 (1943).

69. W. H. Redit, Agr. Handbook No. 195, U. S. Department of Agriculture, U. S. Government Printing Office, Washington, D. C., 1969.

70. Refrigeration Research Foundation, The Commodity Storage Manual. The Foundation, Washington, D. C. (continually updated).

71. A. W. Ruff and G. F. Sainsbury, in ASHRAE Guide and Data Book — Applications 1971, American Society of Heating, Refrigerating and Air-Conditioning Engineers, New York, 1971, Chapter 36, pp. 447-458.

72. A. L. Ryall and W. J. Lipton, Handling, Transportation and Storage of Fruits and Vegetables; Vol. 1, Vegetables and Melons, AVI Publ. Co., Westport, Conn., 1972.

73. K. J. Scott and E. A. Roberts, Aust. J. Exp. Agr. Anim. Husb., 7, 87 (1967).

74. P. F. Sharp, Food Res., 2, 477 (1937).

75. W. H. Smith, Advan. Food Res., 12, 95 (1963).

76. W. L. Smith, Jr., Bot. Rev., 28, 411 (1962).

77. M. C. Smith, Jr., M. D. Judge, and W. J. Stadelman, J. Food Sci., 34, 42 (1969).

78. R. M. Smock, Agr. Ext. Bull. No. 759, Cornell University, Ithaca, New York, 1958.

79. R. M. Smock, in Agr. Ext. Bull. No. 422 (W. J. Lord, ed.), University of Massachusetts, Amherst, 1964, pp. 61-64.

80. R. M. Smock, in Frontiers of Food Research, Symposium Proceedings, Cornell University, Ithaca, New York, 1966, pp. 30-35.

81. R. M. Smock and G. D. Blanpied, Ext. Bull. No. 440, Cornell University, Ithaca, New York, 1969.

82. R. M. Smock and L. Yatsu, Proc. Amer. Soc. Hort. Sci., 75, 53 (1960).

83. W. J. Stadelman, ASHRAE J., March, 61 (1970).

84. M. E. Stansby and T. P. Griffiths, Ind. Eng. Chem., 27, 1452 (1935).

85. J. K. Stewart and M. J. Ceponis, Mktg. Res. Report No. 832, U. S. Department of Agriculture, U. S. Government Printing Office, Washington, D. C., 1968.

86. J. K. Stewart and M. Uota, J. Amer. Soc. Hort. Sci., 96, 27 (1971).

87. A. Stradelli, in Storage of Fruit and Vegetables; Cold Stores, Vol. 1966-1, International Institute of Refrigeration, Paris, 1966, pp. 185-190.

88. J. E. Thomson and L. A. Risse, J. Food Sci., 36, 74 (1971).

89. R. G. Tomkins, Food Investigation Board Report 1932, Great Britain Department of Sci. Ind. Research, 1933, pp. 48-50.

90. S. A. Trout, F. E. Huelin, and G. B. Tindale, Tech. Paper No. 14, Division of Food Preservation and Transport, CSIRO, Australia, 1960.

91. W. M. Urbain, in The Science of Meat and Meat Products (J. F. Price and B. S. Schweigert, eds.), 2nd ed., W. H. Freeman, San Francisco, 1971, Chapter 9, pp. 402-451.

92. United States Department of Agriculture, Regulations Governing the Inspection of Poultry and Poultry Products, 7 CFR, Part 81.66 (b) (1) (2), U. S. Department of Agriculture, Consumer and Marketing Service, U. S. Government Printing Office, Washington, D. C., May 27, 1971.

93. J. E. Van Der Plank and R. Davies, J. Pomol. Hort. Sci., 15, 226 (1937).

94. C. C. Welling and J. J. Smoot, in ASHRAE Guide and Data Book — Applications 1971, American Society of Heating, Refrigerating and Air-Conditioning Engineers, New York, 1971, Chapter 28, pp. 371-380.

95. J. M. Wells, Mktg. Res. Report No. 908, U. S. Department of Agriculture, U. S. Government Printing Office, Washington, D. C., 1971.

96. B. G. Wilkinson, in The Biochemistry of Fruits and Their Products (A. C. Hulme, ed.), Vol. 1, Academic Press, New York, 1970, Chapter 18, pp. 537-554.

97. J. N. Yeatman, R. E. Hardenburg, R. E. Anderson, P. H. Heinze, W. H. Redit, W. E. Tolle, and L. P. McColloch, Agr. Res. Serv. Publ. ARS 51-4, U. S. Department of Agriculture, U. S. Government Printing Office, Washington, D. C., 1965.

98. R. Zacharias, Ernährungsumschau, 9, 33 (1962).

GENERAL REFERENCES

American Society of Heating, Refrigerating and Air-Conditioning Engineers, ASHRAE Guide and Data Book — Applications 1971, American Society of Heating, Refrigerating and Air-Conditioning Engineers, New York, 1971.

R. B. Duckworth, Fruits and Vegetables, Pergamon Press, London, 1966.

J. C. Fidler, in Low Temperature Biology of Foodstuffs (J. Hawthorn and E. J. Rolfe, eds.), Pergamon Press, London, 1968, pp. 271-283.

J. C. Fidler, in Current Trends in Cryobiology (A. U. Smith, ed.), Plenum Press, New York, 1970, Chapter 2, pp. 43-60.

International Institute of Refrigeration, Recommended Conditions for Land Transport of Perishable Foodstuffs, 2nd ed., the Institute, Paris, 1963.

International Institute of Refrigeration, Recommended Conditions for Cold Storage of Perishable Produce, 2nd ed., the Institute, Paris, 1967.

J. M. Lutz and R. E. Hardenburg, Agr. Handbook No. 66, U. S. Department of Agriculture, U. S. Government Printing Office, Washington, D. C., 1968.

W. H. Redit, Agr. Handbook No. 195, U. S. Department of Agriculture, U. S. Government Printing Office, Washington, D. C., 1969.

Refrigeration Research Foundation, The Commodity Storage Manual, The Foundation, Washington, D. C. (continually updated).

W. H. Smith, Advan. Food Res., 12, 95 (1963).

Chapter 6

FREEZING PRESERVATION

O. Fennema

CONTENTS

I. INTRODUCTION

Freezing is unquestionably the most satisfactory method currently available for long-term preservation of foods. Properly conducted freezing is effective for retaining the flavor, color, and nutritive value of food, and is moderately effective for preservation of texture. In spite of the superiority of freezing as compared to other methods of long-term preservation, it almost invariably produces some detrimental effects, the severity depending on the product and the nature of the freezing process. In order to minimize these detrimental effects, it is necessary that the freezing process be well understood. For this reason a discussion of fundamental aspects of freezing precedes the discussion of freezing practices.

II. FUNDAMENTAL ASPECTS OF FREEZING PRESERVATION

Freezing consists of a reduction in temperature, generally to -18° (0° F) or below, and crystallization of part of the water and some of the solutes. Although two or more of these events generally occur simultaneously, it is instructive to consider their effects independently.

A. Supercooling

When the temperature of water is lowered below its freezing point, and crystallization does not occur, it is said to be supercooled or undercooled. Supercooling provides a means of determining the independent effect of a reduction in temperature below the initial freezing point. Numerous studies have been conducted in which samples have been supercooled well below their initial freezing points, returned immediately to an above-zero temperature, then compared to like samples which have been allowed to freeze while undergoing the same pattern of temperature change. These studies indicate that temporary supercooling generally does not impair the quality of foods whereas freezing does. Thus, the undesirable side effects which often accompany preservation by freezing are attributable mainly to water-ice transformations rather than to a reduction in temperature, as such.

B. Crystallization of Water

Crystallization is the formation of a systematically organized solid phase from a solution, melt, or vapor. The concern here is with crystallization from a melt (ice from water) or from solution (solute from water). Fortunately, the principles underlying these two types of crystallization are quite similar so they can be dealt with simultaneously.

The crystallization process consists of nucleation and crystal growth. Nucleation is the association of molecules into a tiny ordered particle of a size sufficient to survive and serve as a site for crystal growth. Crystal growth is simply the enlargement of the nucleus by the orderly addition of molecules. Although the two steps of the crystallization process usually overlap with respect to their times of occurrence, it is possible to control their relative rates and thereby control the characteristics of the final crystalline system. An understanding of how this can be done is best gained by considering nucleation, then crystal growth, and finally the combined events.

1. Nucleation

If an aqueous system is devoid of ice crystals, the temperature of this system, even if it is pure water, invariably falls temporarily below the initial freezing point (temperature at which a tiny crystal of ice can exist in equilibrium with the liquid phase) of the system before nucleation and further solidification can proceed. This temporary supercooling occurs because nucleation is difficult.

Two types of nucleation are possible: homogeneous and heterogeneous. Water, if exceedingly pure, is limited to nucleation of the homogeneous type. A homogeneous nucleus forms by chance orientation of a suitable number of molecules into a tiny ordered particle. Homogeneous nucleation is impossible at 0° and does not become probable until the temperature is reduced many degrees below 0°, particularly if the sample is very small. Homogeneous nucleation is of no concern in most practical situations.

Heterogeneous or catalytic nucleation is the type that occurs in foods and living specimens. Heterogeneous nucleation involves the formation of nuclei adjacent to suspended foreign particles, surface films, or on the walls of containers. Although heterogeneous nucleation necessitates some supercooling, it is more probable, at any given temperature and fixed sample size, than homogeneous nucleation. Temporary supercooling of as much as 10° is not uncommon prior to heterogeneous nucleation in relatively large biological samples [56].

As the temperature is lowered to some critical value characteristic of the sample, nucleation begins, and further decreases in temperature result in an abrupt increase in rate. This behavior is depicted schematically in Figure 6-1.

2. Crystal Growth

Growth of a crystal nucleus constitutes the second step in the crystallization process. Unlike nucleation, crystal growth can occur at temperatures just below the melting point of the system. At temperatures near the melting point, water molecules add to existing nuclei (if present) in preference to forming new nuclei.

The rate of crystallization in a complex aqueous system is governed by the rates of mass and heat transfer. During the course of crystallization, water molecules must move from the liquid phase to a stable site on the crystal surface, and solute molecules must diffuse away from the crystal. Since water molecules

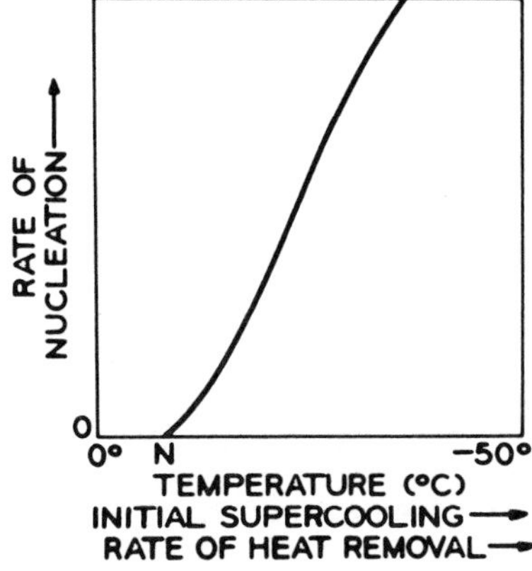

FIG. 6-1. The rate of water nucleation as influenced by supercooling and rate of heat removal (schematic) [16].

are small, highly mobile, and usually present in abundance, in most instances movement of water molecules is unlikely to limit the rate at which ice crystals grow. A possible exception may be during the final stages of freezing when the sample temperature is low, viscosity is high, and little unfrozen water remains.

Foods usually contain an abundance of dissolved solutes, suspended matter, and sometimes cellular components. Small solutes are able to migrate short distances and eventually are concentrated among the growing ice crystals or between segments of a single crystal. Since suspended matter and cellular components are less able or are unable to migrate, ice crystals either form around them or push them aside. Although the need to transfer solutes and other noncrystallizing matter away from enlarging ice crystals can significantly reduce the rate of crystal growth as compared to pure water, this kind of mass transfer is not believed to be rate limiting, except possibly during the late stages of freezing. It therefore can be concluded that the rate of ice crystal growth in high-moisture samples (foods) frozen at commercially realistic rates is generally not limited by mass-transfer processes (with the exception of the late stages of freezing).

It follows that heat transfer is the process which usually limits the rate of crystallization. This is reasonable considering water's large latent heat of crystallization. Tissues and other nonfluid aqueous samples are especially troublesome in this respect since heat transfer must occur primarily by conduction. That growth rates of ice crystals increase greatly as the rate of heat removal is increased, can be demonstrated at a constant temperature, by freezing like specimens of differing size. In this type of experiment the smaller specimen exhibits more rapid crystallization.

The effect of rate of heat removal (heat transfer) on rate of crystallization is depicted schematically in Figure 6-2. Noteworthy points are (1) crystal growth can occur at temperatures very near the melting point and (2) growth rate increases moderately with increasing rates of heat removal until at some very low sample temperature mass-transfer difficulties (high viscosity) cause the growth rate to decline.

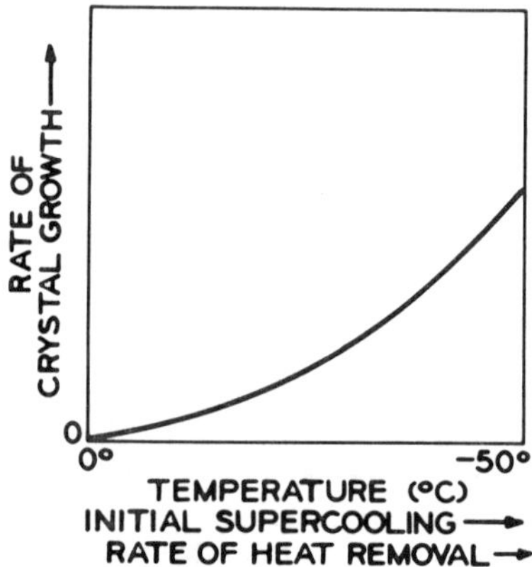

FIG. 6-2. Growth rates of ice crystals as influenced by supercooling and rate of heat removal (schematic) [16, 30].

3. Crystal Size

By comparing the crystal growth-heat removal (supercooling) relationship of Figure 6-2 with the nucleation-supercooling relationship of Figure 6-1, a logical explanation can be derived for the fact that rapidly frozen foods contain ice crystals that are small and numerous, whereas similar samples which have been slowly frozen contain ice crystals that are large and few in number. To aid this comparison Figures 6-1 and 6-2 have been combined in Figure 6-3. In an unseeded sample maintained at a temperature between the melting point and point N, only a few nuclei form and each grows extensively. This corresponds to slow freezing (slow removal of heat energy), during which the sample temperature remains at or near the solid-liquid equilibrium temperature. However, if an unseeded liquid sample is cooled to a temperature well below point N (rapid removal of heat energy and rapid freezing), many nuclei form and each grows to only a limited extent. The mean size of the completed crystals therefore varies inversely with the number of nuclei, and the number of nuclei can be easily controlled in practical situations simply by controlling the rate of heat removal.

It should not be inferred that samples of equal dimensions and frozen in the same manner must contain ice crystals of the same size. Under constant freezing conditions, large differences in crystal size are often observed among different kinds of samples, among different samples of the same kind, and even at various locations within the same sample [82]. For example, prerigor and postrigor fish, when frozen at the same rate, will exhibit a difference in the mean size of crystals [48].

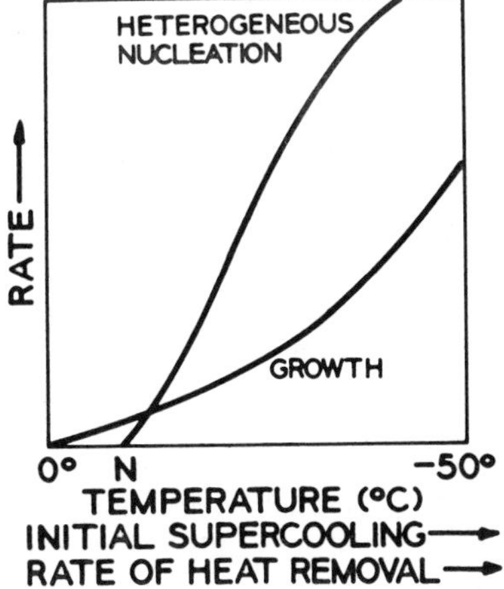

FIG. 6-3. Comparative rates of nucleation and crystal growth of water as influenced by supercooling and rate of heat removal (schematic) [14, 30, 90].

4. Recrystallization

Control of crystal size and shape would be relatively easy if crystals once formed, were stable. Unfortunately, they are relatively unstable; during frozen storage they undergo metamorphic changes which are collectively referred to as recrystallization. The term "recrystallization," as used here, encompasses any change in the number, size, shape, orientation, or perfection of crystals (grains) following completion of initial solidification.

Recrystallization has been observed in a wide variety of substrates including pure ice, ice cream, frozen juices, and frozen plant and animal tissues. The rate at which it occurs is dependent on temperature and the nature of the sample. The rate of recrystallization of ice decreases as the temperature is lowered or the content of nonaqueous components is increased. For example, recrystallization can occur readily in ice at -96°, but it occurs at a slow rate in tissues at -18°.

Recrystallization occurs because systems tend toward a state of equilibrium wherein free energy is minimized and the chemical potential is equalized among all phases. The free energy of a crystalline phase is minimized when its short- and long-range structures are perfect and its size is infinite (no boundaries). Since real crystals are imperfect and all, of course, have boundaries, the question arises as to the equilibrium size and shape of real ice crystals. This can be answered only qualitatively and is perhaps best approached by considering the types of recrystallization which are known to occur. Since standardized terminology is lacking, the following classification of recrystallization is offered to facilitate the present discussion: (a) isomass, (b) migratory, (c) accretive, (d) pressure induced, and (e) irruptive.

a. Isomass Recrystallization. Isomass recrystallization refers to any change in surface (shape) or internal structure of an individual crystal of fixed mass as it moves to a lower energy level. Surface isomass recrystallization occurs when a crystal of irregular shape and large surface-to-volume ratio (dendritic crystals, such as snow flakes) assumes a more compact structure with a smaller surface-to-volume ratio and a lower surface energy. In ice, this "rounding off" process is thought to occur by means of surface diffusion of water molecules.

Polymorphic transformations represent a type of internal isomass recrystallization. This type of recrystallization has been reported in aqueous specimens (e.g., a transformation from cubic to hexagonal symmetry during warming), but only under extreme conditions which are not encountered commercially.

A reduction of defects within a crystal represents another type of internal isomass recrystallization.

b. Migratory Recrystallization. This type of recrystallization, which is also referred to as "grain growth," is the tendency for large crystals in a polycrystal system to grow at the expense of small crystals. The result is an increase in average crystal size, a decrease in the number of crystals, and a decrease in surface energy of the entire crystalline phase. The equilibrium form is a single crystal.

At constant temperature and pressure, migratory recrystallization is the result of differences in the surface energies of large and small crystals. More simply, smaller crystals, with their very small radii of curvature, cannot bind their surface molecules as firmly as can larger crystals. Small crystals therefore exhibit lower melting points (or greater solubilities) than large ones.

When temperature is constant, migratory recrystallization occurs at a significant rate only when the specimen contains numerous crystals with diameters less than about 2 μm. Work by Woodroof [93] with frozen fruits and vegetables and by Arbuckle [7] with ice cream indicates that the mean crystal size in these products is much greater than 2 μm even when freezing is accomplished by direct contact with solidified carbon dioxide. This does not eliminate the possibility that some crystals of less than 2 μm may exist in frozen foods, but it does minimize the importance of migratory recrystallization in frozen foods stored at a constant temperature and pressure.

Fluctuating temperatures and associated vapor pressure gradients in the specimen, enhance migratory recrystallization even when the crystals have diameters in excess of 2 μm. Crystals in areas of low temperature (low vapor pressure) grow at the expense of those in areas of higher temperature (higher vapor pressure). In areas of high vapor pressure, some of the smallest crystals, S, may completely disappear, whereas larger crystals, L, simply decrease in size. Reversal of the vapor pressure gradient causes crystals L to grow but former crystals S do not reappear because of the energy barrier to nucleation.

c. Accretive Recrystallization. Contacting crystals can join together resulting in an increase in crystal size, a decrease in the number of crystals, and a decrease in the surface energy of the entire crystalline phase. "Accretive recrystallization" is an appropriate name for this process although other terms, such as sintering (inappropriate since it involves heating), regelation (inappropriate since it involves pressure), and aggregation, also have been used.

Since frozen foods contain a large amount of ice, with ample contact between ice crystals, it is probable that accretive recrystallization does occur in these specimens.

d. Pressure-induced Recrystallization. If force is applied to a group of crystals, those crystals with their basal planes aligned with the direction of force grow at the expense of those in other orientations. This type of recrystallization can result in an increase in crystal size, a decrease in the number of crystals, and a reorientation so that more crystals have their c axes normal to the direction of the force. It is not known whether this type of recrystallization can occur in foods, but its likelihood is greatest in large specimens where considerable pressure can develop during freezing [15].

e. Irruptive Recrystallization. When very small samples, particularly those with low water contents, are frozen at ultrarapid rates some solidify in a partially noncrystalline state. Upon warming to some critical temperature which is characteristic of the solutes present, crystallization resumes abruptly. This occurrence was referred to by Luyet [49] as "irruptive recrystallization." Irruptive recrystallization apparently does not occur in commercially frozen foods since they contain an abundance of water and are frozen at rates much slower than that required to produce partially noncrystalline ice.

f. Conclusions. Migratory recrystallization appears to be the major type of recrystallization that occurs in frozen foods and the stimulus is likely provided by fluctuating temperatures. Recrystallization, regardless of type, can be minimized by maintaining a low, constant storage temperature.

C. Location of Ice Crystals in Cellular Matter

The location of ice crystals in tissue and cellular suspensions is a function of freezing rate, specimen temperature, and nature of the cells. Slow freezing (cooling at a rate of less than about 1°/min) of plant tissues, animal tissues, or cellular suspensions (microorganisms, spermatazoa, red blood cells, etc.) generally causes ice crystals to form exclusively in extracellular locations. An example of extracellular ice formation is shown for slowly frozen (much slower than 1°/min) postrigor cod muscle in Figure 6-4c. The sequential development of extracellular ice crystals during slow freezing is illustrated schematically in the top four photographs of Figure 6-5.

Although uncommon, intracellular ice crystals have been observed in some slowly frozen specimens, e.g., prerigor cod muscle [48], and tissue frozen for a second time [62]. Slowly frozen blanched vegetables also are not expected to exhibit a preference for extracellular ice formation.

Conditions (slow freezing) leading to preferential extracellular freezing result in large crystals, maximum dislocation of water, a shrunken appearance of cells in the frozen state, and usually something less than maximum attainable food quality.

All kinds of tissue and cellular suspensions, without known exception, exhibit a uniform distribution of ice crystals (both intra- and extracellular) when frozen very rapidly to a low temperature. The rate of freezing needed to produce uniform crystallization generally increases as cell size decreases. An example of uniform

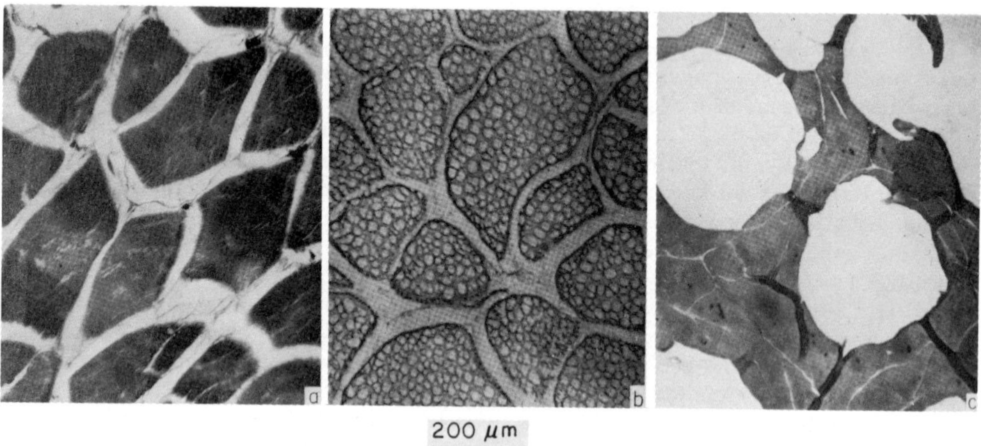

FIG. 6-4. Effect of freezing rate on the location of ice crystals in postrigor cod muscle. (a) Unfrozen, (b) rapidly frozen, (c) Slowly frozen. From Love [47], courtesy of Academic Press.

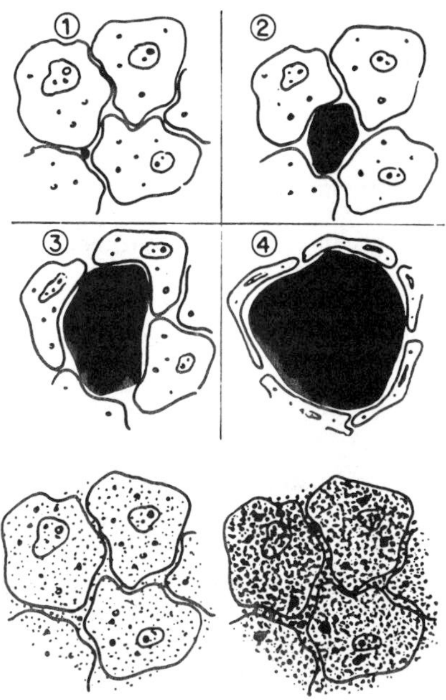

FIG. 6-5. Portrayal of the development of ice crystals in tissue. Upper sequence, slow freezing; lower sequence, rapid freezing. From Meryman [61]; courtesy of the Federation of American Societies for Experimental Biology.

ice formation (mainly intracellular because extracellular spaces are small) in rapidly frozen postrigor cod muscle is shown in Figure 6-4b. The sequential development of uniform crystallization during rapid freezing is illustrated schematically in the lower two photographs of Figure 6-5.

The conditions (rapid freezing) which produce uniform crystallization result in numerous small ice crystals, minimum dislocation of water, a frozen appearance which is similar to the original unfrozen appearance, and food quality which is usually superior to that obtained by slow freezing.

Most evidence favors the view that crystallization, regardless of freezing rate, is initiated primarily in the extracellular fluid [56]. If this assumption is accepted, the observed behavior of samples exposed to different rates of freezing can be easily explained. Following the onset of extracellular crystallization, further crystallization: (1) can continue exclusively in extracellular regions or (2) can occur in intracellular regions by means of (a) intracellular nucleation or (b) growth of extracellular ice crystals through the cell membrane. Exclusive extracellular freezing can be visualized in the following way. As ice forms in extracellular regions during slow freezing, the solute concentration of the unfrozen phase gradually increases and the vapor pressure gradually declines. Since crystals apparently cannot penetrate cellular membranes at relatively high subfreezing temperatures, the intracellular fluid remains in a supercooled condition and its vapor pressure at any given temperature exceeds that of the extracellular fluid and ice crystals. This vapor pressure difference causes water to diffuse from the cells (cellular dehydration) and deposit on extracellular ice crystals. By this means, supercooling of the cell contents is minimized and so is the likelihood of intracellular nucleation. Continued freezing results in considerable shrinkage of the cells and formation of large exclusively extracellular ice crystals.

Intracellular (uniform) freezing is favored by rapid cooling to a low temperature so that the opportunity for cellular dehydration is minimized. In this situation, there is increased probability for either intracellular nucleation or for growth of extracellular ice crystals through the cell membrane.

D. Other Occurrences Associated with Ice Formation

1. Volume Changes

Volume changes during freezing result because pure water at 0° expands approximately 9% when transformed into ice at the same temperature. Most foods also expand on freezing but to a lesser extent than pure water. A few exceptions can be found to the general rule that aqueous systems expand on freezing, e.g., highly concentrated sucrose solutions exhibit a small net contraction upon freezing. Considerable variability exists in the volume changes observed for different aqueous specimens during freezing. The following factors contribute to this variability:

a. Composition. Of special concern is the weight percent of water and the volume percent of air. The presence of solutes or suspended matter has a "substitution" effect; i.e., these materials reduce the fraction of water in the system. Since water is responsible for essentially all of the expansion that occurs during freezing, any decrease in water per unit volume decreases the observed expansion. Intracellular air spaces, which are common in plant tissue, can probably accommodate growing crystals and thereby minimize changes in the specimen's exterior dimensions.
b. Fraction of water that fails to freeze. Any water that is bound or supercooled does not contribute to expansion during freezing. The kind and amount of solutes present influence the amount of bound water and the tendency for a system to supercool.
c. Temperature range under study. Observed changes in volume are dependent on the temperature range to which the sample is exposed. This is not surprising since the following events, each affected differently by temperature, can contribute to changes in volume: (1) cooling of the specimen prior to freezing (contraction), (2) ice formation (expansion), (3) cooling of ice crystals (contraction), (4) solute crystallization (contraction). (5) cooling of solute crystals present in eutectics (contraction), and (6) crystallization and cooling of nonsolutes such as fat (contraction).

Additional complications associated with volume changes during freezing of complex aqueous systems have been discussed by Haynes [33].

The major point to keep in mind with respect to volume changes during freezing is that ice crystals are relatively pure even when they originate from a complex system. It is therefore proper to assign an expansion value of approximately 9% to the quantity of water that freezes. Since most other constituents contract as the temperature is lowered, it is apparent that the volume change is not uniform throughout the system. There are localized areas of expansion (ice crystals) and localized areas of contraction, with a likelihood of stresses and a possibility of mechanical damage. Mechanical damage to the texture of food tissues during freezing is undoubtedly more likely in plant tissue, with its rigid structure and poorly aligned cells, than in muscle, with its pliable consistency and parallel arrangement of cells (fibers).

2. Concentration of Nonaqueous Constituents

During freezing of aqueous solutions, cellular suspensions, or tissues, water from solution is transferred into ice crystals of a variable but rather high degree of purity. Nearly all of the nonaqueous constituents are therefore concentrated in a diminished quantity of unfrozen water. The net effect is similar to conventional dehydration except in this instance the temperature is lower and the separated water is deposited locally in the form of ice. As a result the unfrozen phase changes significantly in such properties as pH, titratable acidity, ionic strength, viscosity, freezing point (and all other colligative properties), surface and interfacial tensions, and oxidation-reduction potential. In addition, oxygen and carbon dioxide may be expelled from solution; water structure and water-solute interactions may be drastically altered; macromolecules are forced closer together, providing a better opportunity for detrimental interactions; and eutectic formation may occur as described below.

As the freeze concentration process progresses, various solutes eventually reach or exceed their respective saturation concentrations and simultaneous crystallization of ice and solute becomes possible. The temperature at which a crystallized solute can exist in equilibrium with ice and the unfrozen phase is known as the "eutectic point" or "eutectic temperature" of the solute. (When water is the solvent, the term "cryohydric" is sometimes used in place of the term "eutectic.") Each solute or combination thereof has a characteristic eutectic temperature (e.g., HCl, -86°; $CaCl_2$, -55°; K_2CO_3, -36.5°; NaCl, -21.13°; sucrose, -14°; glucose, -5°; Na_2CO_3, -2.1°; and the sodium and potassium phosphates, alone and in combination, range from -0.5° to -17.2°). Maximum ice formation is not possible until the eutectic temperature of that component with the lowest eutectic temperature has been reached. This temperature is often referred to as the "final" eutectic temperature. For ice cream and egg white, the final eutectic temperature is reportedly -55° [39, 81]; for meat, it is between -50° and -60° [75]; and for bread, it is apparently near -70° [53]. Other natural foods can be expected to have final eutectic temperatures in the same general range. Consequently, it is safe to say that nearly all frozen foods stored at conventional commercial temperatures contain some unfrozen and potentially freezable water. A sizeable quantity of bound and unfreezable water is also present.

The extent of concentration that occurs during freezing is influenced mainly by the final temperature and to a lesser degree by the eutectic temperatures of the solutes present, agitation, and the rate of cooling. If a solution is frozen slowly over a temperature range wherein eutectics do not form, then the ice crystals are quite pure, equilibrium conditions are approached, and the post-freezing molal concentration of the unfrozen solution (if it is a perfect solution) depends only on temperature. For example, an initially dilute solution attains solid-liquid equilibrium at a given subfreezing temperature by forming an extensive amount of ice, whereas a solution of high initial concentration attains solid liquid equilibrium at the same temperature by forming considerably less ice. According to Mazur [55, 56], the unfrozen fraction of a partially frozen specimen treated as described can be estimated from the expression

$$\alpha \approx \frac{1.86 v m_i}{T_f - T}$$

where

α is the fraction of unfrozen material
v is the number of ions formed per molecule
m_i is the initial molality
T_f is the initial freezing point of the system (°C)
T is specimen temperature (° C)

Thus, at a given subfreezing temperature, the unfrozen phases of all solutions meeting the above qualifying assumptions contain the same ratio of solute particles (ions, molecules, etc.) to water molecules, but the ratio of ice to unfrozen liquid can differ greatly. This principle, which is illustrated in Table 6-1, is believed to apply with fair accuracy to real specimens under commercially realistic conditions.

The formation of eutectics, of course, interferes with the relationship just described. Nevertheless, as the temperature is decreased in the subfreezing range, the unfrozen phase of any solution becomes more concentrated, at least until the final eutectic temperature is obtained. In the not uncommon event of solute supersaturation, concentration can continue at temperatures well below the final eutectic point.

Agitation of the fluid phase during freezing aids formation of pure ice crystals by minimizing accumulation of solutes at the solid-liquid interface [68]. As a result, concentration of solutes in the unfrozen phase is maximized at any given subfreezing temperature.

Slow removal of heat (slow freezing) results in a smooth, continuous, solid-liquid interface; maximum crystal purity; and maximum concentration of solutes in the unfrozen phase [1, 65, 68]. Rapid removal of heat (rapid freezing) results in an irregular or a discontinuous interface, considerable entrapment of solutes by growing crystals, and less than maximum concentration of solutes in the unfrozen phase.

The fact that freeze concentration has been used commercially for the removal of water from fruit juices emphasizes that solutes in the unfrozen phase can be concentrated substantially by freezing. Less obvious is the fact that solutes in tissue also undergo concentration during freezing [89].

Although there is some uncertainty concerning the importance of freeze concentration as a cause of freezing damage, it is likely that some products are susceptible to damage by this means.

TABLE 6-1

Effect of Initial Concentration on the Decrease in Volume and the Increase in Molality of the Unfrozen Phase During Freezing[a]

	Initially dilute	Initially concentrated
Unfrozen (1 liter)	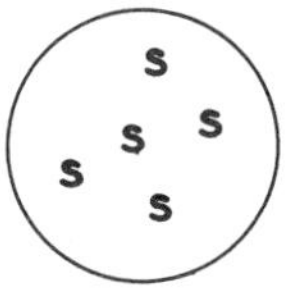	
Frozen		
Concentration unfrozen	5 S per liter	15 S per liter
Concentration in unfrozen phase of frozen sample	30 S per liter	30 S per liter
Increase in concentration by freezing	6X	2X
Volume of unfrozen phase after freezing	1/6 liter	1/2 liter
Amount of ice formed	5/6 liter	1/2 liter

[a] S = solutes; it is assumed that samples behave ideally and differ only in water content.

E. Relative Resistances of Various Tissues to Freezing Damage

Resistance to freezing damage varies greatly depending on the nature of the sample and how it is handled prior to freezing (e.g., pre- versus postrigor freezing of animal tissue). Particularly striking differences are apparent between animal and plant tissues and among various kinds of plant tissues (species, varieties, maturity, growing conditions, etc.). Although plant and animal tissues possess many features in common, their differences are more worthy of attention since these may account for observed differences in resistance to freezing damage.

Plant food tissue is normally semirigid because of the middle lamella, cellulosic cell walls, cell turgidity, and in certain tissues, such cellular components as starch grains. Plant cells are frequently polyhedral and they may or may not be aligned in an orderly fashion. Only partial contact exists among neighboring plant cells and a pectinaceous middle lamella serves as an intercellular cementing substance. Those areas in which the cells are not in perfect contact are occupied by an interconnected series of air spaces which frequently comprise a substantial fraction of the total volume of the tissue. The nutrients for plant cells are provided largely by diffusion from neighboring cells. Numerous tubules, known as plasmodesmata, connect the cytoplasms of adjacent cells and they are thought to have a role in the transfer of nutrients. A vacuole is usually present in plant cells and it may occupy as much as 90% of the volume of a mature cell. Numerous plastids (chloroplasts, chromoplasts, leukoplasts, amyloplasts) are present within the cytoplasm located between the vacuole and the cell wall. Whenever amyloplasts (plastids containing starch) are present, as in potatoes, they commonly dominate the cellular contents.

Muscle fibers (cells) are enveloped in a membrane (sarcolemma) but they do not possess a cell wall. These fibers are flexible, highly elongated cylinders with tapered ends, and they are invariably arranged in an orderly parallel fashion. Muscle fibers are separated by strands of collageneous connective tissue and by the interstitial fluid which is the source of nutrients. The contents of muscle fibers consist largely of proteinaceous contractile myofibrils and these, in turn, are composed of actin and myosin filaments. Muscle is devoid of counterparts of the air spaces, vacuoles, plasmodesmata, and plastids which are common in the cells of plant food tissues. Plant tissues, in contrast, contain nothing comparable to contractile myofibrils.

Cell breakage, cell separation, and impaired texture are frequently observed in frozen-thawed plant tissues and this is probably attributable to the semirigid, poorly aligned nature of many plant cells. Muscle fibers are more likely to separate than to break during freezing, and the texture of muscle is usually not damaged substantially by freezing and thawing. This behavior is at least partly attributable to the flexible nature of muscle fibers and to their parallel arrangement. Resistance to freezing injury is therefore often related to known characteristics of the samples involved.

III. THE FREEZING PROCESS — TECHNOLOGICAL ASPECTS

The freezing process consists of prefreezing treatments, freezing, frozen storage, and thawing. Each phase must be properly conducted if food products of maximum quality are desired.

A. Prefreezing Treatments for Major Classes of Natural Food Tissues

1. Fruits and Vegetables

The quality of fruits and vegetables following the freezing process depends to no small extent on the variety chosen, growing conditions, and stage of maturity at harvest. Optimum quality is usually achieved when maturity is near that which is customary for fresh consumption (mature for fruits, immature for vegetables).

Aside from such obvious operations as washing, grading, sorting, cutting, and proper packaging, some means usually must be employed to control enzyme activity in fruits and vegetables. If this is not done, fruits and vegetables undergo undesirable changes in flavor, color, texture, and nutritive value (degradation of vitamins) during frozen storage. Control of enzyme activity in vegetables is generally achieved by blanching. Blanching or scalding consists of a mild heat treatment accomplished by exposing the product to hot water or steam at 80°-100° for a period of several minutes. This treatment is quite acceptable since most vegetables are cooked prior to consumption.

One of the most troublesome reactions in fruits is enzyme-catalyzed oxidative browning (catalyzed by polyphenoloxidases). This reaction is important because it readily causes discoloration of light colored fruits. Control of polyphenoloxidases usually cannot be achieved by heating since heat damages desirable organoleptic properties of fruits. Chemical additives therefore are generally utilized to retard enzyme-catalyzed oxidative browning (enzymatic browning). These chemicals serve to inhibit enzymes, alter enzyme substrates, or limit entrance of oxygen.

Sulfur dioxide (also sulfites or sulfurous acid) and acids, such as citric and malic, are sometimes used to control enzymatic browning, and they do so by inhibiting the activity of polyphenoloxidases or by reacting with the substrate. Sulfur dioxide interacts chemically with the enzyme or the substrate, whereas acids function by lowering the pH to a value less suitable for the activity of enzymes [35, 69]. Use of sulfur dioxide is frequently avoided in fruits which are not cooked prior to consumption, since an objectionable chemical flavor is sometimes detectable. High levels of sulfur dioxide also can result in decreased firmness.

Ascorbic acid, although relatively expensive, is an effective agent for controlling enzymatic browning and is as well a valuable nutrient. Isoascorbic (erythorbic) acid and dihydroxy maleic acid are less effective than ascorbic acid. The anti-browning capability of ascorbic acid apparently arises from its ability to maintain phenolic substances in a reduced and colorless state [69]. Ascorbic acid has only a slight influence on pH. Concentrations of about 0.3% ascorbic acid in the sugar syrup are usually required for complete effectiveness; however, lower concentrations (about 0.1%) are frequently used.

Sucrose or sugar syrups are often added to fruits prior to freezing and their presence accomplishes several purposes: (1) to contribute sweetness, (2) to help retain volatile aromas, (3) to decrease the amount of water frozen at any given subfreezing temperature, and (4) to lessen enzymatic browning by acting as a barrier to the entrance of oxygen.

Cut fruits, particularly those with light colors, must be protected against enzymatic browning during the interval between cutting and freezing. Immersion in a dilute solution of sodium chloride (1-3%) is a method often used. Polyphenoloxidases are apparently inhibited by chloride ions, but the mechanism is unclear. An alternate and highly effective method for inhibiting polyphenoloxidases involves immersing the fruit for 45 sec in a 0.25% $NaHSO_3$ solution followed by 5 min in a 0.2% K_2HPO_4 solution (pH 8.8). The $NaHSO_3$ solution serves to avoid initial enzymatic browning before the slower acting K_2HPO_4 can exert its effect [12]. Use of $NaHSO_3$ results in no detrimental effects since its concentration is low enough to avoid softening of the tissue and it is removed sufficiently during subsequent treatment with the K_2HPO_4 solutions to avoid flavor defects. Treatment with K_2HPO_4 apparently decreases the susceptibility of polyphenolic substances to enzyme-catalyzed oxidation [64].

Prior to freezing, overly soft fruits (e.g., soft apples) can be firmed to some extent by brief immersion in a solution containing calcium salts. This effect probably occurs because divalent calcium ions form crosslinks between negatively charged polymers of pectic substances.

2. Fish

The quality of frozen-thawed-cooked fish is influenced by a variety of factors, including species, composition (especially fat content), size, how and where caught, elapsed time in the live state between harvest and freezing, the state of rigor when frozen, quality when frozen, and the details of the freezing process.

The major problems encountered during freeze processing of fish are oxidative deterioration, dehydration, toughening, loss of juiciness, and excessive thaw exudate (drip). In addition, fish are especially subject to deterioration by microorganisms and by autolysis, so great care must be taken to follow sanitary practices, to promptly cool to near 0° for prefreezing operations, and to freeze without undue delay. Effective and practical prefreezing or freezing techniques are available for controlling all of these problems except toughening and loss of juiciness. Reasonable control of toughening and loss of juiciness can be accomplished only by storing fish for a minimal time and/or at temperatures of -18° (0° F) or lower.

The state of rigor at the time of freezing has an important bearing on the ultimate quality of fillets. (See Volume 1 of this series for a discussion of the properties of pre- and postrigor muscle.) In order to minimize contraction and lessen increases in toughness of fillets during the freezing process, it is desirable that rigor mortis be completed at a low nonfreezing temperature prior to freezing. Alternatively, fillets may be frozen prerigor provided precautions are taken to avoid contraction during thawing (thaw contraction is less likely when thawing is slow, frozen storage is imposed for at least 2 months at -18°, and the pieces are large). Freezing during rigor is generally undesirable. The state of rigor at the time of freezing is less important with whole fish than it is with fillets and smaller pieces. For this reason, freezing promptly after harvest (prerigor) is sometimes recommended for whole fish [2]. The subject of prefreezing time-temperature treatments for fish is receiving considerable study and better and more detailed procedures are likely to be developed.

Undesirable oxidative changes in fish can be lessened by the following general approaches: (1) eliminating oxygen (vacuum, or flushing with inert gases) and/or providing a barrier to the entrance of oxygen (protective coatings); (2) avoiding contamination with oxidative catalysts (e.g., heavy metals); (3) adding antioxidants and antioxidant synergists; (4) avoiding irradiation, including light; and (5) using very low storage temperatures and/or short storage times.

Dehydration can be avoided simply by applying a suitable protective coating to the fish (ice glaze, molded gelatinous coating, or conventional package). Thaw exudate can be reduced by controlling the rate and extent of glycolysis prior to freezing, by applying suitable chemicals prior to freezing (e.g., polyphosphates), and by following recommended practices for freezing, frozen storage, and thawing.

3. Poultry

High quality frozen poultry can be obtained only if the live birds are of high quality, of proper maturity, have not been recently fed a diet containing fish oil, and are handled in a manner that avoids bruising. In general, tenderness decreases as birds become more mature.

Slaughter and dressing operations (scalding, removal of feathers, evisceration) should be conducted in a manner that does not unnecessarily stimulate the rate of postmortem glycolysis, since rapid glycolysis sometimes results in severe contraction of the muscles and this, in turn, can cause increased toughness.

To minimize growth of microorganisms, eviscerated poultry should be cooled promptly to 10° (50° F) or less. Chilling is usually accomplished by immersing the poultry directly in a mixture of crushed ice and water. This method provides a rapid rate of cooling without loss of moisture from the poultry (actually moisture is gained). Moisture losses can be sizeable if air cooling is used.

Chilling and aging of poultry for 1-6 hr (depending on age and size of birds) at temperatures somewhat above 0° also serve to promote desirable tenderness. The chilling-aging treatment (1) slows the rate of glycolysis; (2) allows gradual completion of rigor mortis, thereby avoiding excessive muscle contraction and the associated increase in toughness; and (3) allows changes (poorly understood) which over a period of several hours enhance the tenderness of poultry as measured in the cooked state [18-20, 37].

Prior to freezing, whole birds are frequently packaged in skin-tight plastic films. This provides significant protection against dehydration and oxidation during frozen storage.

4. Mammalian Meat

The ultimate quality of thawed-cooked meat is influenced by such preslaughter factors as species, interanimal variability, type of muscle, age, nutritional status (degree of finish), glycogen content of the muscle, and chemicals injected just prior to slaughter (e.g., enzymes for tenderization, chemicals to control the rate of glycolysis).

Slaughtered animal carcasses should be chilled promptly to a relatively low nonfreezing temperature (10° - 20°) for a period involving at least 20 hr so that glycolysis occurs slowly prior to freezing. This treatment reduces microbial growth and lessens toughness-inducing muscle contraction. Prerigor excised muscle generally should not be cooled to less than 10° or "cold shortening" occurs. At a temperature near 0° glycolysis is relatively rapid, contraction can be locally severe, and the muscle generally exhibits increased toughness and poor water-holding capacity. Meat used for the manufacture of sausage can be advantageously frozen prerigor. In this instance, rapid glycolysis can be avoided nearly completely during the freezing-thawing-manufacturing process, thus maintaining a high pH and a large capacity to bind water. Prefreezing time-temperature schedules for meat are receiving considerable study, and improved procedures are likely to be forthcoming.

Some kinds of mammalian muscle are improved significantly by aging prior to freezing. Aging (conditioning) consists of holding animal carcasses at cool non-freezing temperatures for a period of several days. Although the mechanisms are still not clear, aging of beef is known to enhance flavor; increase tenderness, juiciness, and water-holding capacity (as compared to meat at the conclusion of rigor mortis); and to decrease thaw exudate and lipid stability. Although beef is improved by aging several days prior to freezing, this is usually not done. Lamb requires little or no aging and pork should not be aged because of its unstable lipid components.

Mammalian meat is often cut into relatively small portions prior to freezing. This operation increases the surface-to-volume ratio of the product, thereby increasing thaw exudate and susceptibility to dehydration and oxidative deterioration.

Chemicals are sometimes added to meat prior to freezing to retard undesirable reactions or to promote desirable changes. Mammalian meat, especially pork, is susceptible to lipid oxidation during frozen storage, and various chemicals have been tested for their antioxidant effectiveness. Inconsistent results have been reported and none of the chemicals are approved for use in unmodified meat. Several of the common antioxidants can, however, be used in sausage. Some flavoring materials, such as smoke, have significant antioxidant properties. Heme pigments and sodium chloride (e.g., in cured meat) act as pro-oxidants at certain concentrations, however.

Proteolytic enzymes, especially papain, are sometimes used prior to freezing to tenderize less desirable cuts of meat. These enzymes can be applied by injection prior to slaughter or by dipping, spraying, or injection after slaughter.

To avoid freezer burn and lessen oxidative deterioration during frozen storage, it is essential that meat be packaged in a close-fitting material that is durable and is an effective barrier to oxygen and water vapor. Oxidative changes are even more effectively reduced if air is excluded by means of vacuum or inert-gas packaging.

B. Freezing

1. Methods

a. Air Freezing. Packaged or unpackaged nonfluid foods can be frozen in air at temperatures ranging from -18° to -40° (0° to -40°F). If "sharp" freezing is employed, air is circulated slowly or not at all and the rate of freezing is exceedingly slow (3-72 hr or more depending on the conditions and the size of the product). Sharp freezing is uncommon in modern freezing operations.

Vigorous circulation of cold air enables freezing to proceed at a moderately rapid rate and this common method is referred to as "air-blast freezing." Air-blast freezing is usually accomplished by placing the product on trays or on a mesh belt and passing it slowly through an insulated tunnel containing air at -18° to -34° (0° to -30° F) or lower. The air moves counter-current to the product at a velocity of 100-3500 lineal ft/min. Air at about -29° (-20° F) and 2500 lineal ft/min is usually satisfactory. Equipment for air-blast freezing has been described by Tressler [84].

Air-blast freezing is economical and is capable of accommodating foods of a variety of sizes and shapes. It can, however, result in (1) excessive dehydration of unpackaged foods if conditions are not carefully controlled, and this in turn necessitates frequent defrosting of equipment; or (2) undesirable bulging of packaged foods which are not confined between rigid surfaces during freezing.

Fluidized-bed freezing is a modification of air-blast freezing. Solid food particles ranging in size from peas to strawberries can be fluidized by forming a bed of particles 1-5 in. deep on a mesh belt (or mesh tray) and then forcing air upward through the bed at a rate sufficient to partially lift or suspend the particles in a manner somewhat reminiscent of a boiling liquid. If the air used for fluidization is appropriately cooled, freezing can be accomplished at a rapid rate. An air velocity of at least 375 lineal ft/min is necessary to fluidize suitable particles, and an air temperature of about -34° (-30°F) is common. Bed depth depends on the ease with which fluidization can be accomplished, and this in turn depends on size, shape, and uniformity of the particles. A bed depth of slightly more than 1 in. is suitable for easily fluidized particles, such as peas and whole kernel corn; a depth of 3-5 in. is used for partially fluidizable particles, such as green beans; and a depth of 8-10 in. can be used for nonfluidizable products, such as fish fillets (since fluidization is not involved in this instance, a more proper name is "through-flow air freezing"). Although freezing time varies with conditions, some representative times for various products are listed in Table 6-2.

Fluidized-bed freezing has proved successful for many kinds and sizes of unpackaged food tissues, although the best results are obtained with products that are relatively small and uniform in size (e.g., peas, limas, cut green beans, strawberries, whole kernel corn, brussels sprouts).

The advantages of fluidized-bed freezing as compared to conventional air-blast freezing are [80] (1) more efficient heat transfer and more rapid rates of freezing and (2) less product dehydration and less frequent defrosting of equipment. For example, dehydration losses of about 1% have been reported during fluidized-bed freezing of shrimp [4]. The short freezing time is apparently responsible for the small loss of moisture [36].

A major disadvantage of fluidized-bed freezing is that large or nonuniform products cannot be fluidized at reasonable air velocities.

b. Plate Freezing. Food products can be frozen by placing them in contact with a metal surface cooled either by cold brine or vaporizing refrigerants, such as refrigerant-12, refrigerant-22, or ammonia. Packaged food products may rest on, slide against, or be pressed between cold metal plates. Unpackaged food tissues, such as shrimp, can be frozen by freeze adhesion to a slowly rotating cold drum. Fluid products, such as ice cream mix or fruit juices, can be frozen in cylindrical scraped-surface heat exchangers.

Double-contact plate freezers are commonly used for freezing foods in orthorhombic packages (retail packages). This equipment, which may be batch, semiautomatic, or automatic, consists of a stack of horizontal cold plates with intervening spaces to accommodate single layers of packaged product. The filled unit appears much like a multilayered sandwich containing cold plates and product in alternating layers. The plates are arranged so that they can be opened or closed in a vertical accordian-like manner, enabling product to be added or removed. When closed, the plates make firm contact with the two major surfaces of the packages, thereby facilitating heat transfer and assuring that the major surfaces of the packages do not bulge during freezing.

TABLE 6-2

Typical Freezing Times in a Fluidized-Bed Freezer[a]

Product	Time to reduce temperature from 21° to -18° (70° to 0° F)[b] (min)
Peas, whole kernel corn	3-4[c]
Lima beans, blueberries	4-5
Diced carrots	6
Cut green beans	5-12
French fried potatoes	8-12
Strawberries	9-13
Large shrimp and fish sticks	12-15
Fish fillets (up to $1\frac{1}{4}$ in. thick)	30

[a]Ref. 3.

[b]Freezing conditions were not stated, but an air temperature of -34° (-30° F) and a velocity in the range of 500-1000 lineal ft/min are common.

[c]Unpackaged peas frozen in a conventional air-blast freezer require 15-20 min; Shipton [80].

Contact-plate freezing is an economical method that minimizes problems of product dehydration, defrosting of equipment, and package bulging. Two notable disadvantages do however exist: (1) packages must be of uniform thickness and (2) freezing occurs at a moderately slow rate as compared to other modern methods. For example, a compact product in a well-fitted package 1-1.5 in. thick, when cooled by plates at -33° (-28° F), requires about 1-1.5 hr to freeze. Freezing times are extended considerably when the package contains a significant volume of void space.

It should be mentioned that plate freezing can, in some instances, result in rapid freezing. This is true, for example, when solid unpackaged items are frozen on a rotary drum or when fluid products are frozen in a scraped-surface heat exchanger.

c. Liquid-immersion Freezing. Liquid-immersion freezing (usually referred to as "direct-immersion freezing") is accomplished when a food product, either packaged or unpackaged, is frozen by immersion in or by spraying with a freezant that remains liquid throughout the process. Aqueous solutions of the following substances have been used as freezants: propylene glycol, glycerol, sodium chloride, calcium chloride, and mixtures of sugars and salt. This technique, although not common, is used commercially for canned citrus juice concentrate (cans of juice are passed continuously through a chamber containing liquid freezant); for poultry, especially during the initial stages of freezing (to impart a uniform, white color to the surface); and occasionally for fish and shrimp. Liquid-immersion freezing can result in moderately rapid freezing as indicated by the fact that 6 oz. cans of orange juice concentrate can be cooled from 4° to -18° (40° to 0° F) in 10-15 min using commercially available equipment.

The major advantages of liquid-immersion freezing are that it results in rapid freezing (especially for foods which are unpackaged or are packaged in skin-tight films) and it is easily adapted to continuous operations. It is difficult, however, to find freezants with suitable properties.

d. Cryogenic Freezing. Cryogenic freezing refers to very rapid freezing achieved by exposing food items, unpackaged or thinly packaged, to an extremely cold freezant undergoing a change of state. The fact that heat removal is accomplished during a change of state by the freezant is used here to distinguish cryogenic freezing from liquid-immersion freezing. The most common food-grade cryogenic freezants are boiling nitrogen and boiling or subliming carbon dioxide. Boiling nitrous oxide also has been considered but at present it is not being used commercially. Boiling CCl_2F_2 (refrigerant-12) does not have a sufficiently low boiling point to qualify as a true cryogenic fluid but it should nevertheless be included in this category since it can provide, by the change of state principle, rates of freezing comparable to those obtained commercially with true cryogenic freezants. Some pertinent properties of these four freezants are listed in Table 6-3. The large usable heat capacity of solid carbon dioxide is especially noteworthy.

The rate of freezing obtained with cryogenic methods is much greater than that obtained with convention air-blast or plate freezing but is only moderately greater than that obtained with fluidized-bed or liquid-immersion freezing. For example, shrimp freeze in about 9 min in a commercial liquid nitrogen freezer and in about 12 min in a fluidized-bed freezer (see Table 6-2).

Liquid nitrogen (LN) is used in many cryogenic food freezers. The schematic diagram of Figure 6-6 represents the main features of major types of LN freezers available in the United States. The product is placed on a conveyor belt and moved into zone A of the insulated chamber, where it is cooled with moderately cold gaseous nitrogen moving counter-current to the product. In zone B, LN is sprayed or dribbled on the product, or alternatively, very cold gaseous nitrogen (at its bp temperature) is brought into contact with the product. None of the current commercial LN freezers employ the common laboratory technique of direct immersion. Following the desired exposure time (governed by belt speed), the product passes into zone C, where it is allowed to equilibrate to the desired final temperature (-18° to -30°) before it is discharged. It should be noted that the final product temperature is usually no different than that obtained during conventional methods of freezing.

Advantages of LN freezing are as follows: (1) Dehydration loss from the product is usually much less than 1%, (2) Oxygen is excluded during freezing, and (3) individually frozen pieces of product undergo minimal freezing damage. (This can be important for freeze-sensitive products, such as tomatoes, avocados, mushrooms, melon balls, bananas, strawberries, and raspberries. Meats frozen cryogenically exhibit minimal thaw exudate and damage to texture. These quality

TABLE 6-3

Properties of Some Cryogenic Freezants Suitable for Use with Foods[a]

Property	Freezant				
	CO_2		N_2O	CCl_2F_2	N_2
Molecular weight	44.01		44.02	120.93	28.016
Boiling point at 1 atm, °F (°C)	-109.3 (-78.5)		-129.1 (-89.5)	-21.62 (-29.79)	-320.46 (-195.81)
Latent heat at bp Btu/lb[b]	Solid (subl.) 246.1	Liquid (vap.) 149.6 at -69.9°F	(vap.) 161.8	(vap.) 71.04	(vap.) 85.8
Specific heat, gas, C_p at 1 atm, Btu/lb °F[c]	0.1988 at 60°F	0.1988 at 60°F	0.2004 at 60°F	0.145 at 86°F	0.2447 at 60°F
Usable heat capacity (Btu/lb)[d]	264	150	184	71	161

[a]From Matheson Gas Data Book [54].

[b]Btu/lb ÷ 1.8 = cal/g.

[c]Btu/lb °F = cal/g °C.

[d]Latent heat + (specific heat x ΔT), where ΔT = (-20° F) - (temperature of boiling or sublimation).

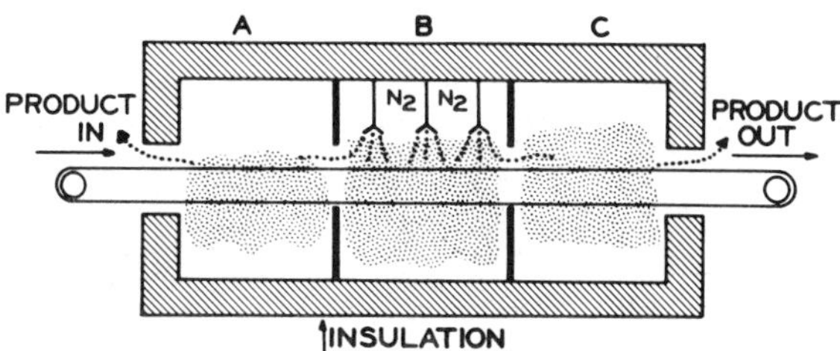

FIG. 6-6. Schematic diagram of a cryogenic food freezer. Zone A, cooling: incoming product is cooled by moderately cold nitrogen gas which is circulated by a fan and is moving counter-currently to the product. Most of the partially warmed nitrogen gas is eventually discharged to the atmosphere at the product entrance. Zone B, freezing: liquid nitrogen is sprayed or dribbled on the product, or nitrogen gas at its boiling point temperature can be brought into contact with the product. Zone C, tempering: Product is brought to a uniform final temperature of -18° to -29° (0° to 20° F). A small volume of nitrogen gas discharges with the product to avoid the entrance of moisture-laden air.

advantages are more likely to be retained if the time of frozen storage is minimized and/or the temperature of frozen storage is -23° (-10° F) or lower). Moreover, (4) the surface of poultry assumes a uniform white appearance, which is regarded as superior to the appearance achieved by slower methods of freezing; and (5) the equipment is simple, suitable for continuous flow operations, adaptable to various production rates and product sizes, of relatively low initial cost, and capable of high production rates in a minimal space. The only disadvantage of LN freezing is high operating cost, and this is attributable nearly entirely to the cost of LN.

Freezing with carbon dioxide involves exposing the product to powdered or liquid carbon dioxide. Most of the advantages cited above for LN freezing also apply to carbon dioxide freezing. However, tumbling of food tissues in the presence of powdered carbon dioxide is not a satisfactory procedure for delicate food tissues. An additional disadvantage of carbon dioxide freezing is that unpackaged products absorb or entrain significant amounts of carbon dioxide. If this carbon dioxide is not removed (e.g., by allowing the frozen product to stand unpackaged for several hours) before the product is packaged in an impervious material, undesirable swelling can occur.

Procedures for freezing foods in boiling CCl_2F_2 (refrigerant-12) have been developed by du Pont. Sufficient data have been collected to satisfy health officials of several countries, including the United States, as to the safety of food-grade CCl_2F_2 when it is properly used as a freezant in direct contact with foods. The product to be frozen in CCl_2F_2 is placed on a mesh belt and conveyed through an insulated freezing chamber. Freezing is accomplished either by spraying the product with food-grade CCl_2F_2 or by a combination process consisting of initial immersion of the product followed by spraying. In both procedures vapors are collected for reuse by condensation on a mechanically cooled metal surface. Inadvertent losses of CCl_2F_2 (i.e., by entrainment in crevices in the product) reportedly amount to about 1-1.5 lb CCl_2F_2 per 100 lb of a relatively nonporous product [41]. This is currently the only cryogenic method wherein the freezant is recovered and reused; however, this technique has been tested experimentally with a nitrous oxide freezing system.

Advantages of freezing in liquid CCl_2F_2 are essentially the same as those cited for LN, except the CCl_2F_2 system has the added advantage of lower costs. Even though CCl_2F_2 boils at a much higher temperature than does LN (see Table 6-3), freezing rates in CCl_2F_2 are claimed to equal or exceed those obtained with LN (as used commercially) because of CCl_2F_2's more favorable film coefficient for heat transfer.

The CCl_2F_2 process apparently is suitable for a wide variety of food products, including peas, corn, asparagus, french fried potatoes, cantaloupe balls, shrimp, meat patties, chicken parts, diced poultry, and packaged fluids and semisolids. Porous products, such as unpackaged cakes, cause a problem because of excessive entrainment of the freezant; however, for some products, this problem can be partially circumvented by dipping them in water immediately prior to freezing. The resulting thin crust of ice serves as a barrier to absorption of CCl_2F_2.

2. Costs of Freezing by Various Methods

When dealing with large, efficient operations, total freezing costs per pound of product reportedly are approximately the same for air-blast, plate, fluidized-bed, and CCl_2F_2 freezing. LN freezing is substantially more expensive than any of the methods just mentioned, and its high cost is attributable to the LN which is used and discarded [8, 13, 41, 71, 83]. The cost of freezing with carbon dioxide is apparently intermediate between these two extremes. There are, however, circumstances that can alter this cost pattern. If unpackaged products are frozen by the air-blast method and the equipment is poorly designed or operated, water loss from the product can be substantial. This occurrence is of particular economic significance if a product of great value (meat, shrimp) is being frozen, since the lost water has the same value per unit weight as the product. In this situation the cost of air-blast freezing can easily exceed the cost of LN freezing, since water losses during the latter method are small. It also should be noted that situations do exist where LN or carbon dioxide freezing can be economically advantageous. For example, when the processing operation is small, is subject to large variations in production rate, or is of uncertain duration, installation of any freezing equipment involving a large capital investment (air blast, plate, fluidized bed) and requiring a large, uniform production rate for efficient operation is inadvisable.

3. Selecting a Method for Freezing Foods

A decision concerning a process for freezing foods should be based on cost and food-quality considerations. It is quite possible that a product processed so as to yield the lowest retail price may be rejected by the consumer for a product of better quality which has been achieved by using a superior but more costly processing method. However, it is possible that some products processed by the most economical methods may have qualities which are only slightly inferior, indistinguishable from, or perhaps even superior to products processed by more expensive methods.

To properly assess the effects of freezing methods on food quality, the product must be evaluated following a treatment similar to that which it is to receive commercially. Food quality as perceived by the consumer is thereby determined, and this is the most useful kind of quality information.

To assess the quality of frozen foods as viewed by the consumer, the product must be evaluated after the following sequence of events:

1. Prefreezing treatments, such as blanching of vegetables; chilling of meats, fish and poultry; aging of beef; or the addition of normally used protective chemicals,
2. Freezing.
3. Frozen storage for a commercially realistic time and temperature. Since frozen storage, as now conducted commercially, is the most detrimental phase of the freezing process, careful thought should be given to selection of storage conditions. Commercial conditions of frozen storage are almost certain to continue to improve. If so, the quality of products frozen by methods which result in minimal damage will, in the future, be more apparent to the consumer. It therefore may be wise when conducting tests, to utilize storage conditions which are somewhat less detrimental to product quality than those which currently prevail.
4. Thawing, and also cooking for those products which are normally consumed in a cooked state.

Quality information can be combined with data on freezing costs to determine which method of freezing is best for the product under consideration.

4. Freezing Methods for Various Classes of Food

Vegetables can be satisfactorily frozen by any of the commercial methods. Plate freezing is common for retail packages of vegetables, and fluidized-bed or air-blast freezing is satisfactory for individually quick frozen (IQF), reasonably nonfragile vegetables.

The situation with respect to fruits is similar to that for vegetables except that fragile fruits must be handled more gently than most vegetables. Very rapid freezing is advantageous for minimizing damage to fruits and vegetables with especially delicate structures; however, the quality advantages observed immediately following freezing tend to disappear during frozen storage as now conducted commercially.

The quality of thawed fish can be influenced somewhat by the rate of freezing but this factor generally is of secondary importance. Attainment of a product temperature of -18° (0° F) or lower in a period of 2 hr or less generally produces satisfactory results. Any of the freezing methods previously discussed can be applied to fish.

Freezing rate within the range of good commercial practice has little effect on water-holding capacity and organoleptic attributes of thawed-cooked poultry. The rate of freezing, however, does have a profound effect on the appearance of the frozen product and it is for this reason that rapid freezing is employed, at least for the surface layer. Slow freezing results in frozen poultry with a natural or sometimes reddish coloration, whereas very rapid freezing produces a uniform, chalky-white frozen appearance that is regarded as desirable. Intermediate chalkiness is produced with intermediate rates of freezing. The chalky-white appearance produced by rapid freezing persists only in the frozen or freeze-dried states (disappears during thawing) and is apparently related to the number and size of ice crystals present and to the thickness and nature of the tissue matrix.

Plate freezing is common for packaged cut-up poultry. Air-blast freezing is common for whole, ready to cook birds packaged in skin-tight plastic films. Air for this purpose should be no warmer than -34° (-30°F) and should be moving at a minimum velocity of 600 lineal ft/min. Freezing by immersion in or by spraying with cold calcium chloride brine or glycol solutions (e.g., an aqueous solution of 46 wt % propylene glycol) is also common for whole birds packaged in skin-tight films. This method and cryogenic freezing are particularly effective for imparting a chalky-white appearance to frozen poultry. Cryogenic freezing, or immersion-freezing using a freezant at -18° (0°F) or lower, produces the whitest possible appearance. The desired white appearance can be obtained equally well if very rapid freezing (liquid immersion or a cryogenic method) is used only during the early stages of freezing and normal air-blast freezing (moderately rapid freezing) is used for the remainder.

Rate of freezing within the range of good commercial practice, has little effect on flavor, juiciness, or tenderness of meat. Thaw exudate, however, is increased significantly if very slow freezing is employed (when the freezing boundary moves at a rate of less than 0.2 cm/hr), particularly with small pieces of meat. Since meat cuts are marketed in a variety of sizes and shapes, air-blast freezing is a suitable and common technique. Acceptable conditions are an air temperature of -29° (most common) to -40° (-20° to -40°F) and a velocity such that the air is changed at least two times a minute. Plate freezing is suitable for meat cuts of uniform thickness. Cryogenic freezing also can be applied successfully to meat. The whitish appearance of rapidly frozen meat can be overcome, if desired, by warming the surface momentarily with infrared heat and then allowing it to refreeze.

5. Freezing Curves (Time-Temperature Plots)

Familiarity with the pattern of temperature changes that food materials undergo during freezing is basic to an understanding of the freezing process. Shown in Figure 6-7 are curves typical of a wide variety of foods when freezing is conducted at various rates. The most striking feature of these curves is their simplicity. Before these curves are considered in detail it should be noted that time-temperature data derived from natural food tissues during moderately slow freezing (curve with the long plateau in Fig. 6-7) will at best only approximate true conditions of solid-liquid equilibrium. Furthermore, when very rapid freezing is employed, the time-temperature data represent nonequilibrium conditions.

Consider now, the slow freezing curve of Figure 6-7. Portion A-S represents simple cooling (removal of sensible heat) with no water-ice transformation. Point S represents supercooling and this event is believed to occur to some extent (usually not more than 10°) in all specimens, although it is not always apparent. Detection of supercooling depends on the sensitivity, response time, and location of the temperature-measuring device.

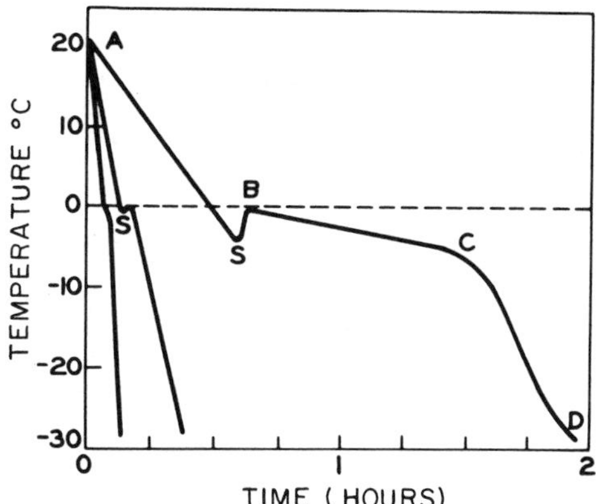

FIG. 6-7. Schematic freezing curves typical of foods and biological matter frozen at different rates. S = supercooling. See text for explanation.

Following onset of crystallization at point S, the released heat of crystallization causes the temperature to rise promptly to the apparent initial freezing point of the sample, point B. The apparent initial freezing point as determined by this means is always somewhat lower than the true initial freezing point, especially if supercooling exceeds more than $1^\circ - 2^\circ$ [27]. The initial freezing point, which is determined by the number of dissolved particles in solution (molecules, ions, dimers, etc.), differs surprisingly little among various types of natural foods. Data illustrating this point are shown in Table 6-4.

Section B-C of Figure 6-7 represents the period during which a major portion (often three-fourths) of the water is crystallized. Since heat which is generated from ice formation constitutes the bulk of the heat energy which must be removed during freezing, and since the ice formed during section B-C results in only a moderate increase in the concentration of solutes in the unfrozen phase (slight depression of freezing point), section B-C exists as a long plateau with a slight negative slope. During the early stages of section B-C, water separates as pure or nearly pure ice crystals, whereas during the late stages of section B-C and beyond, the possibility exists for formation of eutectic mixtures and other types of complex solids of largely unknown structures and compositions. (However, many solutes may persist in a state of supersaturated metastable equilibrium.) The late stages of section B-C and beyond are especially complex in tissue because growing ice crystals exist in and/or among cells, thereby competing with each other for space and for the diminishing supply of freezable water.

Each successive gram of ice formed beyond point C is responsible for a substantial and successively larger increase in molality (lower freezing point) of the unfrozen phase. Furthermore, since the specimen at point C contains far less freezable water than it had initially, removal of a given amount of heat energy can effect a much greater reduction in specimen temperature during section C-D than during B-C.

TABLE 6-4

Initial Freezing Points of Various Classes of Foods[a]

Class of food	Range of freezing points °C (°F)
Nineteen common vegetables	-0.8 to -2.8 (30.5 - 26.9)
Twenty-two common fruits	-0.9 to -2.7 (30.4 - 27.2)
Fresh pork, beef, and lamb	-1.7 to -2.2 (29 - 28)
Milk and eggs	ca. -0.5 (ca. 31)

[a]From Hardenburg and Lutz [32].

Even after cooling to point D, the specimen usually contains some freezable water (unless this temperature is well below the specimen's final eutectic point). Changes in the ice contents of various specimens as a function of temperature are shown in Tables 6-5 and 6-6.

It also should be noted from the curves in Figure 6-7 that increased rates of heat removal cause the various stages of cooling and freezing to become less distinct; and finally, at very great rates of heat removal, they become indistinguishable.

C. Frozen Storage

Frozen foods stored at or near a conventional temperature of -18° (0°F) are not completely frozen, nor are they inert. They deteriorate at a significant rate and the quality loss incurred during a normal period of frozen storage generally exceeds that caused by any other phase of the freezing process (prefreezing treatments, freezing, thawing). Loss of quality during proper frozen storage of foods occurs by chemical or physical means since microorganisms cannot grow under these conditions.

The major physical changes that occur during frozen storage are recrystallization and sublimation. Recrystallization is a common event even in foods stored at -18°. The quality advantages initially apparent in rapidly frozen foods as compared to slowly frozen foods gradually disappear during frozen storage, and recrystallization is at least partially responsible [21, 23, 42-44, 50]. Recrystallization can be effectively controlled by storing products at a constant low temperature and for a minimal time.

Sublimation of ice can occur during frozen storage of improperly packaged frozen food, and this can lead to a defect known as "freezer burn." Freezer burn is particularly noticeable on the surface of poultry, where it is evident as unattractive, dry, brown spots. Sublimation and freezer burn can be minimized by any means which effectively stops loss of moisture from the product, e.g., a

TABLE 6-5

Percent Frozen Water in Ice Cream at Various Temperatures[a]

Water frozen to ice (%)	Temperature °C (°F)	Water frozen to ice (%)	Temperature °C (°F)
0	-2.5 (27.55)	40	-4.2 (24.40)
5	-2.6 (27.35)	45	-4.6 (23.65)
10	-2.7 (27.05)	50	-5.2 (22.62)
15	-2.9 (26.78)	55	-5.9 (21.42)
20	-3.1 (26.40)	60	-6.8 (19.79)
25	-3.3 (26.04)	70	-9.5 (14.99)
30	-3.5 (25.70)	80	-14.9 (5.14)
35	-3.9 (25.03)	90	-30.2 (-22.29)

[a]Composition: fat, 12.5%; serum solids, 10.5%; sugar, 15%; stabilizer, 0.30%; water, 61.7%. From Liska and Rippen [45]. Presumably based on initial water content.

high relative humidity, application of an ice glaze (fish), or packaging with a material which is highly impermeable to water vapor.

Some changes (not necessarily independent) of a chemical or partly chemical nature that occur in foods during frozen storage are (1) degradation of pigments, (2) degradation of vitamins, (3) insolubilization or destabilization of proteins, (4) oxidation of lipids, (5) reactions which lead to diminished differences in quality between rapidly and slowly frozen foods, and (6) reactions which cause an increase in the amount of thaw exudate from tissue as the time of frozen storage is extended. Most of these chemical reactions occur slowly at -18° and decline further in rate as the temperature is reduced. However, during the early stages of freezing, some reactions, including glycolysis, actually increase in rate and others decrease in rate less than expected [27].

The time frozen foods can be maintained in "good condition" depends on the kind of product, how it is processed and packaged, and the storage temperature. In the United States, -18° (0° F) is generally accepted as a satisfactory storage temperature for frozen foods (lower temperatures are common in Europe). Shown in Table 6-7 are expected storage lives for foods that are properly processed and packaged and then stored at -18°. Great differences are evident in the storage lives of the various products.

The effect of temperature on the storage lives of foods is shown in Figure 6-8. It is again apparent that different products held at any given temperature differ greatly in their storage lives. Furthermore, a change in storage temperature affects the storage lives of some products to a much greater extent than it does others.

TABLE 6-6

Percent Frozen Water[a] in Various Food Materials

Temperature °C (°F)	Lean meat[b] (74.5% H_2O)	Haddock[c] (83.6% H_2O)	Egg white[d] (86.5% H_2O)	Egg yolk[d] (50% H_2O)
0 (32)	0	0	0	0
-1 (30.2)	2	9.7	48	42
-2 (28.4)	48	55.6	75	67
-3 (26.6)	64	69.5	82	73
-4 (24.8)	71	75.8	86	77
-5 (23.0)	74	79.6	87	79
-10 (14.0)	83	86.7	92	84
-20 (-4)	88	90.6	93	87
-30 (-22)	89	92.0	94	89
-40 (-40)	—	92.2	—	—

[a]Grams ice/(grams of total initial water) x 100.

[b]Riedel [73].

[c]Riedel [72].

[d]Riedel [74].

In spite of the fact that frozen foods should be maintained at -18° or lower during storage and marketing, sizeable amounts are routinely exposed to much higher temperatures. Most of this mishandling occurs at locations other than the storage warehouse, especially in assembly rooms, delivery trucks, and retail cabinets [40, 63]. Modern, properly operated, home-type freezers or freezer-refrigerator combinations generally provide satisfactory conditions for limited storage of already frozen foods [52].

Temperature fluctuations cannot be completely avoided during commercial frozen storage. The rates at which a variety of products deteriorate have been compared under conditions similar to the following: (1) temperatures fluctuating on the order of ±5° (±9°F) in the vicinity of -18° (0°F) and (2) a comparable steady temperature (not a mean of the extreme temperatures because the rate of deterioration increases exponentially with increasing temperatures [78]). In general, it has been found that the overall quality of fruits, vegetables, meats, poultry, and fish deteriorates at essentially the same rate under both conditions ([22, 25, 29, 31, 38, 70, 88, 92]; also see the numerous studies on time-temperature tolerance of frozen foods, conducted at the United States Department

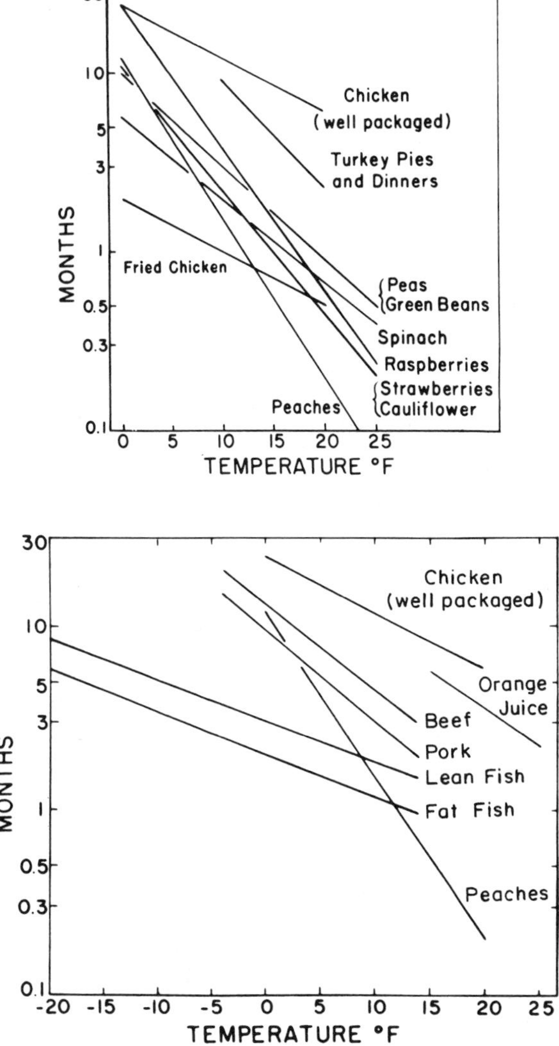

FIG. 6-8. Relative stabilities of frozen foods at various temperatures [79].

of Agriculture's Western Regional Research Laboratory and published in Food Technology, the first article appearing in 1957 under the authorship of Van Arsdel [86]).

In a few instances, fluctuating temperatures can be more detrimental to product quality than a comparable steady-temperature condition. This can occur (1) when the product has a delicate texture which contributes greatly to its overall quality, e.g., bread, manufactured foods thickened with eggs and starch, ice cream, and some fruits; (2) when it is desirable to retain the white surface color of rapidly frozen poultry; (3) when it is desirable to avoid ice crystals on the

TABLE 6-7

Storage Life of Frozen Foods at -18° (0° F)[a]

Products with storage lives of 3 months or less	
Mammalian meats	Pork liver and heart, sweet breads, brains, tripe, special sausage or delicatessen meats, precooked Canadian bacon, precooked sliced ham, precooked sausage
Bakery products	Sponge cake, spice cake
Products with storage lives of 3-6 months	
Mammalian meats	Pork (ground, sausage, bacon); beef, lamb, or veal livers and hearts; tongue; kidneys; precooked whole hams
Poultry	Precooked turkey
Products with storage lives of 6-12 months	
Vegetables	Asparagus, green beans, brussels sprouts, corn on the cob, french fried potatoes
Fish	Fatty types, raw or precooked (6-8 months); lean types, raw (10-12 months); lobsters (8-10 months)
Mammalian meats	Beef (ground), veal (ground, thin cutlets, or cubes), lamb (ground), pork (roasts, chops, smoked ham), precooked meat pies
Poultry	Raw, eviscerated; fried chicken; precooked chicken pies
Bakery products	Cakes (except sponge and spice cakes)
Products with storage lives of 12 months or more	
Fruits (sugared)	Apricots with ascorbic acid, peaches with ascorbic acid, raspberries, strawberries
Vegetables	Lima beans, broccoli, cauliflower, cut corn, carrots, green peas, spinach
Mammalian meats	Beef (roasts, steaks); lamb (roasts, chops)
Fish	Precooked, lean
Eggs	Whole, yolk, white
Bakery products	Precooked bread, rolls and waffles; cookie dough

[a] Average time products can be maintained in "good condition" when processed and packaged in accordance with accepted procedures. ASHRAE Guide and Data Book — Applications [5]; Tressler et al. [85].

surface of products loosely packaged in materials which are essentially impermeable to water vapor and (4) when substantial melting of the product occurs during the temperature fluctuation.

The most serious temperature fluctuations are encountered at locations other than the storage warehouse. The remarkable nonuniformity of temperatures that can occur in a properly operated retail cabinet was demonstrated by Peterson [67]. He recorded time and temperature for a series of small cans of water properly stored in a commercially available retail cabinet, operating at the lowest possible temperature. The cabinet was equipped with an automatic defrost system, preset by the manufacturer to function for a 1 hr period once every 24 hr. The results shown in Figure 6-9 illustrate that marketing conditions for at least some frozen foods are far from ideal.

A means for calculating the "high-quality life" of frozen foods exposed to the irregular temperature conditions encountered during marketing has been described by Van Arsdel and Guadagni [87] and Guadagni [31].

D. Thawing

Thawing is a particularly vexing aspect of the freezing process. First of all, thawing of retail foods is conducted by persons who are unskilled in, and often unconcerned with, proper handling of frozen foods. Furthermore (1) thawing of nonfluid foods (tissue, gels) is inherently slower than freezing when comparable temperature differentials are employed; (2) temperature differentials frequently are less during thawing than during freezing, especially with fruits; and (3) the pattern of temperature change is less desirable during thawing than it is during freezing. During thawing, foods are subject to damage by chemical, physical, and microbial means, although microbial problems are negligible in properly handled foods. In light of these considerations, thawing must be regarded as a greater potential source of damage than freezing.

Further consideration must be given to the fact that tissues, gels, and all other aqueous materials which transmit heat energy primarily by conduction thaw more slowly than they freeze (equal temperature differentials) and that the shape of the time-temperature curve for thawing bears little similarity to a reversed freezing curve [24, 46, 58, 59, 60, 76]. An experiment conducted annually at the University of Wisconsin (laboratory experiment in Food Chemistry) provides information pertinent to these matters. Four 303 x 407 metal cans of a nonflowable starch gel serve as the substrate. The shanks of four thermocouples are mounted in the lid of each can, and the thermocouple junctions are positioned in a plane midway between the top and bottom of the can. One of the four thermocouple junctions is located adjacent to the outer wall, another at the geometric center, and the others at intermediate positions. Two of the cans are equilibrated in a water bath at $+78.5^{\circ}$ and the other two in a dry ice-acetone bath at -78.5°; thus each specimen's initial temperature is approximately 78.5° removed from the freezing point. The two frozen samples are then transferred to the $+78.5^{\circ}$ bath and the two warm samples are transferred to the -78.5° bath. Temperature and time are recorded until each can completes its change of state. Under these conditions, the temperature differential favors neither freezing nor thawing, since the coolant temperature is as much below the specimen's freezing point as the heating medium is above. The temperature gradients existing during freezing

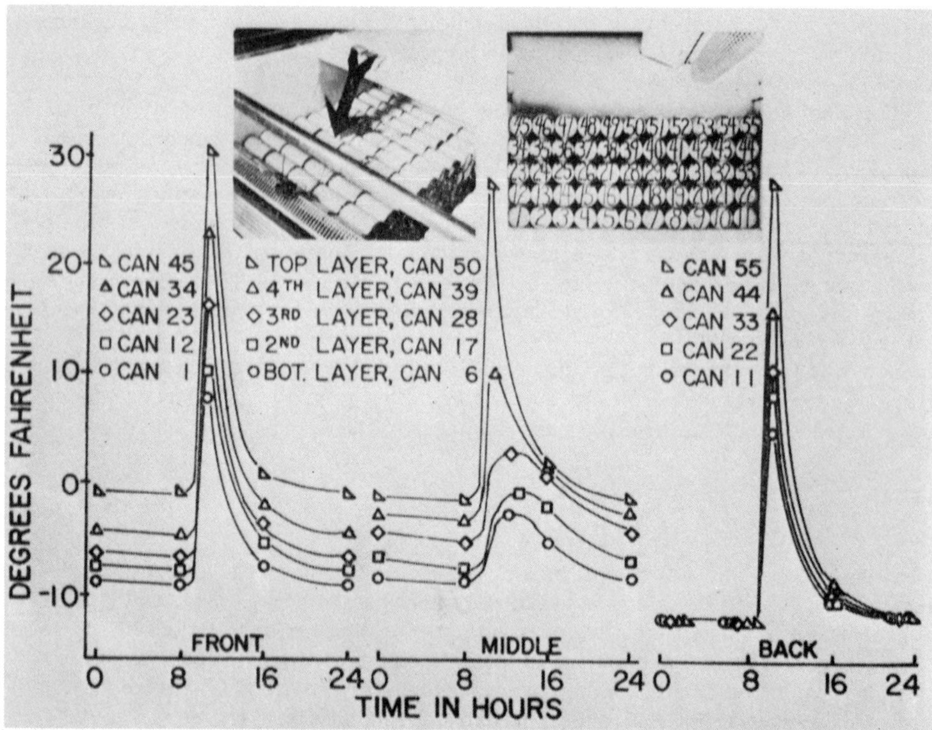

FIG. 6-9. Temperature history of small cans of water stored in a frozen food retail cabinet. Photographs show cabinet, row of cans which contained thermocouples (arrow), and the location numbering system. From Peterson [67]; courtesy of Campbell Soup Company.

and thawing are shown in Figure 6-10. A plot of specimen temperature (geometric center) vs time is shown in Figure 6-11.

Freezing diagrams A, B, and C of Figure 6-10 illustrate that large temperature gradients exist in the outer frozen zone during freezing and that freezing is nearly complete in 26 min. Thawing diagrams D, E, and F of Figure 6-10 illustrate that the temperature of the frozen phase rises rapidly to just below the melting point and remains there during the course of thawing, which is slow and not close to completion after 30 min. Figure 6-11 indicates that the geometric center has frozen after 28 min of cooling, whereas thawing has required 52 min. Also well demonstrated are the different shapes of the freezing and thawing curves. It is of no small importance that the additional time required for thawing is spent just below the melting temperature. This subfreezing temperature zone must be regarded as unfavorable, at least for foods, since (1) recrystallization and growth of microorganisms are both more likely here than at any other subfreezing temperature and (2) many chemical reactions proceed at surprisingly rapid rates.

These differences in the nature and rates of freezing and thawing can be explained on the basis of several properties of water and ice:

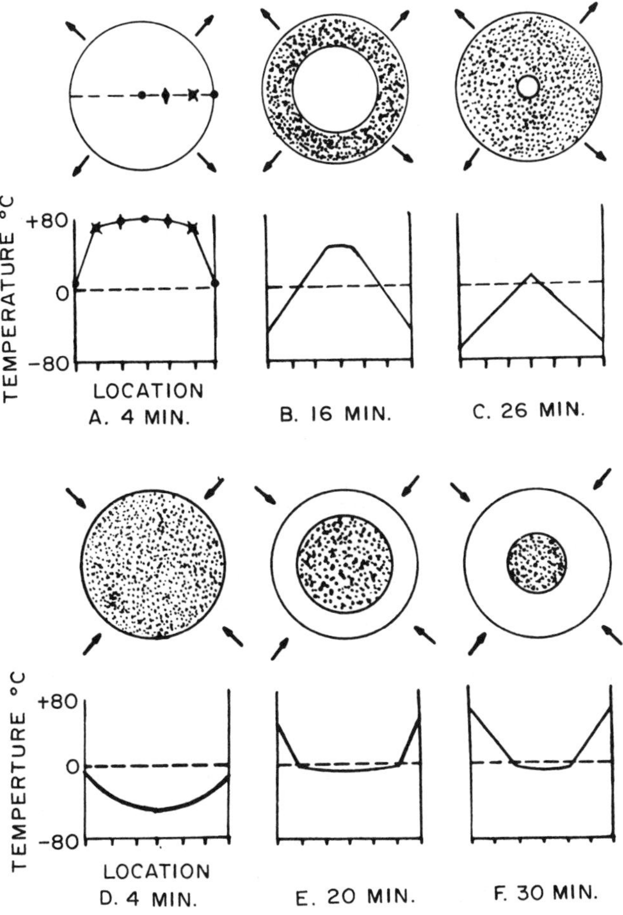

FIG. 6-10. Sequence of freezing and thawing in a cylindrical specimen. A, B, and C, freezing sequence; D, E, and F, thawing sequence. Circles represent the central cross-section of a 303 x 407 can filled with gelatinized starch. The four thermocouple locations are shown in the circular cross section of A. The relationship between the circular cross-sections and the temperature profiles immediately beneath them is also shown in A. Arrows indicate the direction of heat flow and dotted areas represent frozen material. From Fennema and Powrie [26]; courtesy of Academic Press.

1. Latent heat of crystallization (large)
2. Thermal conductivity (ice transmits heat energy at a rate four times faster than water)
3. Thermal diffusivity (ice undergoes a change in temperature at a rate approximately nine times faster than water)

Cooling-freezing or warming-thawing of aqueous substances requires the transfer of considerable heat energy per unit weight, primarily because of water's

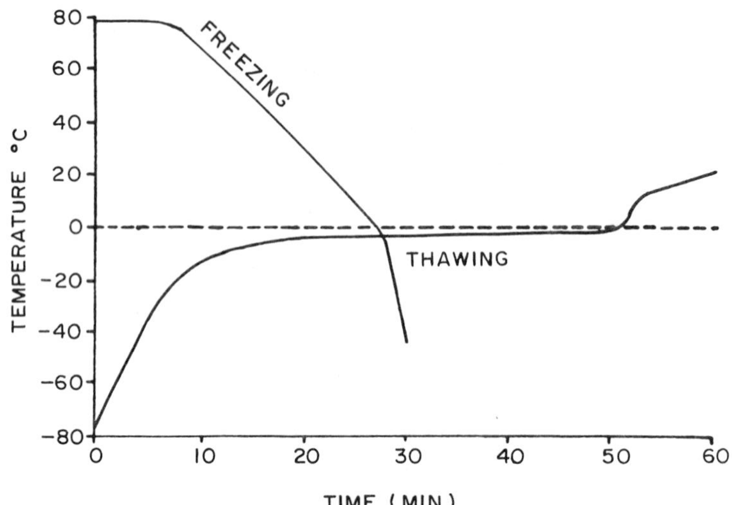

FIG. 6-11. Comparative freezing and thawing curves for the geometric center of a cylindrical specimen (same data as Fig. 6-10). From Fennema and Powrie [26]; courtesy of Academic Press.

large latent heat of crystallization. Freezing involves (1) removal of the latent heat of crystallization through an ice layer that enlarges with time and (2) lowering of specimen temperature. Since ice has a low thermal resistance (high thermal conductivity) and readily undergoes a change in temperature (high thermal diffusivity) freezing can occur quite rapidly. In contrast, thawing involves (1) addition of the latent heat of fusion through a layer of nonflowable water (specimen is a gel or tissue) that enlarges with time and (2) raising the specimen's temperature. Since water has a high thermal resistance (low thermal conductivity as compared to ice) and undergoes changes in temperatures at a rather slow rate (low thermal diffusivity as compared to ice), thawing occurs more slowly than freezing when environmental conditions favor neither process. Put more simply, nonflowable water (in gels or tissues) is a better insulator than ice.

The rapid rise in temperature that the thawing specimen undergoes at the outset (Figs. 6-10 and 6-11), occurs because a significant layer of unfrozen nonflowable water does not yet exist. (The favorable thermal properties of ice account for this occurrence.) However, once the outer surface begins to thaw, heat transfer is significantly impaired, and this occurs when most of the energy required for the change of state has yet to be supplied. The long thawing plateau (Fig. 6-11) is evidence of the extent to which the layer of unfrozen, nonflowable water impedes heat transfer.

It is interesting that the freezing curve of Figure 6-11 does not exhibit a plateau at the freezing point. This behavior is not uncommon for rapidly frozen specimens, especially those (unlike the example shown) in which the thermocouple is located away from the geometric center.

It should be emphasized that these differences in rates of freezing and thawing apply only to situations where heat energy is transmitted primarily by conduction. The principle outlined here would not, for example, apply if dielectric or microwave heating were employed, nor would it apply to normally fluid materials that could be agitated or crushed during thawing.

Differences in the rates of freezing and thawing can be less than shown above if the specimens contain a moderately low water content, as does meat, or a significant air content, as do fruits and vegetables. The effect of low water contents or large air contents, while important in research studies where temperature differentials can be held constant during freezing and thawing, is far less important in practical situations where unequal temperature differentials are nearly always employed. Temperature differentials during rapid commercial freezing can approach 200°, whereas the maximum differential during thawing is on the order of 100° for meat and vegetables and much less for fruits. As a consequence, the comparative slowness of thawing is further accentuated, in spite of the specimen's water or air content.

Although food quality is no doubt damaged at a significant rate during thawing, the time involved is short compared to a normal period of frozen storage. As a result, loss of quality is normally much greater during frozen storage (as conducted commercially) than during thawing. Evidence for this statement comes from the observation that most foods can be rapidly frozen and thawed (no frozen storage) without incurring appreciable damage.

For use in the home, most vegetables can be thawed and cooked by direct immersion in boiling water. Since high temperatures are detrimental to the quality of most fruits, they must be thawed under mild temperature conditions. Suitable techniques consist of placing unopened packages of fruit at room temperature, in the refrigerator, or in cool to slightly warm water [6] until thawing is almost complete. Fruits should be consumed promptly after thawing because their color and texture deteriorate rapidly at this point.

Small portions of frozen meat, fish, or poultry can be cooked directly from the frozen state, if desired. Larger portions of animal tissue are more troublesome to thaw. If a direct cooking technique is used, the oven temperature must be relatively low to avoid overcooking the exterior before the interior has thawed. Other satisfactory techniques for thawing large portions of animal tissue include (1) thawing in a refrigerator, (2) thawing at room temperature while the product is insulated (wrapped in several layers of paper) or uninsulated, and (3) immersion (especially poultry packaged in skin-tight plastic films) in cool to slightly warm water. If either of the last two methods is employed, the period of thawing should be kept to a minimum so that microorganisms have a limited opportunity to multiply.

Radiofrequency heating (dielectric or microwave) has received considerable attention as a possible means of thawing food tissue on a commercial basis [9, 11, 17, 28, 34, 77]. Provided the food tissue is reasonably homogeneous, radiofrequency heating is more rapid and much more uniform than heating by conduction. Unfortunately, frozen foods are not reasonably homogeneous since they always contain frozen and unfrozen phases and a nonuniform distribution of lipids. These components differ greatly in their abilities to absorb radiofrequency energy and this tends to cause localized areas to overheat before other areas have thawed. This problem can usually be overcome by proper control of the following variables: product size, shape, and orientation; the manner in which the product is presented to the radiofrequency field (e.g., immersion in water); frequency, power, or pulsation of the energy source; and the design and location of the electrodes when dielectric heating is used [77].

Other methods which are suitable for thawing foods on a commercial basis include warm (20°) moist air moving at a velocity of at least 1000 lineal ft/min; warm (20°) water moving at about one lineal ft/min and pure steam [51, 57, 66, 91].

REFERENCES

1. H. Akimoto and T. Ogawa, Low Temp. Sci., Ser. B, 11, 47 (1954).

2. M. L. Anderson, Comm. Fish. Rev., 32, 15 (1970).

3. Anonymous, Quick Frozen Foods, 25(6), 36 (1963).

4. Anonymous, Food Proc. Mktg., 28(12), 29 (1967).

5. Anonymous, ASHRAE Guide and Data Book — Applications 1968, American Society of Heating, Refrigerating, and Air-Conditioning Engineers, New York, 1968.

6. Anonymous, Quick Frozen Foods, 30(6), 93 (1968).

7. W. S. Arbuckle, A Microscopical and Statistical Analysis of Texture and Structure of Ice Cream as Affected by Composition, Physical Properties and Processing Methods, University of Missouri Agricultural Experiment Station, Bull. No. 320, 1940.

8. S. Aström, Scand. Refrig., Aug.-Sept., 12 (1971).

9. N. Bengtsson, in First International Congress on Food Science and Technol., London, 1962, Vol 4, Gordon and Breach, 1965, pp. 89-98.

10. N. Bengtsson, Food Technol., 17, 1309 (1963).

11. N. Bengtsson, J. Sci. Food Agr., 14, 592 (1963).

12. H. R. Bolin, F. S. Nury, and B. J. Finkle, Bakers Dig., 38(3), 46 (1964).

13. D. C. Brown, Advan. Cryogenic Eng., 12, 11 (1967).

14. H. E. Buckley, Crystal Growth, Wiley, New York, 1951.

15. E. H. Callow, J. Sci. Food Agr., 3, 145 (1952).

16. B. Chalmers, Principles of Solidification, Wiley, New York, 1964.

17. D. A. Copson, Microwave Heating, AVI Publ. Co., Westport, Conn., 1962.

18. D. deFremery and M. F. Pool, Poultry Sci., 38, 1180 (1959).

19. D. deFremery and M. F. Pool, Food Res., 25, 73 (1960).

20. D. deFremery and M. F. Pool, J. Food Sci., 28, 173 (1963).

21. W. J. Dyer, Food Res., 16, 522 (1951).

22. W. J. Dyer, in Low Temperature Biology of Foodstuffs (J. Hawthorn and E. J. Rolfe, eds.), Pergamon Press, New York, 1968, pp. 429-447.

23. K. G. Dykstra, Refrig. Eng., 64(9), 55 (1956).

24. A. J. Ede, Proc. Intern. Congr. Refrig., 9th Congr., 1, 2029 (1955).

25. C. Emerson, D. E. Brady, and L. N. Tucker, The Effect of Certain Packaging and Storage Treatments on the Acceptability of Frozen Beef, University of Missouri-Columbia Agricultural Experiment Station, Bull. No. 470, 1951.

26. O. Fennema and W. D. Powrie, Advan. Food Res., 13, 219 (1964).

27. O. Fennema, W. D. Powrie, and E. H. Marth, Low-Temperature Preservation of Foods and Living Matter, Marcel Dekker, New York, 1973.

28. S. A. Goldblith, Advan. Food Res., 15, 277 (1966).

29. W. A. Gortner, F. Fenton, F. E. Volz, and E. Gleim, Ind. Eng. Chem., 40, 1423 (1948).

30. C. S. Grove, Jr., R. V. Jelinek, and H. M. Schoen, Advan. Chem. Eng., 3, 1 (1962).

31. D. G. Gaudagni, in Low Temperature Biology of Foodstuffs (J. Hawthorn and E. J. Rolfe, eds.), Pergamon Press, New York, 1968, pp. 399-412.

32. R. E. Hardenburg and J. M. Lutz, in ASHRAE Guide and Data Book — Applications 1968, American Society of Heating, Refrigerating, and Air-Conditioning Engineers, New York, 1971, pp. 459-470.

33. J. M. Haynes, in Low Temperature Biology of Foodstuffs (J. Hawthorn and E. J. Rolfe, eds.), Pergamon Press, New York, 1968, pp. 79-104.

34. A. C. Jason and H. R. Sanders, Food Technol., 16(6), 101 and 107 (1962).

35. M. A. Joslyn and J. D. Ponting, Advan. Food Res., 3, 1 (1951).

36. A. B. Khachaturov, in Progress in Refrigeration Science and Technology (M. Jul and A. M. S. Jul, eds.), Vol. 3, Pergamon Press, New York, 1960, pp. 126-133.

37. A. W. Khan and C. P. Lentz, J. Food Sci., 30, 787 (1965).

38. A. A. Klose, M. F. Pool, and H. Lineweaver, Food Technol., 9, 372 (1955).

39. J. Kuprianoff, in Freeze Drying of Foods (F. R. Fisher, ed.), National Academy of Science-National Research Council, Washington, D. C., 1962, pp. 16-24.

40. J. P. Lane, Food Technol., 20, 549 (1966).

41. F. K. Lawler and L. Trauberman, Food Eng., 41(4), 67 (1969).

42. F. A. Lee, W. A. Gortner, and J. Whitcombe, Ind. Eng. Chem., 38, 341 (1946).

43. F. A. Lee, W. A. Gortner, and J. Whitcombe, Food Technol., 3, 164 (1949).

44. F. A. Lee, R. F. Brooks, A. M. Pearson, J. I. Miller, and F. Volz, Food Res., 15, 8 (1950).

45. B. J. Liska and A. L. Rippen, in ASHRAE Guide and Data Book — Applications 1968, American Society of Heating, Refrigerating and Air-Conditioning Engineers, New York, 1968, p. 356.

46. R. A. K. Long, J. Sci. Food Agr., 6, 621 (1955).

47. R. M. Love, in Cryobiology (H. T. Meryman, ed.), Academic Press, New York, 1966, pp. 317-405.

48. R. M. Love and S. B. Haraldsson, J. Sci. Food Agr., 12, 442 (1961).

49. B. J. Luyet, Ann. N. Y. Acad. Sci., 85, 549 (1960).

50. M. MacArthur, Sci. Agr., 28, 166 (1948).

51. W. A. MacCallum and D. G. Ellis, J. Fish. Res. Bd. Can., 21(1), 115 (1964).

52. E. C. McCracken and F. Churchill, ASHRAE J., 4(11), 27 (1962).

53. H. C. Mannheim, M. P. Steinberg, A. I. Nelson, and T. W. Kendall, Food Technol., 11, 384 (1957).

54. Matheson Gas Data Book, 4th ed., The Matheson Co., Inc., East Rutherford, N. J., 1966.

55. P. Mazur, Biophys. J., 3, 323 (1963).

56. P. Mazur, in Cryobiology (H. T. Meryman, ed.), Academic Press, New York, 1966, pp. 213-315.

57. J. H. Merritt and A. Banks, Supplement au Bulletin de L'Institute International du Froid (Paris), Annexe 1964-1, 65 (1964).

58. H. T. Meryman, Science, 124, 515 (1956).

59. H. T. Meryman, Proc. Roy. Soc., Ser. B, 147, 452 (1957).

60. H. T. Meryman, in Recent Research in Freezing and Drying (A. S. Parkes and A. U. Smith, eds.), Blackwell, Oxford, 1960, pp. 23-39.

61. H. T. Meryman, Fed. Proc., 22, 81 (1963).

62. H. T. Meryman and W. T. Platt, Research Report NM 000 018.01.08, 13, 1, Naval Medical Research Institute, Bethesda, Md. (1955).

63. A. M. Munter, C. H. Byrne, and K. G. Dykstra, Food Technol., 7, 356 (1953).

64. R. F. Nelson and B. J. Finkle, Phytochemistry, 3, 321 (1964).

65. C. S. Pederson and H. G. Beattie, Concentration of Fruit Juices by Freezing, N. Y. State Agricultural Experiment Station, Bull. No. 727, 1947.

66. J. A. Peters, W. A. MacCallum, W. J. Dyer, D. R. Idler, J. W. Slavin, J. P. Lane, D. I. Fraser, and E. J. Laishley, J. Fish. Res. Bd. Can., 25(2), 299 (1968).

67. A. C. Peterson, in Proceedings of the Low Temperature Microbiology Symposium, Campbell Soup Co., Camden, N. J., 1961, pp. 157-195.

68. W. G. Pfann, Solid State Phys., 4, 423 (1957).

69. J. D. Ponting, in Food Enzymes (H. W. Schultz, ed.), AVI Publ. Co., Westport, Conn., 1960, pp. 105-124.

70. S. R. Pottinger, Ice Refrig., 121(2), 13 (1951).

71. C. L. Rasmussen, ASHRAE J., 9(6), 36 (1967).

72. L. Riedel, Kältetechnik, 8, 374 (1956) (in German).

73. L. Riedel, Kältetechnik, 9, 38 (1957) (in German).

74. L. Riedel, Kältetechnik, 9, 342 (1957) (in German).

75. L. Riedel, Kältetechnik, 13, 122 (1961) (in German).

76. A. P. Rinfret, Ann. N. Y. Acad. Sci., 85, 576 (1960).

77. H. R. Sanders, J. Food Technol., 1, 183 (1966).

78. S. Schwimmer, L. Ingraham, and H. M. Hughes, Ind. Eng. Chem., 47, 1149 (1955).

79. A. D. Shepard, in Conference on Frozen Food Quality, U. S. Department of Agriculture, ARS-74-21, 1960, pp. 18-23.

80. J. Shipton, CSIRO Food Preserv. Quart. (Australia), 25(1), 2 (1965).

81. H. H. Sommer, Theory and Practice of Ice Cream Making, 5th ed., Published by the author, Madison, Wisc., 1947.

82. J. L. Stephenson, in Recent Research in Freezing and Drying (A. S. Parkes and A. U. Smith, eds.), Blackwell, Oxford, 1960, pp. 121-145.

83. L. Trauberman, Food Eng., 38(12), 86 (1966).

84. D. K. Tressler, in The Freezing Preservation of Foods, 4th ed. (D. K. Tressler, W. B. Van Arsdel, and M. J. Copley, eds.), Vol. 1, AVI Publ. Co., Westport, Conn., 1968, pp. 120-152.

85. D. K. Tressler, W. B. Van Arsdel, and M. J. Copley (eds.), The Freezing Preservation of Foods, 4th ed., 4 Vols., AVI Publ. Co., Westport, Conn., 1968.

86. W. B. Van Arsdel, Food Technol., 11, 28 (1957).

87. W. B. Van Arsdel and D. G. Guadagni, Food Technol., 13, 14 (1959).

88. W. B. Van Arsdel, M. J. Copley, and R. L. Olson (eds.), Quality and Stability of Frozen Foods, Wiley-Interscience, New York, 1969.

89. L. van den Berg, Food Technol., 15, 434 (1961).

90. A. Van Hook, Crystallization — Theory and Practice, Reinhold, New York, 1961.

91. J. J. Waterman, Fish. News Intl., 2, 113 (1963).

92. J. D. Winter, A. Hustrulid, I. Noble, and E. S. Ross, Food Technol., 6, 311 (1952).

93. J. G. Woodroof, Microscopic Studies of Frozen Fruits and Vegetables, University of Georgia Agricultural Experiment Station, Bull. No. 201, 1938.

GENERAL REFERENCES

ASHRAE Guide and Data Book — Applications 1971, American Society of Heating, Refrigerating and Air-Conditioning Engineers, New York, 1971.

O. Fennema, W. D. Powrie, and E. H. Marth. Low-Temperature Preservation of Foods and Living Matter, Marcel Dekker, New York, 1973.

J. Hawthorne and E. J. Rolfe (eds.), Low Temperature Biology of Foodstuffs, Pergamon Press, New York, 1968.

E. Heen and O. Karsti, in Fish as Food (G. Borgstrom, ed.), Vol. 4, Academic Press, New York, 1965, pp. 355-418.

H. T. Meryman (ed.), Cryobiology, Academic Press, New York, 1966.

P. A. Shumskii, Principles of Structural Glaciology, Dover, New York, 1964.

A. U. Smith (ed.), Current Trends in Cryobiology, Plenum Press, New York, 1970.

D. K. Tressler, W. B. Van Arsdel, and M. J. Copley, The Freezing Preservation of Foods, 4th ed., 4 Vols., AVI Publ. Co., Westport, Conn., 1968.

W. B. Van Arsdel, M. J. Copley, and R. L. Olson (eds.), Quality and Stability of Frozen Foods, Wiley-Interscience, New York, 1969.

G. E. W. Wolstenholme and M. O'Connor (eds.), The Frozen Cell, J. & A. Churchill, London, 1970.

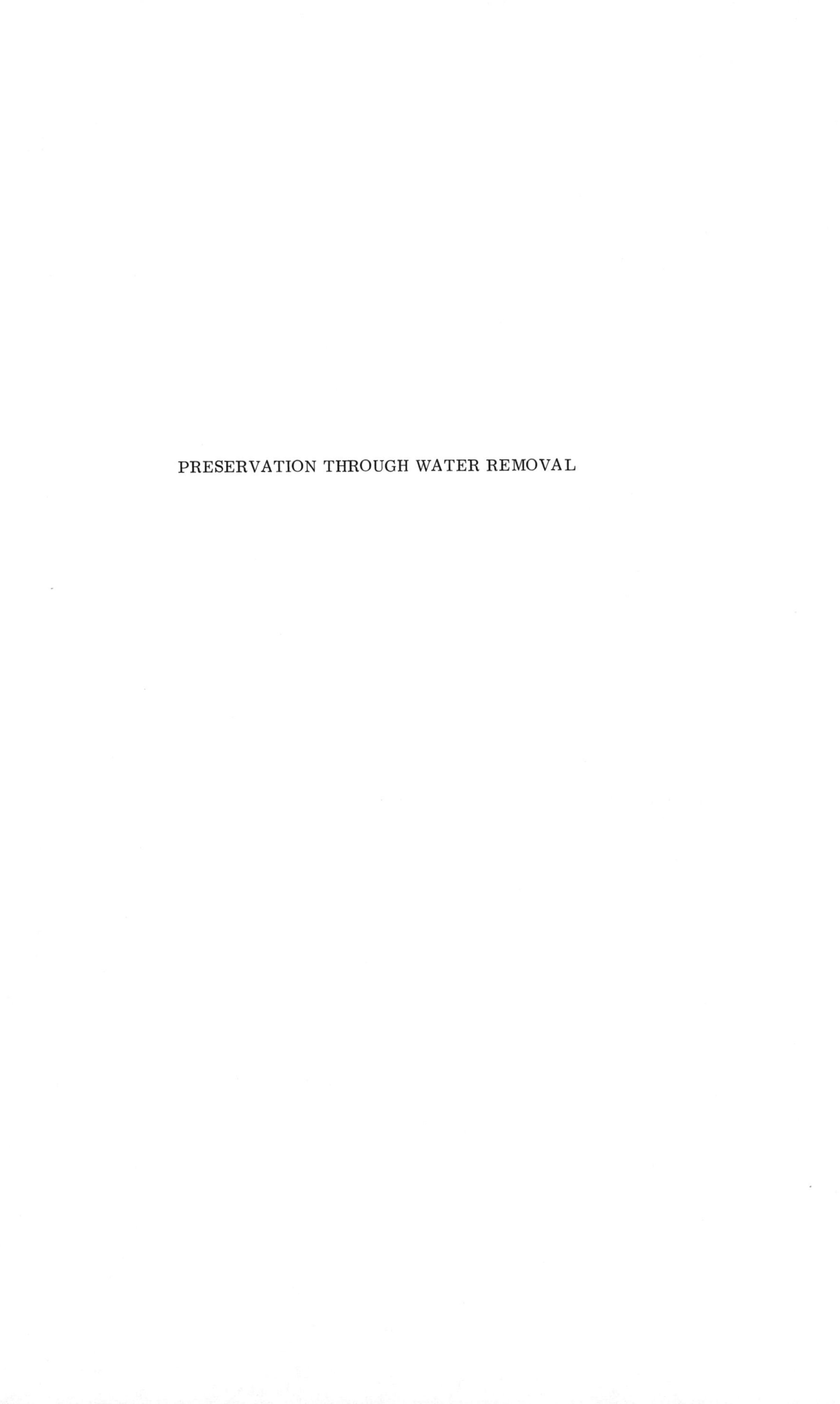

PRESERVATION THROUGH WATER REMOVAL

Chapter 7

EQUILIBRIUM AND RATE CONSIDERATIONS IN PROCESSES FOR FOOD CONCENTRATION AND DEHYDRATION

Marcus Karel

CONTENTS

I. INTRODUCTION

In food concentration and in dehydration, water is removed by one of the several separation processes available for this purpose, including vaporization, crystallization, sublimation, and solvent extraction. In each case, the process is governed by equilibrium as well as rate considerations. In particular, we may write a general equation for separation of water as shown below.

$$\text{Rate of water removal} = \frac{\text{driving force for water removal}}{\text{resistance to water removal}} \qquad (7\text{-}1)$$

This equation is similar to that developed previously for heat transfer and is, in fact, basic to all transport processes (pp. 17, 18). The driving force depends on equilibrium considerations and increases with increasing distance from equilibrium. The closer a given system is to equilibrium, the less the tendency to transfer water. The resistance to transport depends on various transport coefficients.

In this chapter, some of the general principles governing the factors alluded to in Eq. (7-1) are examined.

II. EQUILIBRIUM BETWEEN PHASES

A. The Phase Rule

A useful equation relating the number of intensive variables which can be varied independently (F) to the number of components (Γ) and the number of phases (ϕ) is:

$$F = \Gamma + 2 - \phi \tag{7-2}$$

An intensive variable of one which does not depend on the total amount of each phase (a physically independent homogeneous portion of a system). For instance, temperature, pressure, and composition are intensive variables. The phase rule is useful in predicting the number of variables which can be varied independently.

Consider, for instance, an aqueous solution of sucrose in equilibrium with water vapor above the solution: $\Gamma = 2$ (water and sucrose), $\phi = 2$ (gas and liquid). Accordingly, $F = 2$, and it is possible to vary independently any two of the three pertinent variables of temperature, pressure, and concentration. Thus, if concentration and temperature are fixed, the pressure at which equilibrium is achieved between water vapor and solution is determined as well.

B. Gas-Liquid Equilibria

1. Dalton's Law

A relationship useful in consideration of gas-liquid equilibria is Dalton's law, Eq. (7-3), which states that the total pressure in a gas phase is equal to the sum of partial pressures of the components.

$$P_\theta = p_A + p_B + \cdots \tag{7-3}$$

where

p is partial pressure
P_θ is total pressure
A, B, $\cdots$ are subscripts referring to components A, B, $\cdots$ respectively

2. Ideal Gas Law

In many situations, especially at moderate pressures, it is possible to approximate the behavior of gases by the ideal gas law, Eq. (7-4)

$$PV = NRT \tag{7-4}$$

where

P is pressure
V is volume of the gas phase
N is number of moles of gas phase
R is the gas constant
T is absolute temperature

(Obviously, all variables and constants must be expressed in consistent units. A listing of units of the gas constant R, useful in dehydration processes, is given in Table 7-1.)

When Eqs. (7-3) and (7-4) are combined, a relation is obtained between concentration of a given component in the gas phase and its partial pressure at a given constant temperature. Let

$$n_A \equiv N_A/V = p_A/RT \tag{7-5}$$

where n_A is the molar concentration of component A (moles per unit volume).

Let

$$y_A \equiv n_A/n_\theta \tag{7-6}$$

where

y_A is the mole fraction of A in the gas phase
n_θ is the molar concentration of all components

Then

$$y_A = p_A/P_\theta \tag{7-7}$$

3. Dependence of the Vapor Pressure of a Pure Substance on Temperature

The dependence of vapor pressure on temperature is well approximated by the Clausius-Clapeyron equation:

$$\frac{dP^\circ}{dT} = \frac{\lambda}{T(\Delta V)} \tag{7-8}$$

TABLE 7-1

Selected Values of Gas Constant R

Metric units	
1.986	cal/mole °C
8.314×10^7	ergs/mole °C
82.057	atm cm^3/mole °C
62.36	torr liter/mole °C
English units	
10.73	psi ft^3/mole °F
0.73	atm ft^3/mole °F

where

λ is the latent heat of vaporization
ΔV is the volume change in vaporization
P° is the vapor pressure

If the ideal gas law is assumed applicable, then this equation may be approximated by:

$$\frac{dP^\circ}{P^\circ} = \frac{\lambda}{RT^2} dT \tag{7-9}$$

Integration of this equation between any two given temperatures (T_1) and (T_2) yields Eq. (7-10), which allows the calculation of pressures.

$$2.3 \log_{10} \left(\frac{P_2^\circ}{P_1^\circ}\right) = \left(\frac{P_2^\circ}{P_1^\circ}\right) = \left(\frac{-\lambda}{R} \frac{1}{T_2} - \frac{1}{T_1}\right) \tag{7-10}$$

A plot of the logarithm of vapor pressure vs the reciprocal temperature gives a straight line, provided the ideal relations mentioned above apply and λ does not vary with temperature.

4. Partial Pressure over Solutions

a. Raoult's Law. The partial pressure of water in equilibrium with an aqueous solution is lower than the vapor pressure of pure water. An estimate of the pressure of solvent over a solution is given by Raoult's law:

$$p_A = P_A^\circ x_A \tag{7-11}$$

or

$$y_A = \frac{p_A}{P_\theta} = \frac{P_A^\circ}{P_\theta} x_A \tag{7-12}$$

where x_A is the mole fraction of solvent.

Raoult's law is accurate only in predicting vapor-liquid equilibria for ideal solutions in equilibrium with ideal gases. It is most applicable to dilute solutions. At moderate and high concentrations substantial deviations may occur. The significance of such deviations for food preservation processes is discussed in Chapter 8.

b. Henry's Law. In certain systems where Raoult's law does not apply, the relation between partial pressure and composition of the liquid phase may be estimated by Henry's law:

$$p_A = S_A x_A \tag{7-13}$$

where S_A is a constant. (When $S_A = P_A^\circ$ Henry's law is identical with Raoult's law.)

5. Representation of Equilibrium Relationships for Binary Systems with Two Volatile Components

In a binary mixture, the component with the higher vapor pressure at a given temperature is referred to as the more volatile component, and the ratio of its vapor pressure to that of the less volatile component is called "relative volatility." When Raoult's law is applicable, relative volatility is independent of concentration of the components in the binary mixture.

The relationship between partial pressure, composition, and temperature may be represented graphically in different ways, each of which is useful for particular purposes.

At constant temperature, the partial pressure of one component (usually the more volatile component) may be plotted against its mole fraction in the solution. A directly related plot is that of the "activity" of the component against its mole fraction in solution. The activity (a) is defined in Eq. (7-14) and a typical plot is shown in Figure 7-1.

$$a_A = p_A/P_A^\circ \tag{7-14}$$

This representation is useful for isothermal processes and for evaluating relative humidity in equilibrium with solutions.

In evaluating evaporation and distillation, it is often more convenient to evaluate the relationships between gas phase and liquid phase composition at the boiling point, that is, at a temperature at which the sum of partial pressures of the two volatile components equals atmospheric pressure, usually arbitrarily taken as 760 mm Hg. A diagram relating boiling point to composition of gas and liquid phases of a binary mixture is shown in Figure 7-2. A diagram relating x_A to y_A at the boiling point for a binary mixture is shown in Figure 7-3.

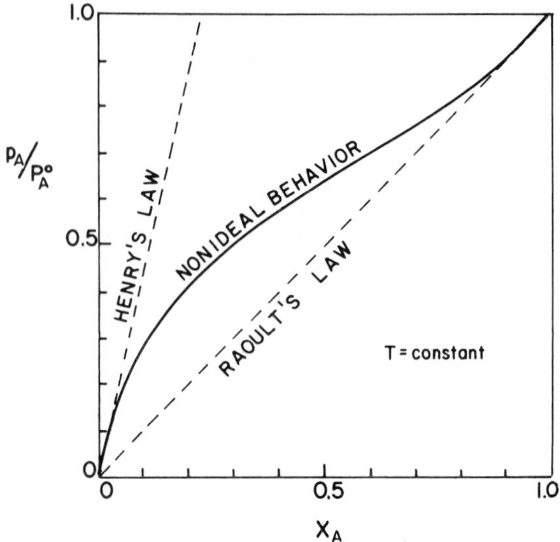

FIG. 7-1. Dependence of partial pressure of a component on its mole fraction in a liquid.

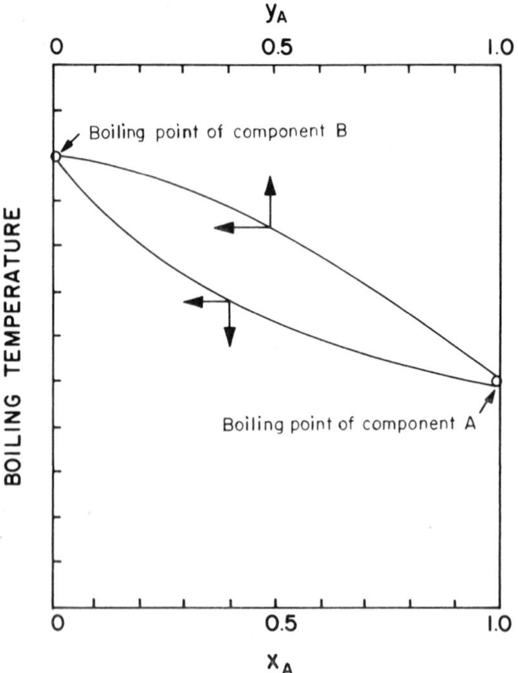

FIG. 7-2. Boiling point diagram for a binary system.

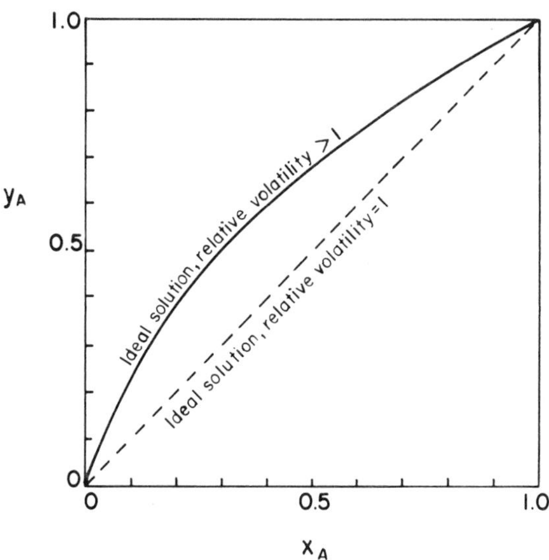

FIG. 7-3. Relation of mole fraction in vapor phase (y) to mole fraction in liquid phase (x) for a binary system.

6. Nonideal Solutions

Most of the solutions of importance in food show nonideal behavior at high concentration and some behave nonideally even in relatively dilute systems. The solid curve shown in Figure 7-1 is typical of deviations from ideal behavior and represents "positive deviation" from Raoult's law.

Equation (7-15) is suitable for correlation of properties of nonideal solutions:

$$a_A = x_A \gamma_A \tag{7-15}$$

where γ_A is an activity coefficient dependent on concentration.

7. Colligative Properties Related to Vapor Pressure

We have noted that the presence of solute results in depression of partial pressure of a solvent below its vapor pressure. Depression of vapor pressure and freezing point and elevation of boiling point and osmotic pressure belong to a group of properties referred to as "colligative." These properties depend on the number of kinetic units in solution (that is, the number of molecules, ions, and particles) and not on the concentration by weight. In the ideal case, all of the above properties can be linearly related to solute concentration.

Thus vapor pressure depression may be written:

$$P^\circ - p = P^\circ (1 - x) = P^\circ x_s \tag{7-16}$$

where

P^{o} is the vapor pressure of pure solvent
p is the partial pressure of solvent over solution
x is the mole fraction of solvent
x_s is the mole fraction of solute

Depression of freezing point and elevation of boiling point are usually approximated by:

$$\Delta T = (\text{const})\ (\text{molality of solute}) \tag{7-17}$$

Osmotic pressure in dilute solutions of nonelectrolytes is well approximated by:

$$\Delta\pi = n_s RT \tag{7-18}$$

where

$\Delta\pi$ is osmotic pressure
n_s is solute concentration

C. Liquid-Liquid Equilibria

Consider a component A soluble in two mutually immiscible liquids L and ℓ. The two liquids, when brought into contact, constitute two phases and component A is transported from one phase to the other until the concentrations of A in phases L and ℓ reach equilibrium levels defined by Eq. (7-19):

$$C_{A(L)} = K_{A(L/\ell)}\ C_{A(\ell)} \tag{7-19}$$

where

$C_{A(L)}$ is the concentration of A in liquid L
$C_{A(\ell)}$ is the concentration of A in liquid ℓ
$K_{A(L/\ell)}$ is the partition coefficient

Equation (7-19) is analogous to the idealized situation for gas-liquid equilibria represented by Henry's law, Eq. (7-13). The partition coefficient may depend on concentration, especially in concentrated solutions, in which case Eq. (7-19) requires modification.

D. Gas-Solid Equilibria

Food dehydration and, in general, the preservation of foods by reduction of water activity depend critically on the equilibrium relationships between water vapor pressure in the atmosphere and the amount of water in the food. Chapter 8 is devoted to a detailed review of these relationships and only a general introduction to gas-solid equilibria, or "sorption" equilibria, is given here.

Gas-solid equilibria are usually presented in the form of relations between partial pressure of a given component in the gas phase and its concentration in the solid phase. Four representations are often made, of which the isotherm is the most common. These representations are as follows.

1. Isotherms

Isotherms are analytical or graphical representations of partial pressure of a component in equilibrium with its concentration in a solid, at constant temperature:

$$p_A = f(C_A), \qquad T = \text{const} \tag{7-20}$$

where C_A is the concentration of component A in the solid phase.

The simplest case is one in which Henry's law applies. In most cases, however, isotherms are nonlinear, and in the case of water held by foods, the isotherms are of type II or of type III of the five types commonly considered in analyses of adsorption [1-3]. Isotherms of type II and III as well as the linear isotherm, which is the form represented by Eq. (7-20), are shown in Figure 7-4.

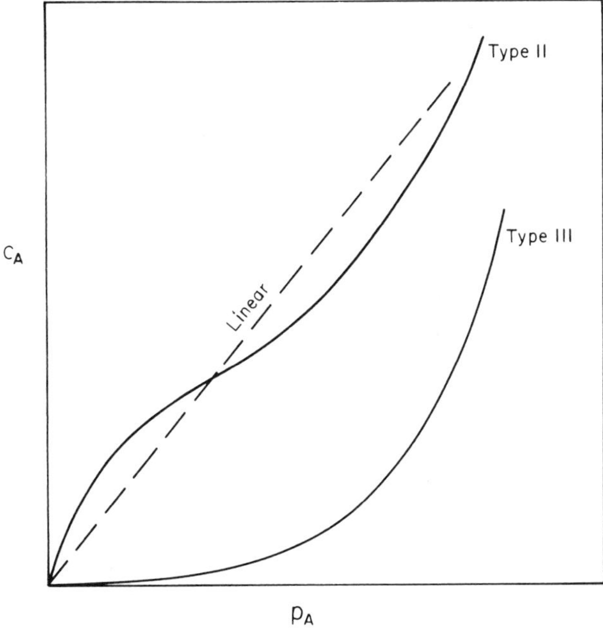

FIG. 7-4. Sorption isotherms. (Isotherms showing linear relation between C_A and p_A are known as linear. Sigmoid isotherms are referred to as type II. Several other isotherms are known and described by Adamson [1], including type III shown here.)

2. Isobars

Isobars represent relations between concentration and temperature while partial pressure of a component is held constant as shown in Figure 7-5 and Eq. (7-21):

$$C_A = f(T), \quad p_A = \text{const} \tag{7-21}$$

3. Constant Activity Curves

Constant activity curves are similar to isobars except that their activity is held constant, as shown in Eq. (7-22):

$$C_A = f(T), \quad p_A/P_A^{\circ} = \text{const} \tag{7-22}$$

4. Isosteres

Isosteres are curves or functions relating equilibrium pressure of an adsorbed component to temperature when the amount adsorbed is held constant, as shown in Figure 7-6 and Eq. (7-23):

$$p_A = f(T), \quad C_A = \text{const} \tag{7-23}$$

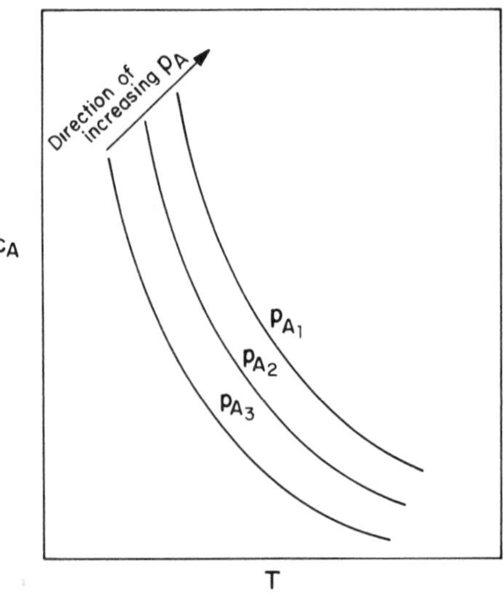

FIG. 7-5. Sorption isobars.

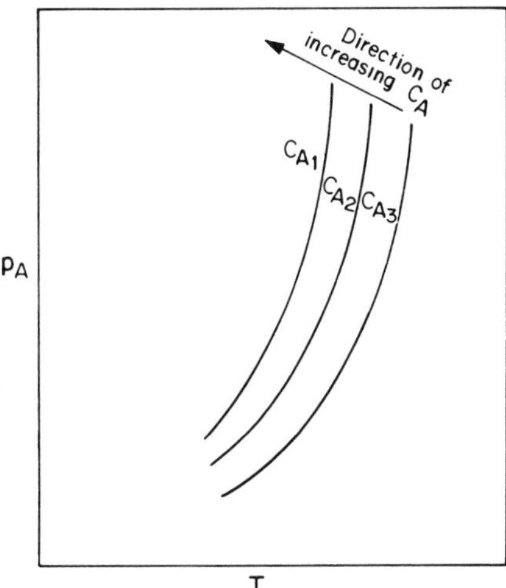

FIG. 7-6. Sorption isosteres.

III. MASS TRANSFER BETWEEN PHASES

A. Interface Between Phases

Phases in contact are usually assumed to be in equilibrium at the plane of contact, that is, the interface. This assumption means that the controlling factor in transfer of components from one phase to another is the rate of transport to and from the interface. The analysis of transfer between phases therefore requires that primary consideration be given to transport in each phase.

The following pages deal with transport of diffusing components in gases, liquids, and porous solids. Water is the substance usually transferred in processes for food preservation, but the same general principles apply to flavor components, preservatives, and other chemical species.

B. Molecular Diffusion in One Direction in Gases and Liquids

When a concentration gradient with respect to a given component exists in only one direction (X), its diffusion may be characterized by Fick's law:

$$J = -D \frac{dC_A}{dX} \tag{7-24}$$

where

J is flux in suitable units (for instance, moles/sec cm^2
D is the diffusion coefficient (cm^2/sec)
C is concentration (moles/cm^3)
X is distance (cm)

When diffusion takes place in steady state, Eq. (7-24) may be integrated to give:

$$J = D\frac{\Delta C}{\Delta X} \tag{7-25}$$

where ΔC is the concentration difference across the distance ΔX.

C. Steady-state Diffusion of a Component Through a Stagnant Gas

In gaseous diffusion, it is often advantageous to express driving forces in terms of pressure differences. When a component can be assumed to diffuse through a stagnant layer of another gas, the following equation applies:

$$J = \frac{DP_\theta}{RT} \cdot \frac{\Delta p}{\Delta X} \cdot \frac{1}{(P_\theta - \bar{p})} \tag{7-26}$$

where

$\bar{p}$ is the mean partial pressure of diffusing gas in layer ΔX
Δp is the partial pressure difference across distance ΔX

When partial pressure differences across ΔX are relatively small, an arithmetic mean may be used for $\bar{p}$. When these become larger, a logarithmic mean should be used.

D. Film Coefficients for Mass Transfer

In analyzing mass transfer, it is convenient to express rates in terms of coefficients incorporating various resistances to mass transfer and relating fluxes to gradients of concentration or pressure.

For instance, consider a liquid phase in contact with a gaseous phase, as shown in Figure 7-7. Assume that a component is transferred from liquid to gas. In the bulk of the liquid, the concentration of the component is uniform and equal to C. The gradient of concentration is assumed to occur in a film adjacent to the gas-liquid interface. At the interface, the concentration (C_i) is in equilibrium with the partial pressure (p_i) of the component at the interface. The partial pressure in the bulk of the gas phase is also considered uniform and equal to p, with all the gradient in partial pressure occurring in a gaseous film adjacent to the interface. Given steady-state conditions, the following equation gives the rate of transfer in the liquid:

$$J = k_L (C - C_i) \tag{7-27}$$

where k_L is a film transfer coefficient in the liquid in units consistent with those of J and C. If for instance J is in moles/sec cm^2 and C is in moles/cm^3, then k_L is in

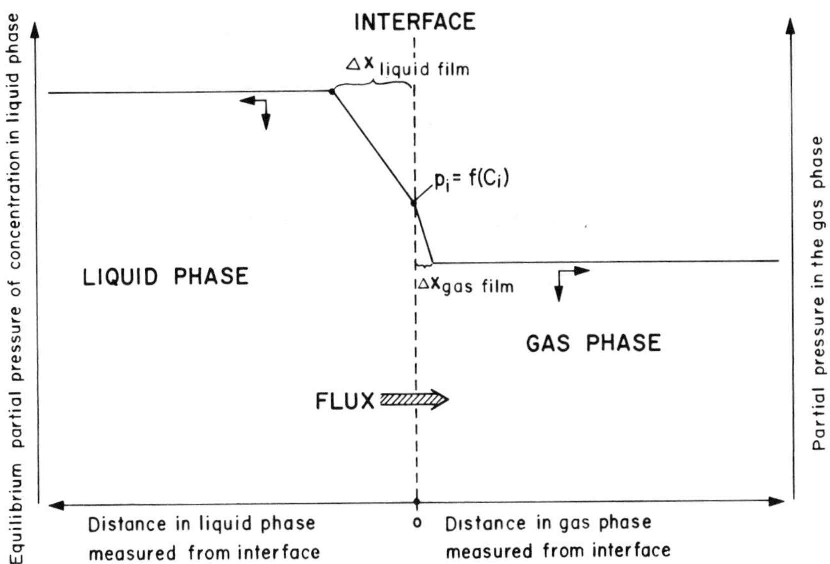

FIG. 7-7. Schematic representation of mass transfer from a liquid to a gas phase.

$$\frac{\text{moles}}{(\text{cm}^2)\ (\text{sec})\ (\text{mole/cm}^3)} = \text{cm/sec}$$

Note that Eq. (7-28) relates k_L to the diffusion coefficient defined by Fick's law in Eq. (7-24):

$$k_L = \frac{D}{\Delta X_{\text{liq film}}} \tag{7-28}$$

where $\Delta X_{\text{liq film}}$ is the thickness of the liquid film. Of course, under usual conditions $\Delta X_{\text{liq film}}$ is not known and k_L cannot be simply related to D.

The rate of transfer in the gas phase is given by:

$$J = k_G(p_i - p) \tag{7-29}$$

where k_G is the film transfer coefficient in consistent units. For instance, if J is in moles/sec^1 cm^2 and p is in atmospheres, then k_G is in moles sec^{-1} cm^{-2} atm^{-1}. Note that k_G may be related to D by Eq. (7-30) if the following assumptions hold: (1) $\Delta X_{\text{gas film}}$ is known, and (2) partial pressure of the component transferred in the gas is very much less than the total pressure.

$$k_G = \frac{D}{RT(\Delta X_{\text{gas film}})} \tag{7-30}$$

where $\Delta X_{\text{gas film}}$ is the thickness of the gas film. Under conditions of steady state, the flux from the bulk of the liquid to the interface equals the flux from the interface to the bulk of the gas phase, and we can write:

$$k_L (C - C_i) = k_G (p_i - p) \tag{7-31}$$

In the usual situation, the actual concentrations and partial pressures at the interface are not measured and the need arises to relate flux to concentrations and partial pressures in the bulk phase. If a function relating gas partial pressures in equilibrium with concentration in liquid is known, as symbolized in Eqs. (7-32) and (7-33), then overall transfer coefficients can be defined as shown in Eqs. (7-34) and (7-35).

$$C' = f(p) \tag{7-32}$$

$$p' = f(C) \tag{7-33}$$

where

C' is the concentration in equilibrium with partial pressure p
p' is the partial pressure in equilibrium with concentration C

$$J = K_L (C - C') \tag{7-34}$$

where K_L is the overall transfer coefficient defined in terms of concentrations in liquid.

$$J = K_G (p' - p) \tag{7-35}$$

where K_G is the overall transfer coefficient defined in terms of partial pressures. It follows that in steady state:

$$K_L (C - C') = K_G (p' - p) \tag{7-36}$$

If Henry's law can be applied to the equilibrium situation, Eq. (7-32) may be rewritten:

$$C = p/S' \tag{7-37}$$

where S' is the Henry's law constant in suitable units. Equation (7-36) can then be written:

$$K_L (p - p') = S' K_G (p' - p) \tag{7-38}$$

and it follows that

$$K_L = S' K_G \tag{7-39}$$

E. Transport in Porous Solids

In many of the operations performed on foods for purposes of preservation, water is removed from porous solids. Details of transport mechanisms in dehydration are discussed in another chapter and here the subject is introduced only in its most general aspects.

Transport in solids may occur by diffusion in the solid itself, in which case Fick's law, Eq. (7-24), can be used.

In porous solids, however, various mechanisms may be involved including:

1. diffusion in the solid itself
2. diffusion in gas-filled pores
3. surface diffusion along the walls of pores and capillaries
4. capillary flow resulting from gradients in surface pressure
5. convective flow in capillaries resulting from differences in total pressure

It is often convenient to establish an overall coefficient for flow in porous solids without attempting to establish the contribution of each of the above factors. One equation often used for this purpose is the following:

$$J = \beta \frac{dp}{dX} \tag{7-40}$$

where

β is an overall permeability coefficient
p is the partial pressure of the transported component
X is distance in the direction of transport

As is to be discussed in a subsequent chapter, β can be related to various properties, such as porosity, internal surface area, and tortuosity of pores.

F. Simultaneous Heat and Mass Transfer

Often, when mass is transferred, heat also must be transferred. When a component is transferred from a liquid to a gas phase, latent heat of vaporization must be supplied. The same amount of heat must be removed when mass transfer occurs in the opposite direction. In every instance the temperature of the interface and transfer rate adjust themselves so that at steady state the rate of heat transfer and rate of mass transfer is balanced.

In some operations, heat transfer is of minor importance in controlling the overall rate of transport from one phase to another. This situation is true of operations in which latent heat effects are small. In liquid-liquid extraction and in leaching, for instance, the overall transfer rate may be considered as limited only by resistances to mass transfer. In other operations, including evaporization and crystallization, heat transfer is the controlling mechanism. Both heat and mass are transferred in large amounts but the rates at which both occur can be determined by considering only the rate of heat exchange with an external source.

In still other operations, particularly humidification, drying, and freeze drying both heat transfer and mass transfer must be considered and balanced in determining overall rates. The methods used for that purpose are discussed in the chapters which follow.

SYMBOLS

(Units are given for some of the quantities. These are intended as examples. Other consistent units in either English or metric systems may be used.)

C	concentration (g/cm^3 or lb/ft^3)
C'	equilibrium concentration (g/cm^3 or lb/ft^3)
D	diffusion coefficient (cm^2/sec or ft^2/hr)
F	number of intensive variables
J	flux ($moles/cm^2$ sec or lb/ft^2 hr or g/cm^2 sec)
$K_{A(L/l)}$	partition coefficient defined in Eq. (7-19)
K_G	overall mass transfer coefficient based on partial pressure difference (lb ft^{-2} hr^{-1} atm^{-1} or moles cm^{-2} sec^{-1} atm^{-1})
K_L	overall mass-transfer coefficient based on concentration difference [lb ft^{-2} hr^{-1} $(lb/ft^3)^{-1}$ or moles cm^{-2} sec^{-1} $(moles/cm^3)^{-1}$]
N	number of moles
P	pressure
P^o	vapor pressure
R	gas constant
S	constant in Henry's law as defined in Eq. (7-13)
S'	constant in Henry's law as defined in Eq. (7-37)
T	temperature
V	volume
X	distance
a	activity
k_L	mass-transfer coefficient in liquid phase (dimensions identical to those of K_L)
k_G	mass-transfer coefficient in gas phase (dimensions identical to those of K_G)
ln	natural logarithm
n	molar concentration (moles per unit volume)
p	partial pressure

$\bar{p}$	mean partial pressure
p'	equilibrium partial pressure
x	mole fraction in liquid phase
y	mole fraction in gas phase
Γ	number of components
β	permeability coefficient defined in Eq. (7-40)
γ	activity coefficient
$\Delta\pi$	osmotic pressure
ΔC	concentration difference
ΔP	pressure difference
ΔT	boiling point elevation or freezing point depression
ΔV	volume change
ΔX	distance interval
$\Delta X_{\text{liq film}}$	thickness of liquid film
$\Delta X_{\text{gas film}}$	thickness of gas film
λ	latent heat of vaporization (cal/mole or Btu/lb)
ϕ	number of phases

Subscripts

A, B, ···	refer to specific components
L, l	refer to specific liquid phases
i	refers to interface between two phases
s	refers to solute
θ	refers to total or sum of all components

REFERENCES

1. A. W. Adamson, Physical Chemistry of Surfaces, Wiley-Interscience, New York, 1960.

2. S. Gal, Die Methodik der Wasserdampf-Sorptionsmessungen, Springer-Verlag, Berlin, Germany, 1967.

3. T. P. Labuza, Food Technol., 22(3), 15, 1967.

GENERAL REFERENCES

A. S. Foust, L. A. Wenzel, C. W. Chimp, L. Mans, and L. B. Anderson, Principles of Unit Operations, John Wiley & Sons, New York, 1960.

W. J. Moore, Physical Chemistry, Prentice-Hall, Englewood Cliffs, N. J., 1955.

T. K. Sherwood and R. L. Pigford, Absorption and Extraction, McGraw-Hill, New York, 1952.

Chapter 8

WATER ACTIVITY AND FOOD PRESERVATION

Marcus Karel

CONTENTS

I. INTRODUCTION

Food-preservation processes have in common their goal of extending the shelf life of foods to allow storage and convenient distribution. The first and most dangerous limitation of shelf life is the activity of microorganisms. Hence the food preservation processes are designed to eliminate the danger of spoilage due to microbes and to avoid their health-threatening activities. Several food-preservation processes achieve this aim by lowering the availability of water to microorganisms. These processes include concentration, dehydration and freeze drying, which are discussed in Chapters 9, 10 and 11. Freezing, in which water is in ice crystals and therefore withheld from microorganisms, has been discussed in Chapter 6. Other preservation processes, such as salting and preservation by adding sugar, also depend on lowering the availability of water by increasing osmotic pressure. These processes are not physical methods of food preservation, however, and are outside the scope of this chapter.

In addition to microbial growth, chemical and physical processes may occur in foods during storage and may damage quality. Control of water content affects the rate of these processes and may contribute to extension of storage life.

A fundamental aspect of processes that lower water content is the relation between the amount and the state of water in foods. The state of water is especially important because of its effect on biological, chemical, and physical processes that limit the shelf life of foods.

II. FREE AND BOUND WATER IN FOODS

There is substantial evidence that a portion of the water in foods is strongly bound to individual sites and that an additional amount is bound less firmly but still not available as a solvent for various soluble food components. There is a variety of methods by which the degree of binding of water within the food can be established, and among these are (1) determination of unfrozen water, (2) measurement of nuclear magnetic resonance, (3) determination of dielectric properties of food, and (4) measurement of vapor pressure. Additional methods of determining water binding have been proposed, but the methods listed above provide adequate insight into the pertinent properties of water in food.

A number of authors have established that upon cooling water-containing foods to low temperatures, well below their freezing points, a portion of the total water remains unfrozen. The most successful methods of analysis of unfrozen water have utilized differential thermal analysis (DTA). Food samples with different total water contents have been cooled to temperatures below -60° C and then rewarmed in the DTA apparatus. When water contents are sufficiently low that only "unfreezable water" is present, there is an absence of DTA peaks. At higher water contents, peaks are observed due to the absorption of latent heat of fusion. By using a sufficient number of different water contents it is possible to determine quite precisely the point at which all of the remaining water is "unfreezable." Using DTA, Duckworth [8] found that amounts of unfreezable water were in the range of 0.13-0.46 g water per gram of solids for food components. Typical foods had values between 0.2 and 0.4.

Wide-line proton magnetic resonance (NMR) is used routinely to determine water in foods. The basis of its application is as follows: protons of molecules in liquid state, as a result of the more or less random motion of these molecules, experience a similar net magnetic field when placed in a magnet. They therefore give sharp and large signals in NMR determinations, because they absorb radio energy at about the same frequency. In solids, however, interactions with neighboring atoms change the proton response and absorptions by the many protons are spread out over a large frequency. Hence NMR signals are broad and shallow and, within the "window" used to record the signal, they may be negligible compared to the liquid signal. A portion of the total water in foods shows signals intermediate in character between liquid and solid water, and if the total content is known, NMR may be used to estimate "bound water content" [39]. The amount of water bound to hydrophilic food components is estimated at about 0.1-0.3 g/g of solids.

The amount of water bound by proteins, polymers, and foods also can be determined by dielectric measurements. Dielectric properties of water molecules depend on the freedom of these molecules to move in response to changes in the applied electric field. "Bound" water shows properties intermediate between the rigidly held dipoles of ice and the much more mobile liquid water molecules. The amount of bound water, however, is difficult to estimate by this method. Brey et al. [4]; working with proteins, reported 0.05-0.1 g/g solids; Roebuck et al. [36] reported 0.3-0.4 g/g bound in starch.

The most successful method for studying properties of water involves preparation of "sorption isotherms," that is, curves relating the partial pressure of water in the food to its water content at constant temperature. In some cases it is more convenient to study the curves relating water activity, a, to water content. Water activity is defined as the ratio of partial pressure of water in the food (p) to the vapor pressure of pure water (P^{o}) at a given temperature.

III. WATER ACTIVITY AND SORPTION BEHAVIOR OF FOODS

In Chapter 7, it has been seen that the equilibrium relation between a component present in both a solid and a gaseous phase can be represented by several types of curves. In particular, isotherms, isobars, or constant activity curves and isosteres were discussed. In each case these curves represent relations among three key variables: (1) partial pressure of the component in the gas phase, (2) concentration of the component in the solid phase, and (3) temperature. In the case of water we shall be concerned with the relations between water activity, moisture content of the food, and temperature.

Isotherms of food are usually obtained at constant temperature by one of two basic methods (indicated schematically in Fig. 8-1).

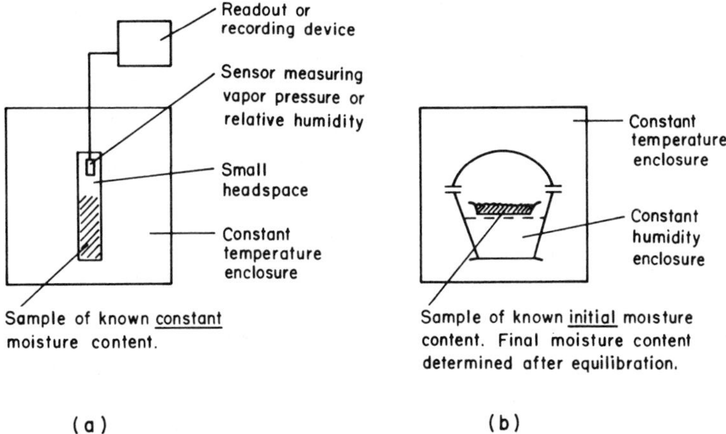

FIG. 8-1. Schematic representation of principles involved in two basic methods for measuring sorption isotherms.

In the first method, (Figure 8-1a) a food of known moisture content is allowed to come to equilibrium with a small headspace in a tight enclosure and partial pressure of water activity is measured manometrically, or relative humidity is measured. Water activity is equal to equilibrium relative humidity divided by 100:

$$a = \frac{\text{ERH}}{100} \tag{8-1}$$

where ERH is the equilibrium relative humidity (%).

Relative humidity sensors of great variety are available for this purpose, including electric hygrometers, dewpoint cells, hair psychrometers, and others. A good review of instrumentation is given by Gal [12].

A simple manometric method of measuring equilibrium partial pressure was reported by Taylor [41] and an electric hygrometer method was reported by Karel et al. [19]. An indicator paper strip method was developed for this purpose [24].

A second basic method (Figure 8-1b) for preparing isotherms is the exposure of a small sample of food to various constant humidity atmospheres. After equilibrium is reached the moisture content is determined gravimetrically or by other methods. A number of saturated salt solutions are available for this purpose and some of these are listed in Table 8-1. Saturated salt solutions have the advantage of maintaining a constant humidity as long as the amount of salt present is above saturation level. Solutions of glycerol or sulfuric acid also can be used to control humidity, provided their concentration is known (or determined) at the end of the equilibration period. Solutions providing constant humidity have been listed by Rockland [35] and O'Brien [32].

There is an additional method, based on approach to equilibrium. When it is not feasible to allow samples to come to equilibrium an interpolation procedure may be used in which weighed samples of food of known moisture content are exposed for a short period of time to various relative humidities and weight gain or loss is determined. By interpolation it is possible to establish the relative humidity at which samples neither gain nor lose weight. This humidity is regarded as the ERH. By repeating the process at several moisture contents, a complete sorption isotherm can be determined [27]. Recent reviews of methods for determining sorption isotherms include those by Gal [12] and by Toledo [42].

Typical isotherms in foods are S shaped, as shown in Figure 8-2, and they may be approximated by a variety of mathematical relationships, some of which are listed in Table 8-2.

A portion of the total water content present in food is strongly bound to specific sites. These sites include the hydroxyl groups of polysaccharides, the carbonyl and amino groups of proteins, and others on which water can be held by hydrogen bonding, by ion-dipole bonds, or by other strong interactions. The most effective way of estimating the contribution of adsorption at specific sites to total water binding is the use of the BET (Brunauer-Emmet-Teller) isotherm [25]. This mathematical relation, shown in Eq. (8-2), is based on oversimplified assumptions but is extremely useful as an estimate of the so-called "monolayer value," which can be considered as equivalent to the amount of water held adsorbed on specific sites. Figure 8-3 shows how sorption data may be plotted to determine the monolayer value m_1 and the constant C.

TABLE 8-1

Some Constant Relative Humidity Solutions

Salt	Approximate solubility in hot water (g/100 ml)	Constant relative humidity at room temperature (%) [a]
LiBr	240	7
$LiCl \cdot H_2O$	100	11
$K(CH_2COO)$	500	22
$MgCl_2 \cdot 6H_2O$	300	32
K_2CO_3	150	43
$NaBr \cdot 2H_2O$	120	58
$NaNO_2$	150	65
NaCl	40	75
KCl	50	85
$BaCl_2$	50	90
K_2SO_4	20	97

[a] Constant relative humidity varies slightly with temperature. For details see Rockland [35].

$$\frac{a}{m(1 - a)} = \frac{1}{m_1 C} + \frac{C - 1}{m_1 C} a \qquad (8\text{-}2)$$

where

a is water activity
m is water content (g H_2O/g solids)
m_1 is the monolayer value
C is a constant

Figure 8-3 is a BET plot for dehydrated orange crystals at 25° C [21]. The monolayer value for this material can then be calculated as follows. From Figure 8-3 we obtain:

$$m_1 C = 1/0.022 = 45.45$$

and

$$C - 1 = 0.89\, m_1 C$$

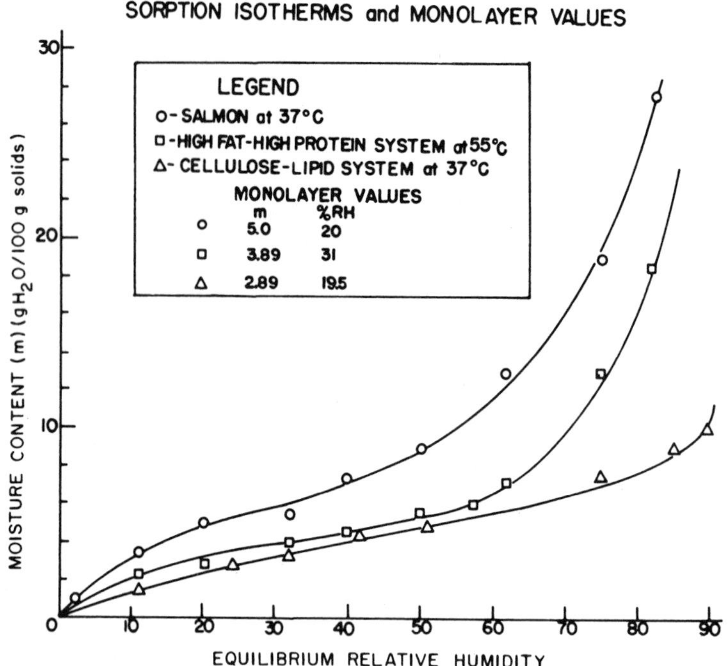

FIG. 8-2. Typical isotherms for water in foods and in model systems representing foods.

Therefore

$$C = 1 + (0.89)(45.45) = 41.45$$

$$m_1 = 45.45/C = 45.45/41.45 = 1.1 \text{ g } H_2O/\text{g solids}$$

In addition to water adsorbed on specific sites there is a portion of the total water which has its vapor pressure depressed because it is located in small capillaries. The relationship between water activity and radius of a capillary is shown in Eq. (8-3).

$$\ln(a) = \frac{-2\gamma}{r}(\cos\theta)\,C_1 \qquad (8\text{-}3)$$

where

γ is the free surface energy of water
r is the capillary radius
θ is the contact angle
C_1 is a constant

TABLE 8-2

Some Equations[a] Reported as Describing Food-Water Isotherms[b]

Henderson	$(1 - a) = e^{-cm^n}$
Fugassi	$m = \dfrac{ca}{(ca)(1 - a) + ca}$
Kuhn	$m = \dfrac{c_1}{(\ln a)^n} + C_2$
Oswin	$m = c\left(\dfrac{a}{1 - a}\right)^n$
Mizrahi	$a = \dfrac{c_1 + m}{c_2 + m}$
Linear isotherm	$m = c_1 a + c_2$

[a] a = water activity; m = water content; c = constant; n = exponent.

[b] From Karel [18].

The extent to which capillarity depresses water activity in foods is not clear. In order for activity to be lowered to a value of 0.9, capillaries must have radii of less than 0.000001 cm.

Finally, water capable of acting as a solvent is subject to depression of vapor pressure by small solutes. Ideally, this vapor depression is given by Raoult's law, Eq. (8-4), but in foods there are usually substantial deviations from the ideal relationship.

$$a = x_w \tag{8-4}$$

where x_w is the mole fraction of water in solution.

In determining x_w the total number of kinetic units in solution is counted. One mole of NaCl, for instance, counts as 2 moles of kinetic units (Na^+ and Cl^-).

Table 8-3 contains data relating the activity of water to concentrations of various solutes. If ideal relations are followed, then water activity should be numerically equal to the mole fraction of water in solution. Deviations from ideality may arise from several causes:

a. Not all water in food is capable of acting as a solvent.
b. Not all of the solute is in actual solution. (Some may, for instance, be bound to other insoluble food components, as in the case of salts bound to proteins.)
c. Interactions between solute molecules may cause deviations from ideality.

Minimum activities achievable by common solutes present in foods are shown in Table 8-4.

TABLE 8-3

Water Activity of Selected Solutions

Water activity	Molality (moles solute/liter pure H_2O)			
	Ideal solute	NaCl	Sucrose	Glycerol
0.90	6.17	2.83	4.11	5.6
0.80	13.9	5.15	—	11.5

Not all of the water present in food acts as a solvent. Water bound on specific sites (monolayer water) does not act as solvent, and Duckworth [9] has recently demonstrated that an additional fraction of water, as well, does not dissolve solutes. He used NMR techniques and noted that proton-containing solutes, such as sugars, gave in solution NMR signals which differed from those given in the precipitated state. Each solute had a specific water activity at which it went into

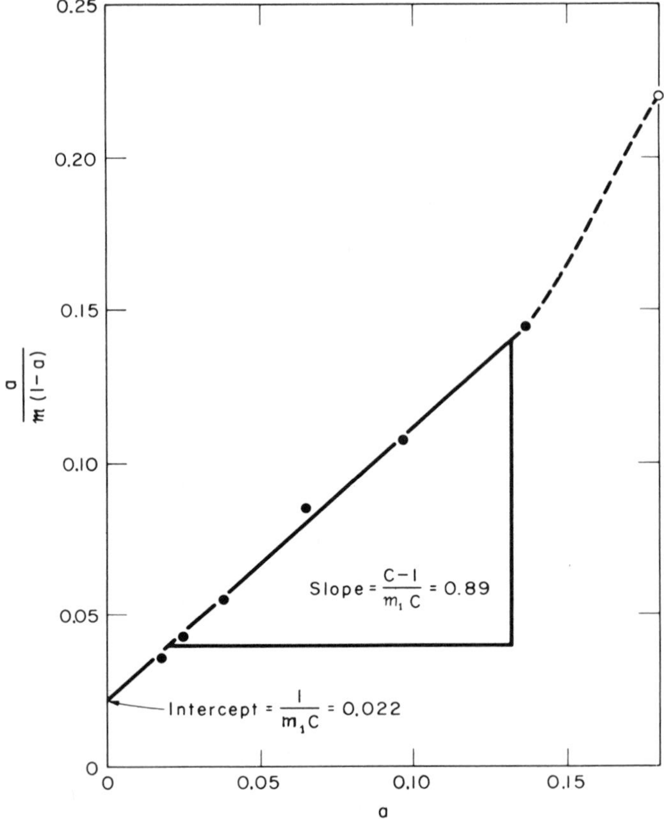

FIG. 8-3. BET plot for determination of the "monolayer value" for water in foods.

TABLE 8-4

Minimum Activities of Solutions (at Room Temperature)

	Solubility limit (% w/w)	Minimum activity
Sucrose	67	0.86
Glucose	47	0.915
Invert sugar	63	0.82
Sucrose + invert sugar (Sucrose, 37.6%; invert sugar, 62.4%)	75	0.71
NaCl	27	0.74

solution, and the presence of other insoluble components in the food did not alter this activity. Thus he found that sucrose went into solution at an activity of 0.82 in several sucrose-polymer-water systems. The amount of water to which this activity corresponded varied from 0.11 to 0.34 g/g solids, depending on what polymer was used in the system. The monolayer value was well below this amount, corresponding to 20-45% of the nonsolvent water. Figure 8-4 shows schematically the major types of water in food.

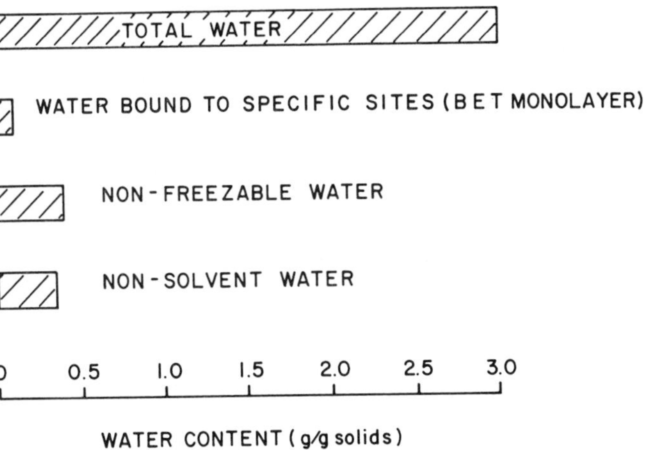

FIG. 8-4. Schematic representation of the types of "bound" water in foods.

Soluble substances present in foods may also undergo phase transformations and the occurrence of these phase changes can depend on the presence of water. Sorption of water in such systems is complicated and can require consideration of kinetics as well as equilibrium of sorption. The considerations involved are illustrated in Figure 8-5, which shows the sorption behavior of sucrose in several states. Crystalline sucrose sorbs very little water until water activity reaches approximately 0.8 and the sucrose begins to dissolve. When the drying procedures are sufficiently rapid to produce amorphous sucrose, however, exposure to increasing humidities results in water uptake, reaching sorption levels far higher than those of crystalline sucrose. This difference in behavior is due to the greater internal area available for water sorption in the amorphous material and to the greater ability of water to penetrate into the H-bonded structure, which is less

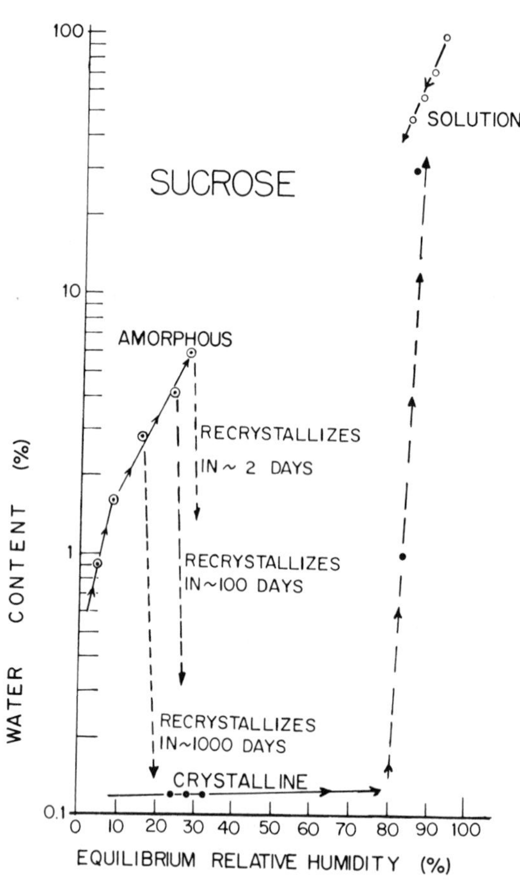

FIG. 8-5. Sorption behavior of water in sucrose.

regular in amorphous sucrose than in sucrose crystals. However, adsorption of water imparts mobility to the sugar molecules; this mobility results eventually in the sucrose transforming from the metastable amorphous state to the more stable crystalline state. In this process the sugar first gains some water but then loses practically all of its water content, because the recrystallization that occurs extremely slowly at low humidities accelerates with transient water pickup at higher humidities [18].

Another complicating factor in sorption behavior of foods, is the so-called "sorption hysteresis." This phenomenon is illustrated in Figure 8-6. The water content of a food sample showing hysteresis is lower during equilibration by adsorption (pickup of water by an initially dry sample) than in desorption (drying of an initially "wet" sample).

Additional properties of water are discussed in Chapter 2 of Part I of this series.

IV. ENERGY OF BINDING OF WATER

By determining sorption isotherms at several temperatures it is possible to obtain "isosteres," curves relating temperature to partial pressure of water or water activity at constant moisture contents. When the natural logarithm of water activity for a given moisture content is plotted against the reciprocal of absolute temperature the slope of the resultant straight line is equal to $\Delta H_A/R$. ΔH_A is the heat of adsorption, (above and beyond the latent heat of vaporization of normal, nonadsorbed, water) and R is the gas constant. If, on the other hand, one plots on the abscissa the reciprocal of absolute temperature and on the ordinate the natural logarithm of the partial pressure of water in food, then the slope of the resultant straight line gives $\Delta\bar{H}/R$, where $\Delta\bar{H}$ is the total heat of vaporization plus adsorption.

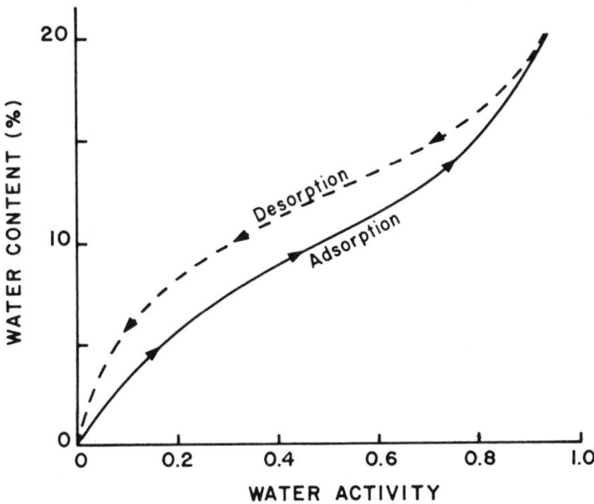

FIG. 8-6. Water sorption hysteresis.

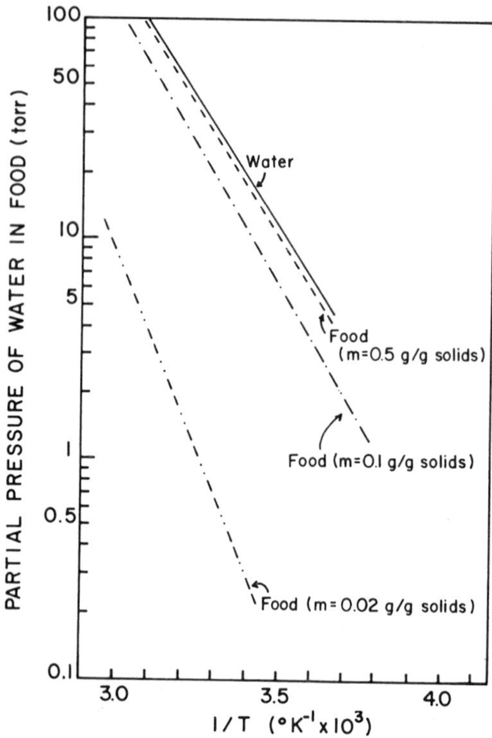

FIG. 8-7. Plot of partial pressure of water against the reciprocal of absolute temperature.

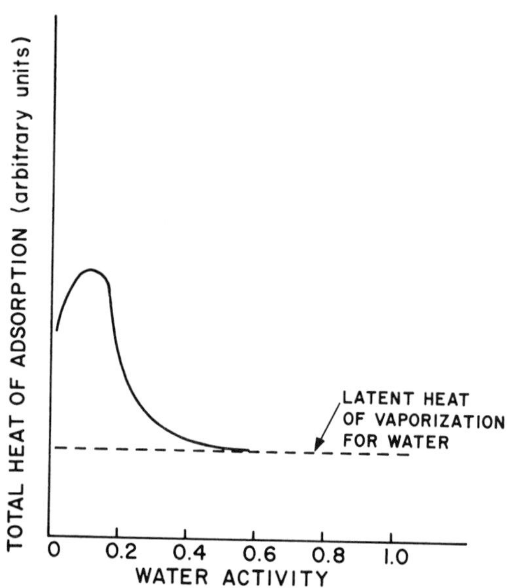

FIG. 8-8. Typical dependence of total heat of sorption on water activity.

TABLE 8-5

BET Monolayer Values and Maximum Total Heats of Adsorption for Selected Foods and Food Components

Substance	Approximate BET monolayer (g H_2O/g solids)	Maximum $\overline{\Delta H}$ (kcal/mole)
Starch [9]	0.11	~14
Polygalacturonic acid [3]	0.04	~20
Gelatin [9]	0.11	~12
Lactose, amorphous [11]	0.06	~11.6
Dextran [11]	0.09	~12
Potato flakes [15]	0.05	
Spray-dried whole milk [15]	0.03	
Freeze-dried beef [17]	0.04	~12

Figure 8-7 shows a typical semilogarithmic plot of partial pressure of water vs the reciprocal of absolute temperature, from which ΔH can be calculated, and Figure 8-8 shows the typical dependence of ΔH on water activity. Typically, maximum energy of binding occurs at the BET monolayer value, and Table 8-5 shows typical values for several foods and food components.

V. PRACTICAL APPLICATIONS OF SORPTION OF WATER IN FOODS

Knowledge of sorption behavior of food is useful in concentration and dehydration processes for two reasons:

a. It is of importance in design of the processes themselves, because it has an important impact on the ease or difficulty of water removal, which depends on the partial pressure of water over the food and on the energy of binding of the water in the food.
b. Water activity affects food stability and therefore it must be brought to a suitable level at the conclusion of drying and maintained within an acceptable range of activity values during storage.

In analyzing sorption behavior of complex food mixtures, such as dehydrated soups, cake mixes, and military multi-component rations, it is often desirable to be able to calculate the sorption isotherm of the mixture from sorption isotherms of the components. This is not always possible, and especially with foods containing humectants, such as corn syrup, glycerol, and sorbitol, there have been reported deviations from behavior expected for ideal mixtures. Nevertheless, the behavior of an ideal mixture is often useful as a starting model of behavior and is, for instance, useful in calculation of desired compositions. In the case of confectionery products, a very wide range of water activities is possible by controlling the concentrations of the several soluble components, including sucrose, invert sugar, and glucose. Fondants and jellies, for instance, have water activities in the range of 0.7-0.85 but some candies, such as certain kinds of toffees and hard candies, have activities as low as 0.1-0.2. The usual approach is to calculate the number of moles of solute needed to achieve a given activity using ideal relations [Eq. (8-4) and then to adjust compositions in terms of the major ingredients (i.e., glucose, fructose, and sucrose). Complications arise not only because of deviations from the ideal equation but also because solubility limits for one or more of the solutes are exceeded. In this connection it must be remembered that the solubility of each component in a mixture is depressed by the presence of other components. In the case of mixtures containing insoluble components binding water, the activity of water in the mixture can be approximated on the basis of known sorption behavior of each component.

One of the empirical equations for calculating water activity of sugar confections based on ideal relations is that of Money and Born [30]:

$$a = \frac{1.00}{1 + 0.27n} \tag{8-5}$$

where n is the number of moles of sugar per 100 g of water. Only sugar actually present in solution is considered and amounts in excess of solubility limits are assumed inactive and are not counted in n.

Another empirical approach used in candy formulation is the method of Grover [13]. In this method different candy ingredients (including polymeric components, such as proteins and gums) are assigned a "sucrose equivalent conversion factor," S. This factor is based on experimental vapor pressures measured in solutions of these ingredients; hence in part it incorporates corrections to Raoult's law. (It does so in part only because a true correction factor must vary with composition). Table 8-6 shows the sucrose equivalent factors for common candy ingredients. To calculate water activity one determines the concentration of each ingredient in grams per gram of water and multiplies it by S for the specific ingredient (sugar amounts present in excess of the solubility limit are subtracted from the amount used in calculating the concentration). These products are then entered in Eq. (8-6) and water activity is calculated.

$$a = 1.04 - 0.1\ (\Sigma S_i c_i) + 0.0045\ \Sigma\ (S_i c_i)^2 \tag{8-6}$$

where

c_i is the concentration of component i
S_i is the sucrose equivalent for this ingredient

The method has limitations and does not apply at very low moisture contents [5].

Heiss [14] also used an idealized relation for calculation of the water activity of hard candies and found that it gave reasonable approximations at high water contents but poor results at low water contents.

TABLE 8-6

Sucrose Equivalent Conversion Factors for Various Candy Ingredients

Ingredient	Conversion factor (S)
Sucrose	1.0
Lactose	1.0
Invert sugar	1.3
Corn syrup solids (45 D. E.)	0.8
Gelatin	1.3
Starch and other polysaccharides	0.8
Citric acid and its salts	2.5
Sodium chloride	9.0

[a] After Grover [13].

In solutions, better approximations can be made using more complicated equations. Among these is the one developed for multi-component solutions by Norrish [31]:

$$\ln a = \ln x_w + [(1 - k_1)^{1/2} x_1 + (-k_2)^{1/2} x_2 \cdots]^2 \qquad (8\text{-}7)$$

where

$x_1 \cdots x_n$ are mole fractions of solutes
$k_1 \cdots k_n$ are binary coefficients for each solute as defined in Eq. (8-8)

$$\ln a = \ln x_w + k_1 x_1^2 \qquad (8\text{-}8)$$

In not too concentrated solutions, all of the above relations give reasonable approximations. For the case of a specific corn syrup the Grover equation gives a = 0.825; the Money-Born equation gives a = 0.85 and the Norrish equation gives a = 0.844 [23].

One of the reasons for estimating water activity of mixtures is the need for knowledge about the potential for water transfer between different components of a heterogeneous food. For instance, it is entirely possible to have a situation in which a candy has a fondant cream layer with an initial water content of 12% and an adjacent jelly layer with a water content of 20%, and yet have water being transported from the cream layer to the jelly layer! The reason for this transfer is the higher water activity of the fondant cream layer. Similar transfers between components can occur in filled pastries, in mixed fruits, and in various mixtures of dehydrated foods, e.g., dry soups.

In the case of very dry substances, such as dehydrated foods, depression of water activity is not primarily due to solute depression of vapor pressure and equations based on this principle, such as Raoult's law or the Norrish equation, are of no value. Water activity of mixtures, however, can be approximated by utilizing sorption isotherms of components. A simple graphical approximation has been developed by Salwin and Slawson [37]. In this procedure, portions of the isotherms of components are approximated by straight lines, as shown in Figure 8-9, and the water activity of the mixture is estimated by Eq. (8-9).

$$a = \frac{a_1 s_1 w_1 + a_2 s_2 w_2 + \cdots\cdots a_n s_n w_n}{s_1 w_1 + s_2 w_2 + s_n w_n} \tag{8-9}$$

where

$a_1 \cdots\cdots a_n$ are water activities of components
$s_1 \cdots\cdots s_n$ are isotherm slopes of components
$w_1 \cdots\cdots w_n$ are weights of solids of components
$a \sim$ is water activity of equilibrated mixture

where greater accuracy is desired the isotherms can be approximated by nonlinear equation.

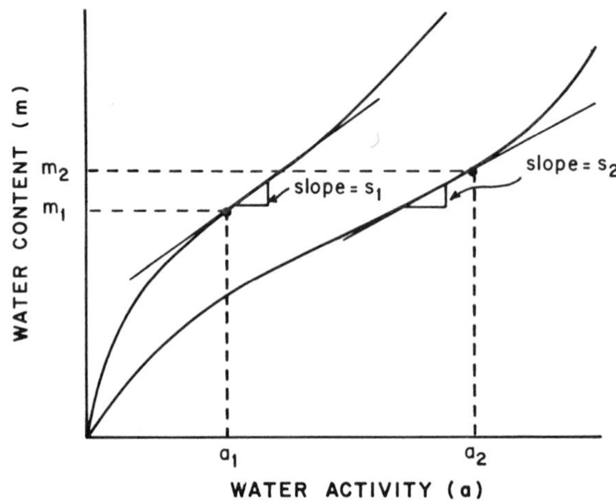

FIG. 8-9. Graphical representation of sorption relations used in calculation of water activity of mixtures, Eq. (8-9).

VI. WATER ACTIVITY AND FOOD STABILITY

A. Water Activity and Microbial Spoilage

In a classic article Scott [38] has summarized the water relations of microorganisms, and most of the principles he suggests are still valid. These include the following:

1. Water activity, rather than water content, determines the lower limit of available water for microbial growth. Most bacteria do not grow below a = 0.91 and most molds cease to grow below a = 0.8. Some xerophilic fungi have been reported to grow at water activities of 0.65, but the range of 0.70-0.75 is generally considered their lower limit. A schematic representation of dependence of microbial growth on water activity is shown in Figure 8-10.
2. Environmental factors affect the level of water activity required for microbial growth. The general principle which often applies is that the less favorable the other environmental factors (nutritional adequacy, pH, oxygen pressure, temperature) the higher becomes the minimum water activity at which microorganisms can grow.
3. Some adaptation to low activities occurs, particularly when water activity is depressed by addition of water-soluble substances (principle of intermediate-

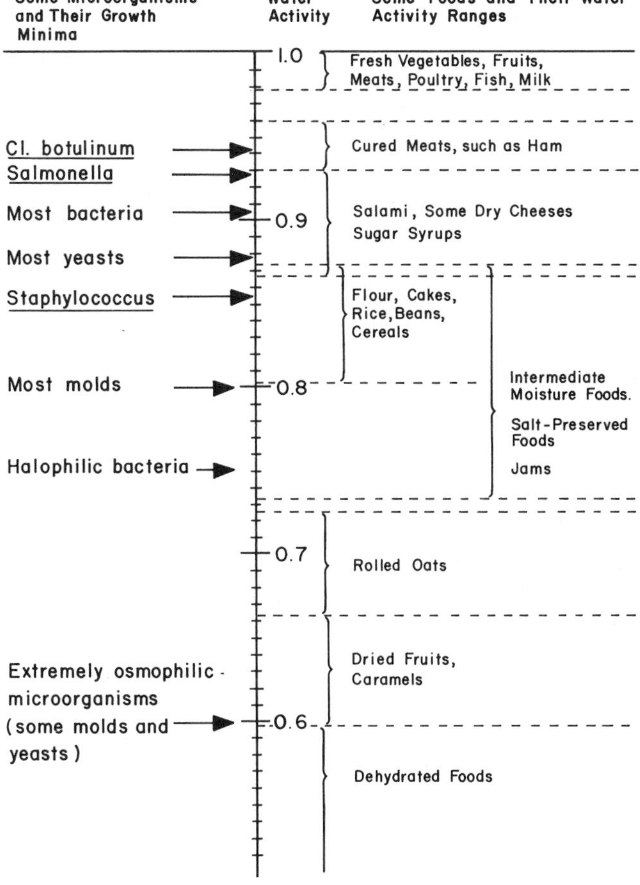

FIG. 8-10. Schematic representation of water activity minima for growth of microorganisms and of typical activity ranges for some foods.

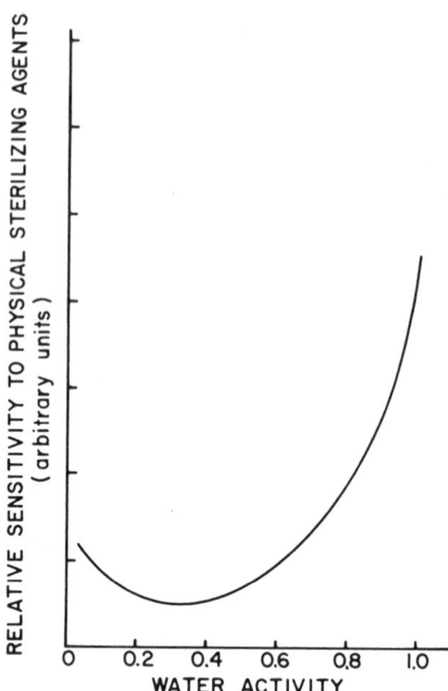

FIG. 8-11. Effect of water activity on sensitivity of microorganisms to physical sterilizing agents.

moisture foods), rather than by water crystallization (frozen foods) or water removal (dehydrated foods).

4. When water activity is depressed by solutes, the solutes themselves may have effects which complicate the effect of water activity per se. For instance, at a given water activity, microbial growth is less effectively depressed by glycerol than by sodium chloride.

More recent researches, especially with freeze-dried and intermediate-moisture foods, have resulted in the following major additional findings [43]:

5. Water activity modifies sensitivity of microorganisms to heat, light, and chemicals. In general, organisms are most sensitive at high water activities (i.e., in dilute solution) and minimum sensitivity occurs in an intermediate-moisture range. A schematic representation of the effect of water activity on microbial sensitivity to sterilization is shown in Figure 8-11.
6. Minimum water activities for production of toxins are often higher than those for microbial growth. This phenomenon may represent an important safety factor in the distribution of dehydrated and intermediate-moisture foods.

B. Effects of Water Activity and Content on the Chemical Deterioration of Foods

The effects of water on chemical reactions in foods are more complicated than are its effects on microbial growth [22, 26]. Water activity is not the only parameter defining the lower limits of chemical activity, and water can act in one or more of the following roles: (a) as a solvent for reactants and for products; (b) as a reactant (e.g., in hydrolysis reactions); (c) as a product of reactions (e.g., in condensation reactions, such as occur in nonenzymic browning); and (d) as a modifier of the catalytic or inhibitory activities of other substances (e.g., water inactivates some metallic catalysts of lipid peroxidation).

1. Enzymatic Reactions in Foods

The occurrence of enzymatic reactions in food at low water contents has been the subject of extensive studies over the past three decades, especially by German workers [2]. While very substantial progress has been made the exact influence of water on enzymatic activity remains to be elucidated. In the region of adsorption of water on specific sites in the foods, that is, below the BET monolayer value, enzymatic reactions occur extremely slowly or cease entirely. Apparently this lack of activity is due to lack of mobility for the substrate to diffuse to the active site of the enzymes. The extreme importance of water as a mobilizer of various changes in low-moisture substrates has led some Russian workers to refer to it as a "mobility catalyzer" [6]. It is interesting to note that even in the case of hydrolazes, which require water as a reactant, the mobilizing activity of water rather than its availability as a reactant is the critical factor in limiting enzyme action. In the case of lipases at low water contents the importance of mobility is further demonstrated by the fact that solid triglycerides undergo enzymatic hydrolysis at a much slower rate than do mobile, liquid triglycerides.

Apparently, the active sites of enzymes function even at very low water activity but the substrates and any other reactants, such as water, must be sufficiently mobile to allow for transport to this site.

2. Nonenzymatic Browning

Most dehydrated foods, and practically all intermediate-moisture foods, are subject to nonenzymatic browning. This reaction is water dependent and invariably shows a maximum rate at intermediate moistures. These effects result from water's dual role as solvent and as a product of the reaction, and hence an inhibitor. At low water activity the limiting factor is inadequate mobility; therefore, addition of water accelerates the reaction.

Since water is also a product of the reaction (condensation reactions involved in browning result in liberation or water) it can therefore inhibit the reaction by "product inhibition." Furthermore, at high water contents the dilution effect becomes important for all reactions having rates that depend on concentration [10] (all reactions which have a reaction order greater than zero).

The exact position of browning maxima depends on the specific food but, generally, concentrated liquids (e.g., unfrozen fruit concentrates) and intermediate-moisture foods (e.g., pastry fillings and so-called evaporated fruit, such as prunes) are in the range of moisture content most advantageous for browning.

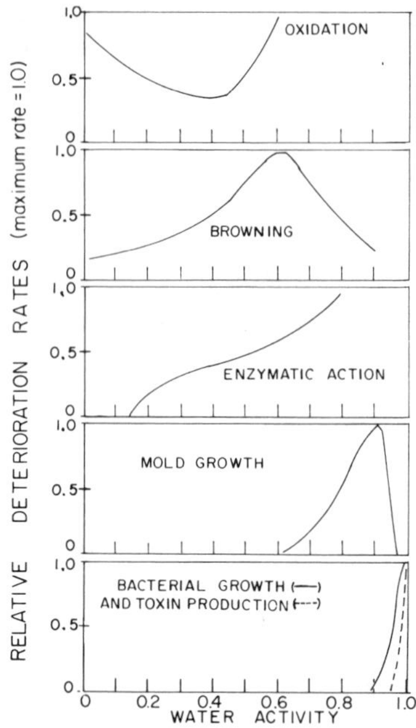

FIG. 8-12. Dependence of relative rates of food deterioration on water activity. Bacterial growth, S. aureus; Toxin production, Enterotoxin B; Troller [43], Mold growth, Xeromyces bisporus, after Christian [7]. Enzymatic activity, oat lipase acting on monoolein in ground oats, after Acker [1]. Browning, nonenzymatic browning of pork bites, after Karel and Labuza [20]. Oxidation, lipid oxidation in potato chips, Quast and Karel [34].

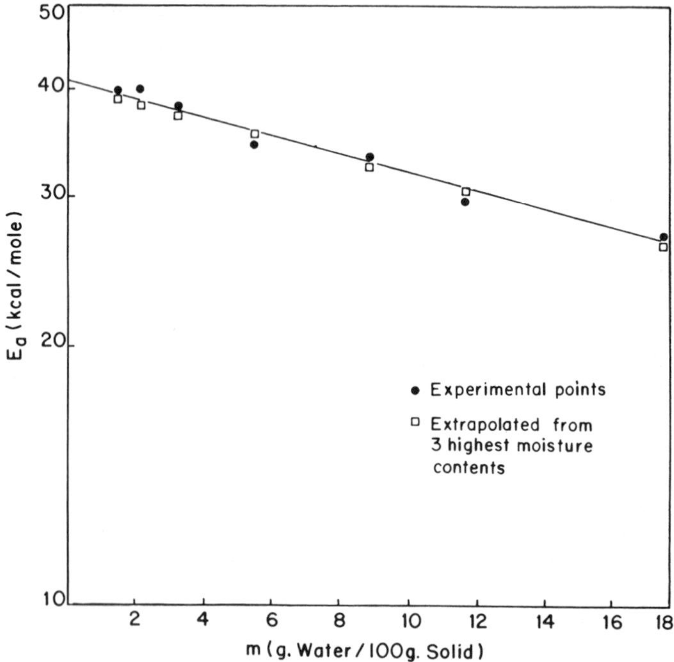

FIG. 8-13. Dependence of activation energy for nonenzymatic browning in dehydrated cabbage on moisture content.

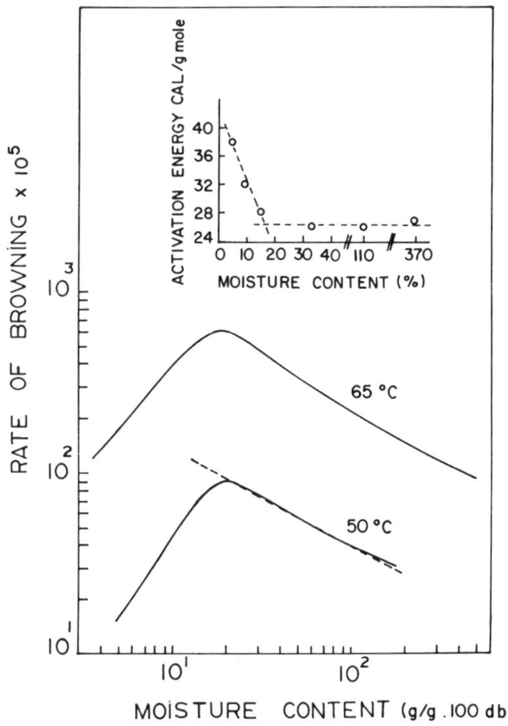

FIG. 8-14. Dependence of rates of browning and of activation energy for browning on the moisture content of potatoes (after Hendel et al. [16]).

Typical effects of water activity on nonenzymatic browning, as well as on other food deterioration mechanisms, are shown in Figure 8-12.

Water activity or water content, however, affect not only the rate at a given temperature but also the activation energy for browning. Figure 8-13 shows the activation energy as a function of moisture for dehydrated cabbage, and Figure 8-14 shows how the known dependence of activation energy for browning of potatoes [16] can be used to calculate the dependence of browning on water content at two different temperatures [28, 29].

3. Lipid Oxidation and Reactions Between Lipids and Proteins

Water affects oxidation of lipids and other free-radical reactions in foods, and the effect is very complicated. Table 8-7 shows the dependence of rate of oxidation of potato chips on water activity. Increasing concentration of water slows the reaction up to a point where the intermediate-moisture range often begins. Peroxide production and oxygen absorption often decrease as the relative humidity is increased above the monolayer of water content.

The mechanisms by which water exerts this protective effect are as follows. (1) On the food surfaces, water hydrogen bonds to the hydroperoxides produced during the free-radical reaction. This hydrogen bonding protects the hydroperoxides from decomposing and therefore slows the rate of initiation through peroxide decomposition. (2) Trace metals catalyze the initiation steps of oxidation, but once these metals are hydrated their catalytic activity is reduced. The extent of the reduction depends on the metal. (3) Water can react with the trace metals to produce insoluble metal hydroxides, taking them out of the reaction phase. (4) The presence of water can also directly affect the free radicals produced during lipid oxidation.

At low moisture levels, added water often gives substantial protection against oxidation. At high moisture levels, however, oxidation increases again. This may be explained by the increased diffusion of those metal catalysts which have not been inactivated and possibly also by the swelling of the porous matrix of dry foods, which allows for faster uptake of atmospheric oxygen. Data on oxidation rates at intermediate-moisture levels, however, are sparse and more studies are needed.

Figure 8-15 shows schematically the dependence of oxidation of potato chips on water activity. It may be noted that water has a strong protective effect early in oxidation (when the oxidation is just beginning and the extent of oxidation, E, is essentially zero) but offers little protection once oxidation is well advanced [33].

In foods which contain many reactive components, reactions initially characterized as "oxidation of lipids" can involve other components as well. Reactions between oxidized lipids and proteins, for instance, may progress at different rates and by different mechanisms, depending on water content and water activity [40]. A more detailed discussion of chemical reactions in foods is given in Part I of this series.

TABLE 8-7

Effects of Water Activity on Oxidation of Potato Chips Stored at 37° C

Water activity	Initial oxygen absorption rate in air (μl O_2/g hr)	Oxidation rate in a closed container[a] (μl O_2/g hr)	Peroxide value after 2000 hr
0.001	2.7	0.15	70
0.11	0.26	0.13	—
0.20	0.16	—	15
0.32	0.15	0.07	—
0.40	0.16	0.06	5
0.62	—	0.19	—
0.75	—	1.6	—

[a]Initial concentration of oxygen 12.2%; rate measured when oxygen content dropped to 10%.

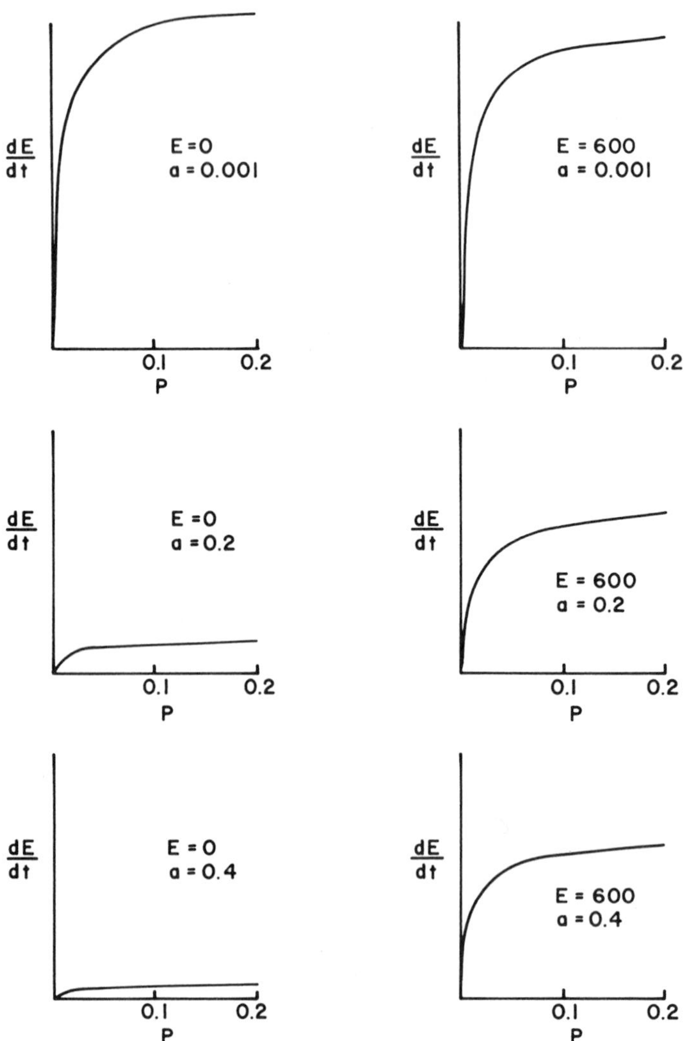

FIG. 8-15. Oxidation of potato chips as affected by water activity and other factors. dE/dt = Rate of oxidation (μl/g hr); E = extent of oxidation (μl/g); a = water activity; P = oxygen pressure, atmosphere.

SYMBOLS

(Units are given for some quantities. These are intended as examples. Other consistent units in either the English or the metric systems may be used.)

E	extent of oxidation, microliters per gram
ERH	equilibrium relative humidity (%)

C, C_1	constants
R	gas constant
S	sucrose equivalent conversion factor
a	water activity
c	concentration (g/g of water)
$k_1 \ldots k_n$	constants for components 1 through n
m	water content (g/g of solids)
m_1	BET monolayer value
n	number of moles per 100 g water
s	slope of a linear portion of the isotherm
$w_1 \ldots w_n$	weights of solids of components 1 through n in a mixture
$x_1 \ldots x_n$	mole fraction of solutes 1 through n in a solution
x_w	mole fraction of water
γ	free surface energy (ergs/cm^2)
ΔH_A	latent heat of adsorption (kcal/mole)
$\Delta\overline{H}$	latent heat of adsorption plus latent heat of vaporization of water (kcal/mole)
θ	contact angle

REFERENCES

1. L. Acker, Z. Ernaehrungswiss, 8 (Suppl.) 45 (1969).

2. L. Acker, Food Technol., 23, 1257 (1970).

3. F. A. Bettelheim, A. Block, and L. J. Kaufman, Biopolymers, 9, 1531 (1970).

4. W. S. Brey, M. A. Heeb, and T. M. Ward, J. Colloid Interface Sci., 30, 13 (1969).

5. S. H. Cakebread, Manuf. Confectioner, Jan. 1970, p. 42.

6. Y. N. Chirgadze and A. M. Owsepayan, Biopolymers, 11, 2179 (1972).

7. J. H. B. Christian, in Recent Advances in Food Science, J. M. Leitch and D. N. Rhodes (eds.), Butterworth, London, 1963, p. 248.

8. R. B. Duckworth, J. Food Technol., 6, 317 (1972).

9. R. B. Duckworth, Proc. Inst. Food Sci. Technol. (U. K.), 5(2), 60 (1972).

10. K. Eichner and M. Karel, J. Agr. Food Chem., 20, 218 (1972).

11. J. M. Flink and M. Karel, J. Food Technol., 7, 199 (1972).

12. S. Gal, Die Methodik der Wasserdampf-Sorptionsmessungen, Springer Verlag, Berlin, 1967.

13. D. W. Grover, J. Soc. Chem. Ind., 66, 201 (1947).

14. R. Heiss, Staerke, 7, 45 (1955).

15. R. Heiss, Haltbarkeit und Sorptionsverhalten wasserarmer Lebensmittel, Springer Verlag, Berlin, 1968.

16. C. E. Hendel, V. G. Silveira, and W. O. Harrington, Food Technol., 9, 433 1955.

17. J. G. Kapsalis, U. S. Army Natick Labs., Natick, Mass., Report 67-87 Fl, AD655488, 1966.

18. M. Karel, CRC Crit. Revs. Food Technol., 3, 329 (1973).

19. M. Karel, Y. Aikawa, and B. E. Proctor, Mod. Packag., 29(2), 153 (1955).

20. M. Karel and T. P. Labuza, MIT Report on Contract Research with U. S. Air Force School of Aerospace Medicine, Brooks AF Base, Contract No. F41-609-68-C-0015, 1969.

21. M. Karel and J. T. R. Nickerson, Food Technol., 18, 194 (1964).

22. T. Krebes and M. Behun, Proc. 11th Intl. Congr. Refriger., Vol. II, 1597 (1963).

23. C. Krueger and W. Steinmetzer, Zucker, 25, 356 (1972).

24. O. Kvaale and E. Dalhoff, Food Technol., 17(5), 151 (1963).

25. T. P. Labuza, Food Technol., 22, 263 (1968).

26. T. P. Labuza, CRC Crit. Revs. Food Technol., 3(2), 219 (1972).

27. A. H. Landrock and B. E. Proctor, Food Technol., 5, 332 (1951).

28. S. Mizrahi, T. P. Labuza, and M. Karel, J. Food Sci., 35, 799 (1970).

29. S. Mizrahi, T. P. Labuza, and M. Karel, J. Food Sci., 35, 804 (1970).

30. R. W. Money and R. Born, J. Sci. Food Agr., 2, 180 (1951).

31. R. S. Norrish, J. Food Technol., 1, 25 (1966).

32. F. E. M. O'Brien, J. Sci. Instrum., 25, 73 (1948).

33. D. G. Quast and M. Karel, J. Food Sci., 37, 584 (1972).

34. D. G. Quast, M. Karel, and W. M. Rand, J. Food Sci., 37, 673 (1972).

35. L. B. Rockland, Anal. Chem., 32, 1375 (1960).

36. B. D. Roebuck, S. A. Goldblith, and W. B. Westphal, J. Food Sci., 37, 199 (1972).

37. H. Salwin and V. Slawson, Food Technol., 13, 715 (1959).

38. W. J. Scott, Advan. Food Res., 7, 84 (1957).

39. S. Shanbhag, M. P. Steinberg, and A. I. Nelson, J. Food Sci., 35, 612 (1970).

40. S. R. Tannenbaum, H. Barth, and J. P. Leroux, J. Agr. Food Chem., 17, 1353 (1969).

41. A. A. Taylor, Food Technol., 15, 536 (1961).

42. R. T. Toledo, Proc. Meat Ind. Res. Conf., American Meat Institute Foundation, Chicago, Ill., 1973, p. 85.

43. J. Troller, J. Milk Food Technol., 36(5), 276 (1973).

Chapter 9

CONCENTRATION OF FOODS

Marcus Karel

CONTENTS

I. INTRODUCTION

Removal of water and the consequent lowering of water activity constitutes an important principle of food preservation. When removal of water is carried to essential completion resulting in food materials with water contents between 0 and 15-20%, the methods involved are called "dehydration processes." In other cases only part of the water is removed, which results in concentrated solutions or dispersions or in semisolid products with water contents in excess of 20%. The methods involved are called "concentration processes." The purposes of concentration can vary. It may be used as an economical preparatory step for subsequent dehydration, as in the cases of spray-dried tea or freeze-dried coffee. It may be used to reduce the bulk of materials to be preserved by freezing or by sterilization, as in the cases of frozen orange juice or evaporated milk. It may also be used as a method of preservation in its own right, for such food products as maple syrup, which do not deteriorate readily at the water activities achievable by concentration.

Methods of concentration which are discussed in this chapter include evaporation, freeze concentration, and membrane separation. Osmotic dehydration and solvent extraction also achieve concentration but are usually carried out in conjunction with dehydration methods. Osmotic dehydration is discussed in Chapter 10.

II. EVAPORATION

A. Principles of Evaporator Operation

Evaporation is the removal by vaporization of part of the solvent from a solution or dispersion of essentially nonvolatile solutes. Evaporation is distinguished from crystallization and drying by the fact that the final product is a concentrated dispersion or solution rather than precipitated solids.

In practice, evaporation is conducted by boiling off the solvent; that is, vaporization is conducted at the boiling point of the solution or dispersion. Where heat sensitivity is a factor, as is often the case with foods, the boiling point may be lowered by lowering the pressure at which evaporation takes place. The process of vaporization involves simultaneous heat and mass transfer, since it depends on the supply of heat required to vaporize the solvent and on removal of vapor from the vicinity of the surface at which vaporization takes place.

Several rate processes are involved in evaporation, but the engineer concerned with the operation can usually consider the entire process in terms of transfer of heat from the heater to the boiling liquid. Heat-transfer analysis, combined with mass and energy balance, is usually adequate for evaporation, provided the boiling point of the liquid is known as a function of pertinent temperatures and concentrations.

An evaporator consists of a heat exchanger capable of maintaining the liquid at its boiling point and a device to separate the vapor phase from the liquid. The mass and energy balance may be developed as shown schematically in Figure 9-1. Figure 9-1 represents an evaporator operating in steady state. The following symbols are used:

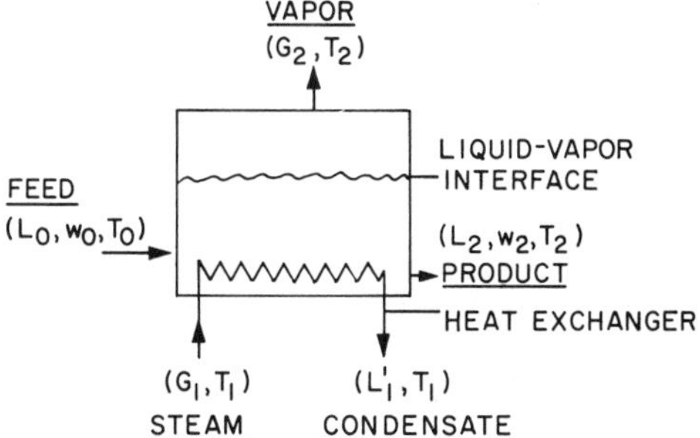

FIG. 9-1. Schematic representation of an evaporator in steady-state operation.

L, L' is liquid streams (lb/hr)
G is vapor stream (lb/hr)
w is solid concentration (lb solid/lb total)
T is temperature (°F)

The subscripts are:

0 = feed stream
1 = steam of condensate in heat exchanger
2 = streams leaving evaporator

It is assumed that the entering feed is brought instantaneously to the boiling point T_2 and that the vapor G_2 and the product L_2 leave the evaporator at this temperature. It is further assumed that steam enters the heat exchanger as saturated vapor (G_1) at temperature T_1 and leaves as liquid condensate (L'_1) at the same temperature.

Mass balance results in the following equations:

overall material balance

$$L_0 = L_2 + G_2 \tag{9-1}$$

Solute balance

$$w_0 L_0 = w_2 L_2 = w_2 (L_0 - G_2) \tag{9-2}$$

Therefore

$$G_2 = L_0 \left(1 - \frac{w_0}{w_2}\right) \tag{9-3}$$

An energy balance may be conducted on the assumption of negligible heat losses to the surroundings:

Heat supplied by condensing steam = heat required to heat and vaporize

$$G_1\lambda_1 = (L_0)(c_{p0})(T_2 - T_0) + G_2\lambda_2 \tag{9-4}$$

where

λ_1 is latent heat of vaporization of water at T_1 (Btu/lb)
λ_2 is latent heat of vaporization of water at T_2 (Btu/lb)
c_{p0} is specific heat of feed solution (Btu/lb °F)

Combination of Eqs. (9-3) and (9-4) allows calculation of the amount of steam required per pound of water removed for a given degree of concentration (w_2/w_0) and a given feed temperature (T_0).

An equally important consideration is the size of heat exchanger capable of supporting the necessary overall rate of heat transfer. The heat transfer rate is:

$$q = UA(T_1 - T_2) \tag{9-5}$$

where

U is overall heat transfer coefficient (Btu/hr^{-1} ft^{-2} $°F^{-1}$)
A is heat transfer area (ft^2)
q is overall heat transfer rate (Btu/hr)

It should be noted that the coefficient U is a function of the sum of resistances to heat transfer:

$$U = 1/r_\theta \tag{9-6}$$

where r_θ is the total resistance:

$$r_\theta = (1/h_1) + r_{wall} + (1/h_2) \tag{9-7}$$

where

$1/h_1$ is the resistance in steam
h_1 is heat transfer coefficient in steam phase (Btu hr^{-1} ft^{-2} $°F^{-1}$)
$r_{wall} = (X_{wall}/k_{wall})$ is the resistance of heat exchanger wall
X_{wall} is wall thickness (ft)
k_{wall} is thermal conductivity of wall (Btu ft^{-1} hr^{-1} $°F^{-1}$)
$1/h_2$ is resistance in liquid phase
h_2 is heat transfer coefficient in liquid phase (Btu hr^{-1} ft^{-2} $°F^{-1}$)

In well-designed and efficiently operated evaporators, the overall resistance r_θ is dominated by resistance in the boiling liquid, and U is often well approximated by h_2. It must be noted here that h_2 depends strongly on the conditions in the liquid, in particular on viscosity, temperature, and type and magnitude of convection. Furthermore, the overall resistance is increased greatly by deposition of a solid or semisolid layer on heat-exchange surfaces (fouling).

B. Boiling Point Estimation

The driving force for heat transfer is the temperature difference ($T_1 - T_2$), which depends on the pressure of steam in the heat exchanger and on the boiling point of the liquid (T_2). The boiling point in turn depends on pressure at the evaporating surface and on concentration of solutes. Boiling point elevation is a colligative property which depends on the number of kinetic units in solution. Under ideal conditions, therefore, the boiling point depends only on the total number of particles, molecules, and ions present per unit weight of solvent. In practice, however, the situation in food materials is quite removed from ideality and the boiling point must be determined empirically. An aid to this determination is the use of the Dühring plot (Fig. 9-2), which is based on the often applicable rule that the boiling point of a solution is a linear function of the boiling point of the solvent. The Dühring plot allows approximation of boiling points of solutions on the basis of measurements of boiling points at only two pressures.

Another consideration in boiling point elevation is the contribution to pressure of the hydrostatic head due to depth of the liquid. It is often negligible, but where it is of importance it is common to base design calculations on a boiling temperature determined from the pressure at half the depth of the liquid in the evaporator.

C. Food Properties and Evaporator Performance

A major consideration in engineering reviews of evaporation is the problem of steam economy, that is, the problem of maximizing evaporation per unit of energy supplied to the evaporator. In food applications, however, a number of considerations pertaining to properties of the foods assumes a dominant role. The following are of particular importance.

1. Viscosity and Consistency

Many of the food products subjected to evaporative concentration become quite viscous when concentrated. In some cases, non-Newtonian behavior is observed

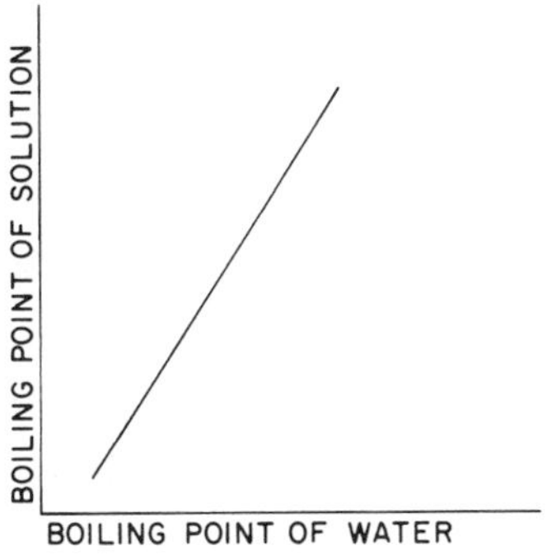

FIG. 9-2. Dühring plot.

and peculiar temperature dependence is encountered due to such effects as denaturation of proteins. Viscosity affects not only the rate of heat transfer but also pumping requirements and other material handling considerations. In processing of very viscous fruit juices, special treatments may be necessary to reduce viscosity. Israeli scientists, for instance [19], have developed an ultrasonic process for reduction of viscosity during concentration of orange juice.

2. Fouling

Fouling is often critical in food evaporation. Proteins and polysaccharides in particular are capable of forming deposits which are difficult to remove and which adversely affect efficiency of heat transfer. The interactions are often complex. In a study concerned with fouling during evaporation of tomato juice, Morgan [20] found that the presence of fiber, pectin, and protein was necessary for fouling, and that removal of any one of these components would greatly improve evaporating efficiency. Practical solutions to the problem involve design of evaporators which minimize development of stagnant layers along surfaces of heaters.

3. Foaming

The presence of various natural surface-active components, including proteins, is often the cause of extensive foaming, and this reduces heat transfer efficiency and may produce problems in removal of product and vapor from the evaporator. Use of antifoam agents can be of some value, but their choice is limited to those acceptable as food additives.

4. Corrosion

Some foods, especially fruit juices, contain components which are corrosive to heat-exchange surfaces. This damages the equipment and results in undesirable transfer of metals to the evaporated product.

5. Entrained Liquid

Separation of entrained liquid from the vapor stream is often difficult. Efficient separation is, however, important in order to avoid economically intolerable losses of solid matter.

6. Flavor

Desirability of food depends on various volatile flavor components which often are lost during vaporization of water. Design of evaporation processes is often directed either toward decreasing such losses or toward recovering volatile flavor components from the vapor phase.

7. Heat Sensitivity

Heat sensitivity is a problem of particular concern since it affects the quality of food. The operational factors, time and temperature, affect the degree of damage caused by heat. A "heat hazard index" has been discussed by Carter and Kraybill [3]. This index assumes the following relationship:

$$\text{"Dh"} = \log_{10} [(P)(t)] \tag{9-8}$$

where

Dh is decomposition hazard index
P is absolute pressure (microns)
t is time of residence (sec)

Carter and Kraybill [3] showed that evaporators have operating conditions corresponding to a "Dh" range of about 1 for molecular stills, which operate in high vacuum, to 12 or more for equipment operating at atmospheric pressure. For reference purposes, these authors indicated that preservation of vitamin D would require a "Dh" value below 1 but that petroleum oils could tolerate "Dh" values above 12.

Unfortunately, heat sensitivity of food is too complicated to be predicted adequately by an index as simple as "Dh." One of the complicating factors is the fact that time of residence is a mean value, and some liquid resides much longer than this mean. Mean residence times in various types of evaporators used for juice concentration are reported by Casimir and Kefford [4] to range from 0.11 min in a centrifugal evaporator to 54 min in a recirculating tube-and-shell evaporator. They have found, however, that the time to replace 97% of a tracer (that is, the minimum time required to remove all but 3% of the original liquid) is three to four times greater than the mean residence time. Furthermore, that portion of the liquid immediately adjacent to heat exchanger surfaces may be at a temperature significantly higher than the mean temperature (boiling point) of the liquid. As should be obvious from the hypothetical temperature gradients shown in Figure 9-3, heat damage at the wall (surface of the heat source) is more likely when the liquid

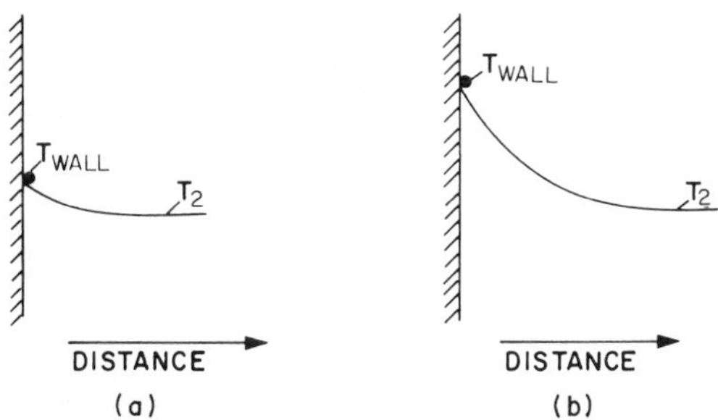

FIG. 9-3. Temperature gradients in evaporating liquids. (a) High h_2; (b) low h_2. h_2, liquid heat-transfer coefficient; T_{wall}, wall temperature; T_2, boiling temperature.

heat-transfer coefficient is low. This condition arises when viscosity of the liquid is high and/or there is inadequate motion of the liquid.

When average changes in food properties are the primary consideration, deviations from mean conditions may not be too important. It may, for instance, be possible to predict average nutrient losses or average increase in optical density caused by browning from an index based on mean conditions. In other situations, however, including production of trace compounds with high potency (very strong off flavors, toxins), such oversimplification may be dangerous.

D. Types of Evaporators

Evaporators used in the food industry may be classified in different ways: by operating pressure; vacuum versus atmospheric; by number of "effects"; single effect versus multiple effect[1]; by type of convection; natural versus forced; and by continuity of operation; batch versus continuous.

Natural convection evaporators can be operated either at atmospheric or at reduced pressure and on either a batch or a continuous basis. Examples are discussed below.

1. Unstirred Open Pan or Kettle Operating at Atmospheric Pressure

These simple batch evaporators are either steam jacketed or heated with closed steam coils and are used for concentration of sauces, jams, and confections when product temperatures in excess of 100° C can be tolerated.

2. Natural Circulation Tube Evaporators

A number of different designs using tubular heat exchangers can be used for continuous evaporation processes. Of these the "standard evaporator" shown in Figure 9-4 is used quite extensively. In this evaporator the solution boils inside vertical tubes with the steam condensing in a chest through which the tubes pass. The steam chest is doughnut shaped, allowing easy downward flow through the large central opening.

Circulation past the heating surface is induced by boiling in the tubes, each of which is usually 2-3 in. in diameter and 4-6 ft in length.

The "standard evaporator" is quite satisfactory for evaporation of relatively low-viscosity, noncorrosive liquids which are resistant to high temperatures. The overall heat-transfer coefficient U can be quite high under some conditions (100-500 Btu hr^{-1} ft^{-2} $°F^{-1}$).

---

[1] As is to be seen in Section II. F, it is often possible to conserve total expenditure of energy by utilizing the vapor from an evaporator to heat and boil the liquid in another evaporator, operating at a lower pressure. In this situation, each evaporator is called an effect and the process is called multiple-effect evaporation.

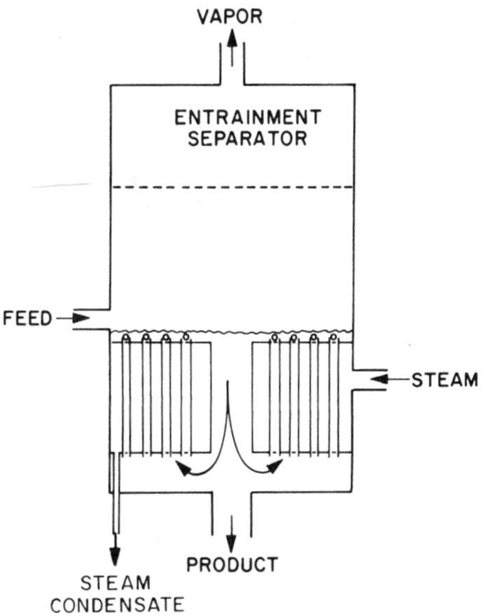

FIG. 9-4. Standard evaporator.

3. Long-tube Evaporators (Climbing Film Evaporators)

A very effective type of evaporator construction consists of a vertical shell and tube heat exchanger. The boiling liquid is inside vertical tubes (1-2 in. in diameter, 10-40 ft in length), and steam condenses on the outside of the tubes.

Good heat-transfer characteristics are imparted by formation of a thin liquid film on internal surfaces of the tubes, due to the lift provided by the vigorous upward flow of vapor through the center of the tube. Under proper operating conditions, high heat-transfer coefficients can be attained and foaming is prevented. Major problems are:

a. Potential for formation of bubbles and entrained liquid flow. (Figure 9-5 shows proper and faulty flow conditions inside tubes.)
b. Susceptibility to formation of deposits on the internal tube surfaces (fouling).

In applications involving viscous liquids it is usually necessary to resort to forced convection. Typical examples of such evaporators include tube-and-shell evaporators similar to those discussed above but utilizing pumps for circulation of the fluid. Forced-convection evaporators can be operated at atmospheric or reduced pressure and on a continuous or on a batch basis.

Of particular importance in the food industry, which often deals with viscous, heat-sensitive fluids, are the vacuum evaporators utilizing forced convection.

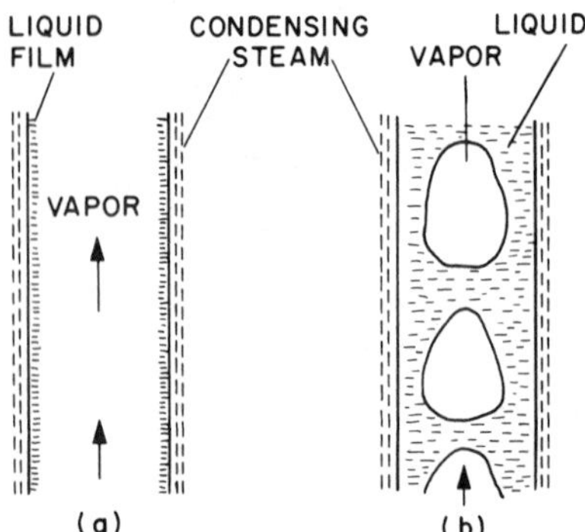

FIG. 9-5. Climbing film evaporator (a) Correct operation; (b) incorrect operation because of bubble formation and liquid entrainment.

4. Stirred Vacuum Kettle

A simple example of a forced-convection evaporator is the stirred vacuum kettle shown in Figure 9-6. This evaporator is used extensively in various batch applications.

For evaporation of liquids with high viscosities a modification of the simple stirred vacuum kettle has been developed by Western Utilization Research Division of the USDA. In this evaporator, called the Wurling evaporator, the heat exchanger, in the form of a steam coil, is rotated through the liquid. Good heat transfer has been reported in concentration of tomato paste to 20-50% solids [2].

5. Mechanically Aided Thin-film Vacuum Evaporators

Continuous processes for heat-sensitive materials often necessitate mechanically aided thin-film evaporators. In these evaporators mechanical devices are used to assure that the boiling liquid flows as a thin film at the heat exchanger surface. Equipment is usually operated at a much reduced pressure which allows high evaporation rates and low operating temperatures.

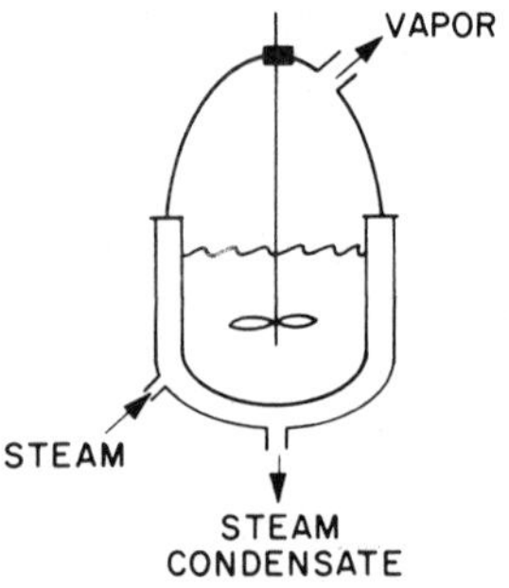

FIG. 9-6. Vacuum kettle.

Evaporators of this type include several in which a thin film of liquid is spread by an internal rotor. Rotors often have a fixed clearance of 0.01-0.03 in., fixed blades with adjustable clearance, or blades which are spring loaded and "wipe" the heat exchange surface. An example of this type of evaporator is the Rototherm E, shown in Figure 9-7. This evaporator operates at very low pressure, has a residence time of as little as 10-15 sec, and can handle fluids of very high viscosities. Heat is applied to the jacket. Centrifugal force holds the liquid on the

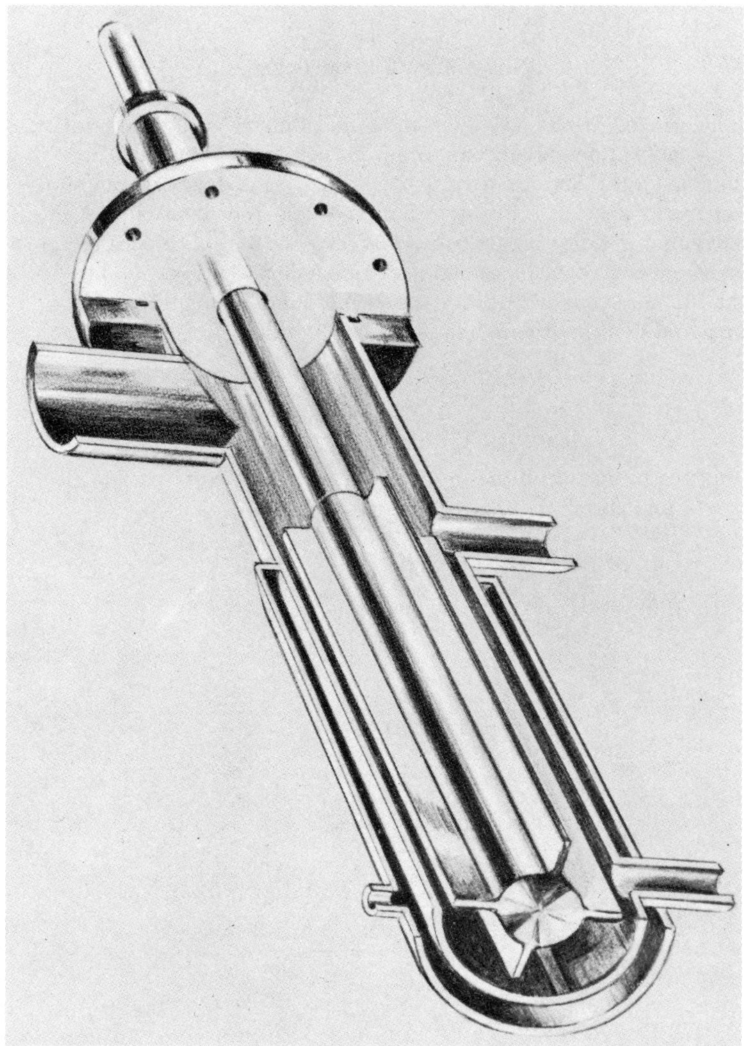

FIG. 9-7. Horizontal thin-film evaporator (Rototherm, courtesy of Artisan Industries, Inc., Waltham, Mass.).

heated wall and produces a turbulent film between the wall and the rotor blade tips. Vapor flow can be either co-current with or counter-current to the liquid.

Another example is the centrifugal evaporator. In this evaporator a film is spread centrifugally on a moving heated surface. A film thickness of 0.004 in. and contact times of 0.5-2 sec are possible for low-viscosity liquids. A diagrammatic representation of a centrifugal evaporator is shown in Figure 9-8.

Mechanical thin-film evaporators are used for heat-sensitive liquids, especially if their viscosities are high (over 1000-10,000 cps) or if there is a tendency to foam. A comparison of performance factors for several types of evaporators is given in Table 9-1.

E. Evaporation with Feed Preheating

It has been mentioned previously that steam economy is a major engineering consideration in evaporation, albeit one which in food processing may have to be subordinated to considerations affecting product quality. Presented below are several types of evaporator designs allowing efficient steam consumption.

First consider the following example. A feed of 10% solids at a temperature of 112°F is to be concentrated in a single-stage evaporator (Fig. 9-1) to 20% solids by boiling at 121°F. (For simplicity's sake, assume no significant rise in boiling point.) Using 100 lb of feed as a base, Eq. (9-3) yields:

$$G_2 = 100\left(1 - \frac{0.1}{0.2}\right) = 50 \text{ lb of vapor}$$

The heat required to evaporate 50 lb of water is calculated from Eq. (9-4), assuming $c_{p0} = 1$ and noting that λ_2 is 970:

$$\text{Total heat required (Btu)} = (L_0)(c_{p0})(T_2 - T_0) + G_2\lambda_2$$

$$= (100)(1)(100) + (50)(970) = 10,000 + 48,500 = 58,500$$

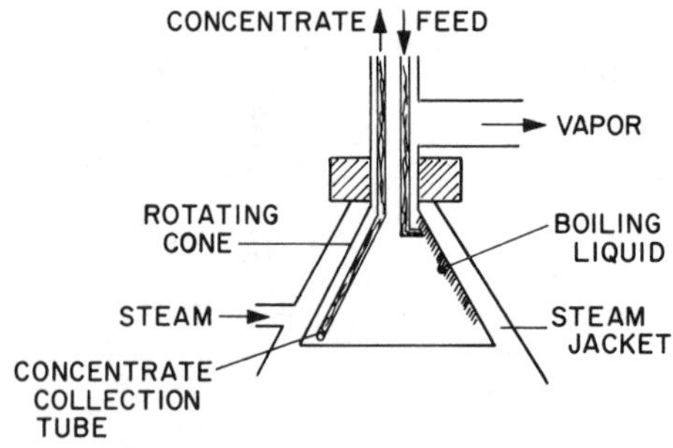

FIG. 9-8. Centrifugal evaporator.

TABLE 9-1

Comparison of Several Types of Evaporators [a]

	Overall heat-transfer coefficient ($Btu\ ft^{-2}\ {}^{\circ}F^{-1}\ hr^{-1}$)		Dh factor	Equipment cost ($\$/ft^2$) for 100 ft^2 size
	Los viscosity	High viscosity		
Standard	30-300	Not suitable	8-10	50
Climbing film	100-400	Below 50	7-8	100
Forced circulation tube and shell (recirculating)	$\simeq$ 500	100	6-8	200
Mechanically aided thin-film evaporator	$\simeq$ 500	300	3-6	500

[a] After Carter and Kraybill [3].

It should be noted that of the 58,500 Btu supplied to the 100 lb of liquid, 48,500 Btu goes into raising the energy of the 50 lb of vapor above that of liquid at the same temperature. That energy cannot be used for evaporation of more liquid in a single-stage evaporator because the vapor is at the same temperature as the boiling liquid and heat transfer cannot be accomplished without a temperature difference. There are, however, several ways in which this energy can be recovered.

In a single-stage evaporator the vapor (G_2) can be used for preheating the entering feed, thus reducing the amount of steam required for evaporation. A schematic diagram of this arrangement is shown in Figure 9-9.

F. Multiple-effect Evaporators

The energy contained in the vapor discharged from an evaporator can be partially recovered by using the vapor as a heat source in another evaporator (effect), operating at a lower pressure (lower boiling point). This is the principle of multiple-effect systems. Three different arrangements are shown in Figure 9-10. Consider the "forward feed" arrangement. Three effects are shown in which boiling occurs at pressures P_2, P_3, and P_4 (where $P_2 > P_3 > P_4$), corresponding to temperatures T_2, T_3, and T_4 (where $T_2 > T_3 > T_4$). Calculations are often greatly simplified by making the following assumptions, which allow a reasonable approximation of the overall operation of multiple-effect evaporators:

1. Feed (L_0) enters at temperature T_2.
2. There are no heat losses to surroundings.
3. Latent heats are independent of temperature ($\lambda_1 \simeq \lambda_2 \simeq \lambda_3 \simeq \lambda_4$).
4. Areas of heat exchangers are the same in all effects.
5. Overall heat-transfer coefficients are the same in all effects.

Natural consequences of the above assumptions are: (1) the total amount of heat transferred and amount of water vaporized is the same in each effect and (2) the number of pounds of water vaporized per pound of steam supplied at the first effect (G_1) is equal to the number of effects. In actual operation the above assumptions are not true and trial-and-error methods must be used to establish the actual operating conditions accurately. The steam economy achievable under real operating conditions is less than that calculated from the ideal assumptions, as shown in Table 9-2.

Different arrangements with respect to direction of feed are possible: cocurrent or forward feed, counter-current or backward feed, and parallel feed. Of these, parallel feed is not common in evaporators but is often used in crystallizers, where precipitation of solids makes continuous flow of feed from one effect to another impractical.

From an engineering point of view, forward feed has the advantage of having the liquid flow with the pressure gradient (the first effect is at the highest pressure). Costs of pumping equipment are minimized and control is less complicated than in backward feed evaporators. Backward feed has two important engineering advantages: (1) when the feed comes in relatively cold, steam consumption is more economical. (2) Viscosity of liquids decreases with increasing temperature but increases with concentration of solids. In backward feed, the tendency to high viscosity in the concentrated liquid is somewhat compensated for by its high temperature. Heat transfer and flow characteristics are thereby improved.

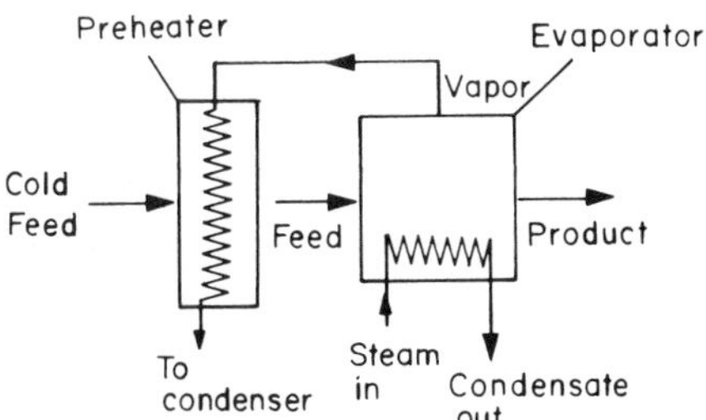

FIG. 9-9. Schematic representation of an evaporating system with a preheater.

The two types of feed flow, however, are also different with respect to their effect on chemical changes in food quality. Many chemical reactions, notably nonenzymatic browning, proceed most rapidly in the range of water contents corresponding to those found in the product (outlet) stream. In backward feed, this

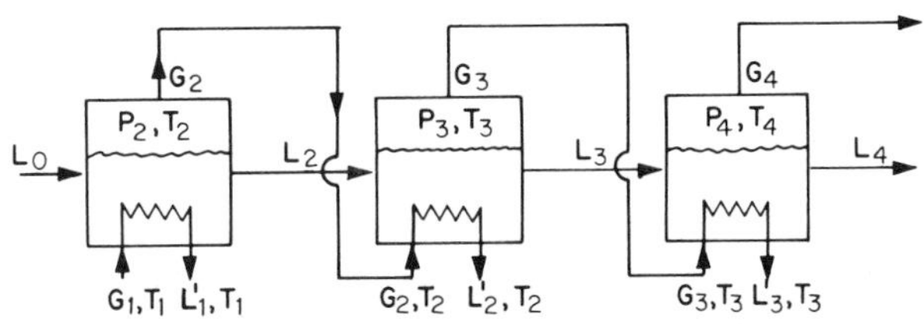

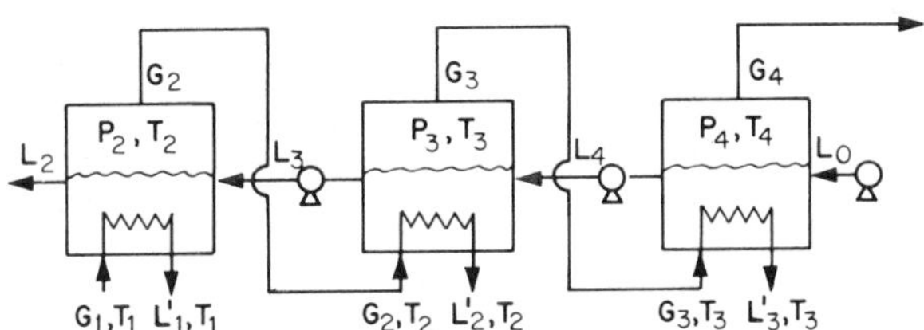

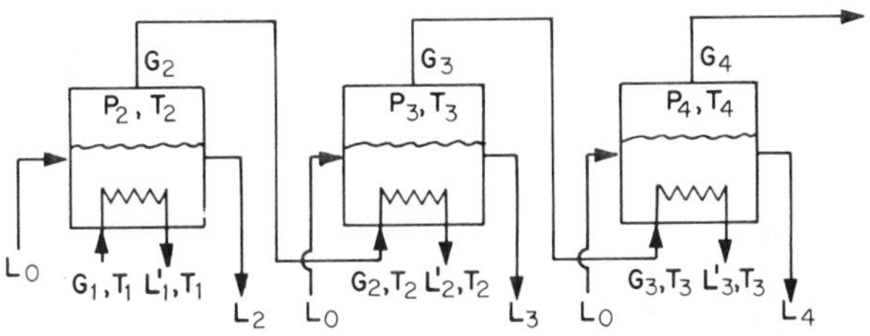

FIG. 9-10. Multiple-effect evaporators with different flow arrangements.

TABLE 9-2

Steam Economy of Multiple-effect Evaporators [a]

Number of effects	Theoretical: lb of water evaporated	Practical: lb of steam to primary effect
One	1.0	0.95
Two	2.0	1.7
Four	4.0	3.2

[a] After Slade [25].

stream is also at the highest temperature and chemical changes may be unacceptably high in this arrangement. Similarly, fouling due to precipitation of denatured proteins may be promoted by the coincidence of high concentration and high temperature in backward feed.

G. Evaporators with Vapor Recompression

Another way of obtaining improved steam economy is by vapor recompression. We may consider the rationale of this method with the following example. Consider an atmospheric single-stage evaporator (Fig. 9-1) with feed (L_0) entering at 212°F and boiling at the same temperature. (There is no boiling point rise.) Let the heat supply be provided by steam at a pressure of 24 psia. The steam (G_1) enters the heat exchanger as saturated vapor and leaves as saturated liquid (L'_1), both at 238° F. The process of vaporization of L_0 to G_2, and of G_1 to L'_1, are shown on a temperature-entropy diagram in Figure 9-11. By consulting the steam table we find that the steam, condensing at 238° F, gives up 953 Btu/lb. The amount of heat required per pound of vaporized liquid at 212° F is 970 Btu. Thus 1 lb of condensing steam vaporizes 0.985 lb of liquid.

In order to obtain 1 lb of steam at 238°F, the distillate L'_1 can be returned to the boiler, in which case 953 Btu must be supplied per pound (Fig. 9-12a). Another possibility is to obtain 0.985 lb by compressing G_2 from 15 psia (atmospheric pressure) to 24 psia and 238° F and then to obtain the remaining 0.015 lb by boiling makeup water (Fig. 9-12b). Heat required is then only 10 Btu/lb of vapor compressed plus the amount to heat and vaporize the makeup water, for a total of only 28 Btu. A representation of the system in the form of a process path line on a temperature-entropy diagram is given in Figure 9-11.

From the point of view of the thermodynamic cost of vaporization, therefore, it appears that recompression is quite attractive, even compared to multi-effect evaporation. Unfortunately, there are major drawbacks and, in particular, the volume of vapor to be compressed presents difficulties. If the operation is conducted at atmospheric pressure, each pound of vapor (at 212°F) to be compressed

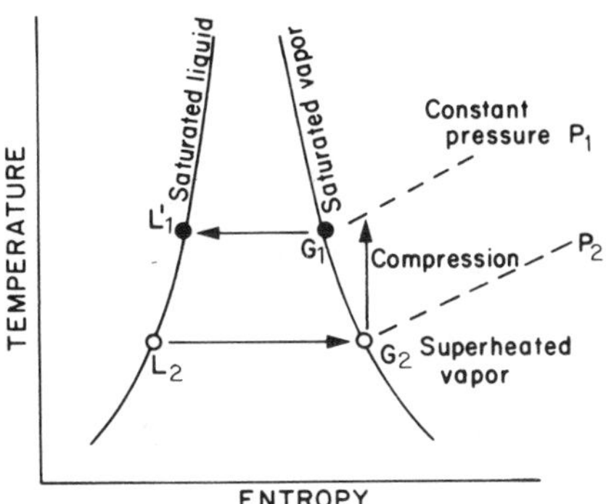

FIG. 9-11. Temperature-entropy diagram representing processes in evaporation and vapor compression.

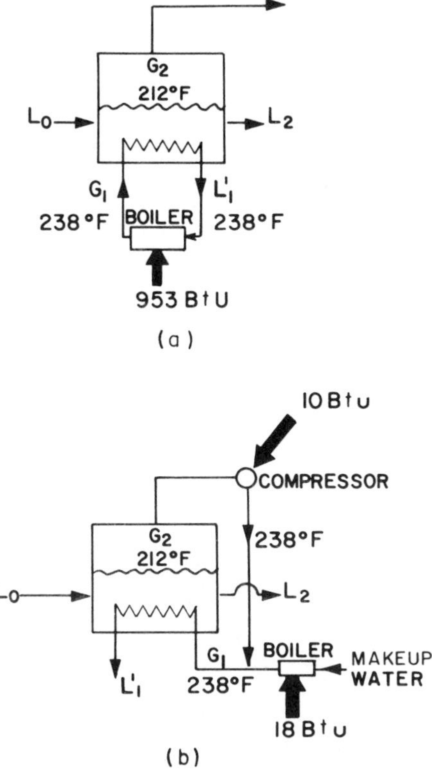

FIG. 9-12. Schematic diagram of two arrangements for regeneration of high-pressure steam needed for evaporator operation. (a) Steam condensate reevaporated in boiler; (b) vapor compression system.

occupies 27 cu ft. The volume to be compressed is, of course, even larger when heat sensitivity of the product requires operation under vacuum. If, for instance, evaporation is carried out at 27 in. Hg vacuum, corresponding to a boiling point of water of 115°F, then 1 lb of vapor occupies 225 cu ft. If constant displacement compression equipment is to be used under these conditions, the cost becomes prohibitive.

It is feasible, at least in the case of a reasonably high evaporation temperature, to use very high-pressure steam and combine it with part of the vapor to produce the steam required for the heat transfer. For instance, in the example shown in Figure 9-13, 1 lb of 140 psia steam is mixed with 0.75 lb of vapor at 15 psia to produce 1.75 lb of steam at 25 psia. In this example, over 40% of the energy potentially lost in the vapor is recovered. This method for improving steam economy requires very high-pressure steam and becomes feasible only in special circumstances.

The volume of gas to be compressed can be greatly reduced by transferring the energy in the water vapor to another gas. This can be done by using a heat pump cycle with a fluid, such as ammonia.

The operation of such a cycle is shown schematically in Figure 9-14. Feed (L_0) enters the evaporator and is partially vaporized at temperature T_2, the heat of vaporization being supplied by ammonia condensing in a heat exchanger in contact with the evaporator surface. The resultant water vapor (G_2) condenses in a heat exchanger supplying the heat for boiling ammonia. Ammonia gas is compressed (E) to a temperature high enough for heat transfer to the boiling aqueous liquid, and the hot ammonia liquid is cooled by expansion at F.

The advantages of this system may be noted from the following typical operating conditions. Feed enters at 80°F and is evaporated at this temperature in a vacuum of 27 in. Hg. The resultant vapor (G_2), in order to provide adequate heat transfer, must be heated to a higher temperature. Assuming that 100° F is a suitable temperature, this requires compression of 633 cu ft/lb to a volume of 350 cu ft/lb; if the latent heat carried per pound of vapor is considered, the volume involved is:

$$\frac{633 \text{ cu ft/lb}}{1050 \text{ Btu/lb}} \simeq 0.6 \text{ cu ft/Btu}$$

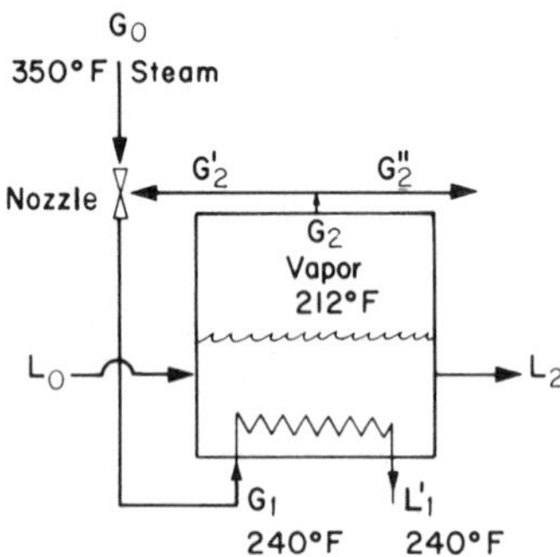

FIG. 9-13. Vapor compression using a nozzle to compress low-pressure vapor with high-pressure steam.

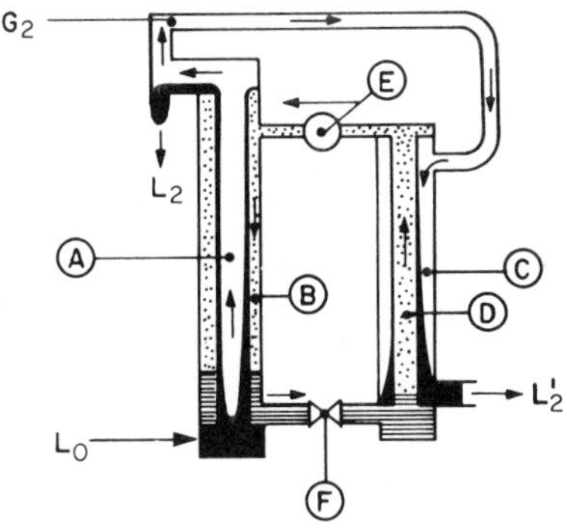

L_0 FEED
L_2 CONCENTRATE
G_2 VAPOR
D_2 CONDENSED VAPOR
A EVAPORATOR AT T_2, P_2
B AMMONIA CONDENSING AT T_B, P_B
C WATER VAPOR CONDENSER AT T_C, P_2
D AMMONIA BOILING AT T_D, P_D
E AMMONIA COMPRESSOR (VAPOR COMPRESSED FROM P_D TO P_B)
F AMMONIA EXPANSION VALVE (LIQUID COOLED FROM T_B TO T_D)

T TEMPERATURE
P PRESSURE

FIG. 9-14. Heat pump cycle using ammonia for vapor recompression.

If, on the other hand, ammonia is used as shown in Figure 9-14, the following operating conditions can be used:

1. Evaporator (A): temperature, 80° F; pressure, 0.5 psia
2. Ammonia condenser (B): temperature, 100° F; pressure, 211 psia
3. Vapor condenser (C): temperature, 100° F, pressure, 0.5 psia
4. Ammonia evaporator (D); temperature, 80° F; pressure, 153 psia

Ammonia gas at 80° F and 153 psia occupies only 1.95 cu ft/lb, and even though it carries less latent energy per pound than water the volume reduction of this system is still enormous. The volume per unit energy content is:

$$\frac{1.95 \text{ cu ft/lb}}{500 \text{ Btu/lb}} \simeq 0.0039 \text{ cu ft/Btu}$$

The disadvantage of the system lies mainly in the capital cost of an ammonia compression system. In the food industry, these systems have been used in citrus juice evaporation and in some other applications in which refrigeration systems involving ammonia are needed in addition to vapor recompression.

H. Principles of Operation of Equipment Used in Evaporators

From an operational point of view the equipment used in evaporators performs the following operations: (1) flow of liquids (pumping of feed and concentrate streams, steam condensate, makeup water); (2) heat transfer (heat exchangers in evaporators, condensers, preheaters); (3) gas and vapor removal (pumps for maintenance of vacuum in evaporators); and (4) gas-liquid separation by mechanical means (entrainment separators for removal of droplets in vapor phase). Of these, vacuum pumps and entrainment separators deserve further discussion.

1. Vacuum Pumps

Removal of vapor and noncondensables (air) from evaporators can be accomplished by one of the following types of pumps:

a. Positive displacement, oil-sealed mechanical pumps for air removal used in conjunction with condensers for removal of water vapor. This type of equipment is suitable for pressures between 1 and 50 torr absolute.
b. Water-sealed mechanical pumps. The degree of vacuum which can be attained with these pumps depends on water temperature. Vacuum attainable in this type of equipment, represented by the Nash pump, is 20-27 in. Hg. (absolute pressure, 75 torr and up).
c. Steam ejectors. These are most useful in applications requiring removal of large quantities of vapor at absolute pressures of 20-100 torr.

Diffusion pumps backed by condensers and by mechanical pumps are used in applications requiring extremely low absolute pressures (1-100 μm Hg or lower). Evaporators operating at these high vacuums are called "molecular stills."

Steam ejectors are very common in the food industry and they can be used not only for producing a vacuum, but also for heating, cooking, mixing, and compressing various fluids. The operational principle is shown schematically in Figure 9-15. A quantity of high-pressure steam (G_1, lb/hr) at pressure P_1 is injected

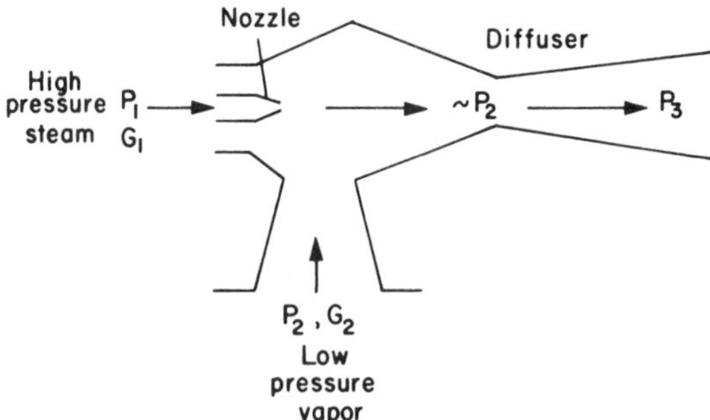

FIG. 9-15. Steam ejector.

into a convergent-divergent nozzle. It draws into the nozzle a quantity (G_2) of low-pressure vapor at pressure P_2. It is usually assumed that mixing of the two streams results in pressure P_2 being attained by the combined stream of fluids ($G_1 + G_2$) at the throat of the nozzle, where the fluids attain a high velocity. In the divergent portion of the nozzle, the kinetic energy is converted in part to energy needed to increase the pressure of the stream from P_2 to P_3. Thus high-energy steam is used to pump and compress low-pressure vapor to higher pressures. Ejectors have in common with Nash pumps the capability of operating without water-vapor condensers. Mechanical pumps sealed with oil, on the other hand, are generally incapable of handling significant water vapor loads and require the use of water-cooled or refrigerated condensers.

Mechanical pumps operate at essentially constant volumetric speed over a range of pressures. Their pumping speed begins to fall off as the pressure in the system being pumped approaches the minimum or "blank-off" pressure which can be attained by the pump. Typical performance curves for single-stage and two-stage pumps are shown in Figure 9-16.

Most pumps are available with a device designed to minimize condensation of vapors in the pump. This device, called a "gas ballast," consists of a valve admitting atmospheric air during the compression stroke to assist in removal of water vapor. Use of gas ballast, however, carries with it the penalty of decreasing the ultimate vacuum which can be attained by the pump. Figure 9-16 shows schematic curves for a single-stage pump operating with the gas ballast valve open or closed.

It should be noted that in the range of pressures at which the volumetric speed of the pump remains essentially constant, the gravimetric speed decreases linearly with decreasing pressure. This is, of course, a direct consequence of the ideal gas law:

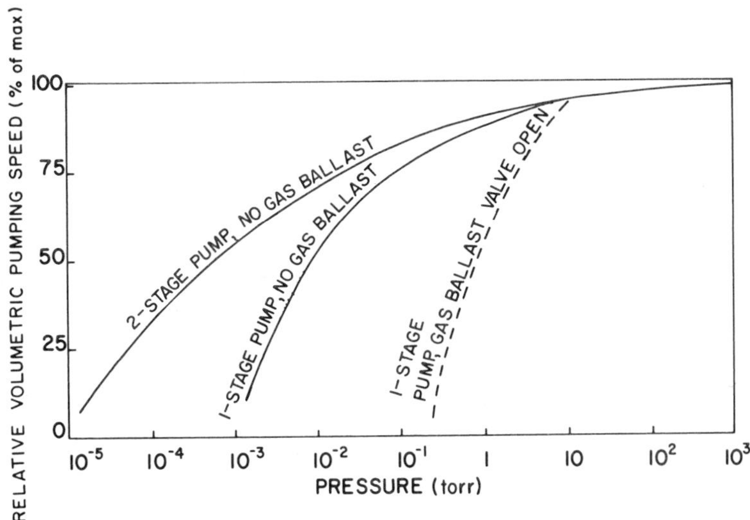

FIG. 9-16. Dependence of volumetric pumping capacity of mechanical vacuum pumps on pressure.

$$PV = nRT \tag{9-9}$$

where

P is pressure
V is volume
n is number of moles
R is the gas constant
T is absolute temperature

Therefore, if the volumetric speed is constant,

$$\frac{dV}{dt} = Sp \tag{9-10}$$

then the molar pumping speed (dn/dt) is:

$$\frac{dn}{dt} = Sp \left(\frac{RT}{P}\right) \tag{9-11}$$

where Sp is the volumetric pump speed, and the gravimetric pumping speed is:

$$G = Sp\ (RT/P)\ (MW) \tag{9-12}$$

where

G is the gravimetric flow of gas (for instance, lb/hr)
MW is the molecular weight of gas

As a consequence, the time to reduce pressure in a closed system from P_1 to P_2, given a constant pump speed Sp, may be calculated from Eq. (9-13):

$$\Delta t = \frac{V_s}{Sp} \ln \frac{P_1}{P_2} \tag{9-13}$$

where V_s is the volume of the closed system.

In practice, however, the time is often longer because Sp drops as the pressure drops and because there is often additional load on the pump due to outgassing of walls, water vapor infiltration from condensers, and system leaks. An empirical equation using a correction factor K_{corr} is often applied:

$$\Delta t = (K_{corr}) \frac{V_s}{Sp} \ln \frac{P_1}{P_2} \tag{9-14}$$

K_{corr} is close to unity at pressures above 10 torr, but increases to 4 or more at pressures between 0.1 and 10 torr. Of course, very much greater corrections are required when condensers do not function properly or in the presence of major leaks.

2. Entrainment Separators

Entrainment separators are devices used in evaporators to avoid incorporation of liquid droplets in the discharging vapor stream. They involve the principle of "momentum separation": a sudden change of velocity of the stream, produced by impingement on some obstacle to flow, causes denser droplets to settle out. Some types of entrainment separators are shown in Figure 9-17.

The food industry often uses combinations of the types of equipment discussed previously. For a more detailed discussion of equipment and practices in evaporation of food products, the reader is referred to Slade [25], Farrall [9], and Armerding [1]. Engineering calculations involved in evaporation are well discussed by Loncin [15], Peters [23], and Coulson and Richardson [6].

III. FREEZE CONCENTRATION

A. Introduction

Evaporation is the most economical and most widely used method of concentration. It has, however, two important limitations: (1) loss of volatile components other than water (flavors) and (2) potential for heat damage to product quality. Freeze

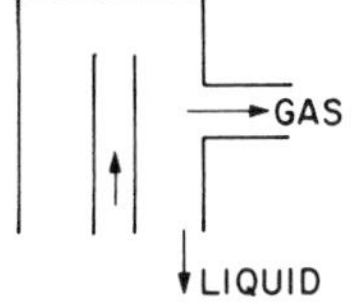

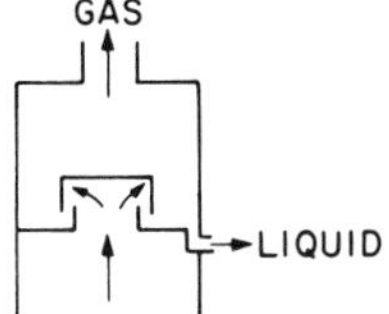

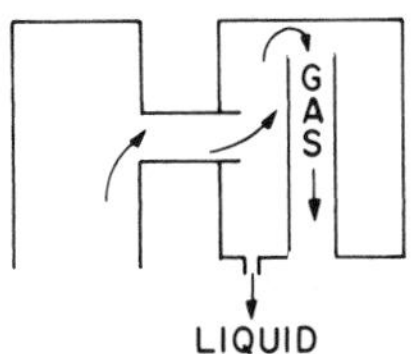

FIG. 9-17. Entrainment separators.

concentration is a process which can sometimes be used to overcome these limitations. This process involves partial freezing of the product and removal of the pure ice crystals, thus leaving behind all of the nonaqueous constituents in a diminished quantity of water. The major limitations of the process include relatively high cost, difficulties in effective separation of ice crystals without loss of food solids, and a relatively low upper limit on total solid concentration in the product.

Freeze concentration has been applied primarily where quality considerations and, in particular, the need for retention of organic volatiles are the dominant considerations, as in concentration of wines and beer (where alcohol and flavors are to be preserved) and concentration of coffee prior to freeze drying (where flavor retention is an important goal).

B. Principle of the Process

1. Equilibrium Considerations

Separation by freezing is based on solid-liquid phase equilibrium. The solution to be concentrated contains the solvent, water, and a large number of soluble components. It is, however, less complicated and usually adequate to consider the system as pseudo-binary, that is, considering all substances dissolved in the water as one component.

The simplest phase diagram of a binary mixture is shown in Figure 9-18. If a binary mixture is cooled under conditions allowing equilibrium to be attained, then pure ice crystals separate out at a point which corresponds to composition w_A and the freezing point of this solution ($T_{A'}$). Further cooling results in more of the

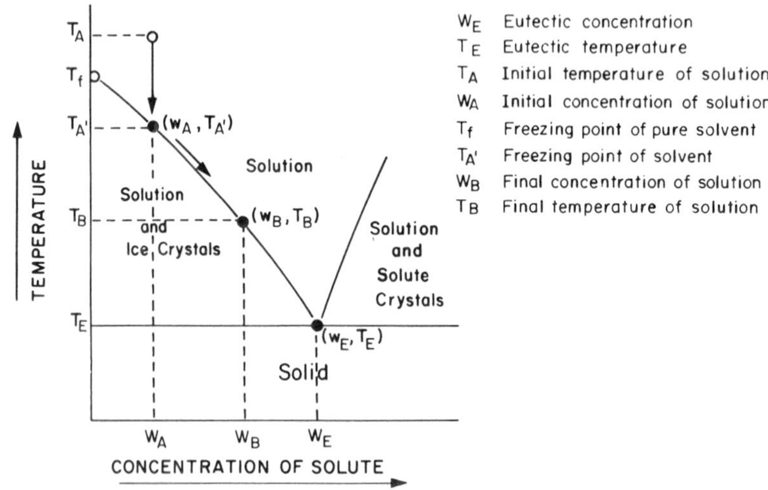

FIG. 9-18. Phase diagram for a simple binary system.

ice crystals separating out, while liquid composition follows the line [$(w_{A'} T_{A'})$ $(w_{B'} T_B)$]. At w_E the crystallizing solid has the same composition as the supernatant liquid.

It is obvious that the concentration process can only be applied up to concentrations below that of the eutectic; and in most cases, because of problems connected with separating ice from a very viscous liquid (both because of concentration and of low temperature), it must stop well below the eutectic. In the example shown in Figure 9-18, the final concentration is w_B. The quantity of water to be removed by freeze concentration can be obtained by a simple mass balance, shown in Figure 9-19.

$$L_A = L_B + L_I \tag{9-15}$$

$$L_A w_A = L_B w_B + L_I w_I \tag{9-16}$$

If we assume that the separation is perfect and no solute is entrapped with the ice stream L_I, then the water removed from the product stream is:

$$L_A w_A - L_B w_B = L_I \tag{9-17}$$

and per pound of feed, the amount of water removed is:

$$w_e \equiv \left(\frac{L_I}{L_A}\right) = 1 - \frac{w_A}{w_B} \tag{9-18}$$

When the amount of solute entrapped in ice is significant, the following equation applies:

$$w_e = \frac{w_B - w_A}{w_B - w_I} \tag{9-19}$$

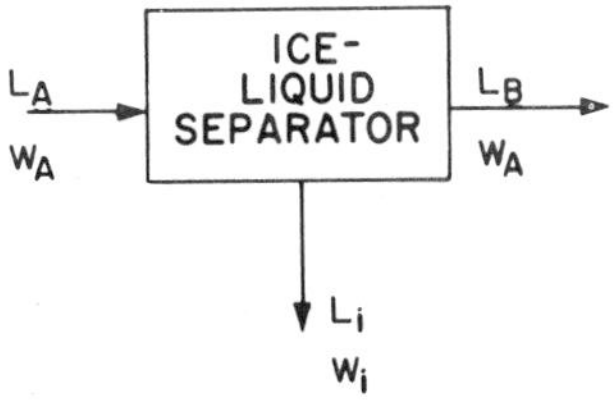

FIG. 9-19. Schematic representation of an ice-liquid separator. L, gravimetric flow (lb/hr); w, lbs solute/lb water (or ice). Subscripts: A, feed; B, concentrate; i, ice.

The above discussion was based on a binary system. In real foods, of course, there are many components. If they all remain in solution, the considerations are identical to the binary process since, from equilibrium considerations, only freezing point depressions need to be considered and eutectic concentrations are never reached.

Initial solid concentrations of food liquids typically range from 5 to 20%, and final concentrations of 35-50% solids are achievable by freeze concentration [22]. The equilibrium freezing temperatures [T_B when referring to the schematic diagram (Fig. 9-18)] corresponding to 40% solids are 24°F for coffee extract, 20°F for orange juice, and 18°F for apple juice. When high concentrations of low molecular weight solutes are present, T_B drops more rapidly with concentration. For instance, at 18° F, the unfrozen phase of an alcohol solution contains 22% alcohol.

2. Rate Considerations

The rate of crystal growth and the average crystal size depend on a number of operational factors, including in particular (a) the rate of nucleation, which in turn is strongly dependent on temperature and composition; (b) the rate of diffusion of water molecules to the surfaces of growing crystals; and (c) the rate of removal of the heat of fusion from the crystal surfaces.

In freeze concentration it is desirable to grow large, symmetrical crystals at a relatively slow rate. Large crystals are easier to separate by mechanical means, and large crystals have fewer points of contact per unit volume, thus minimizing the potential for entrapping liquid and solute. Symmetry of crystals also minimizes entrainment. Good mass-transfer conditions, usually due to turbulent flow in the crystallizers, combined with large heat-transfer areas at relatively high temperatures are most conducive to the type of crystal growth which is desired. Equipment design follows these principles, within limits imposed by economic factors.

Symmetry of crystals is also improved by partial melting and refreezing prior to separation, which is occasionally practiced in freeze-concentration systems.

C. Freeze Concentration and Freeze Desalination

One of the widely discussed applications of freeze separation is desalination. Freeze concentration and desalination are based on the same general principles but differ primarily in their products. In desalination the product is the pure ice crystal stream; in freeze concentration, it is the liquid from which the ice crystals are separated. Entrainment of solute in the ice stream is undesirable in either case. In desalination, however, it is feasible, at least in principle, to obtain high purity of the ice stream by maintaining the degree of concentration at a low level. Since a high degree of concentration is needed in the freeze concentration of foods, this course of action is not feasible in food applications. An additional difference lies in the high viscosity at moderate concentration due to polymers and dispersed particles present in food liquids such as juices. Finally, freezing by direct injection of refrigerant is often more feasible in desalination than in food concentration, because refrigerant removal from the product stream is difficult in the case of foods.

D. Principles of Equipment Used in Freeze Concentration

A freeze-concentration system contains as a minimum the following components: (1) a crystallizer or freezer in which the crystals are produced; and (2) a separator for separation of the crystals and the liquid concentrate. Various additional components may be added to improve operational efficiency. Examples are shown in Figure 9-20, which shows a hypothetical two-stage process utilizing centrifuges for ice separation, and in Figure 9-21, which represents a pulsating unit developed by Phillips Petroleum Company for concentrating beer.

1. Crystallizer

It has been mentioned in Section III. B. 2 that the rate of growth of crystals and the size and uniformity of ice crystals depend on heat- and mass-transfer conditions existing during crystallization. Very large uniform crystals require a very large exchange surface and a high degree of turbulence. When these are provided, crystals as large as 3 mm in diameter can be grown continuously under laboratory conditions. In commercial equipment, cost considerations preclude growing crystals larger than 1/4 to 1/2 mm. This size, however, is usually adequate to achieve separation.

2. Separators

Centrifuges are most commonly used for separation. They provide forces up to 1000 x g, which forces even viscous liquid off crystal surfaces. There remains a layer held by surface forces, however, and this layer may represent a substantial loss of valuable solids or alcohol. One approach to minimizing this loss is the washing of ice crystals prior to their discharge to the melter. The spent wash water is of course returned to the crystallizer if solids are to be recovered (Fig. 9-20).

Filter beds may be used instead of centrifuges and this method is utilized in a design developed by Phillips Petroleum Company (Fig. 9-21).

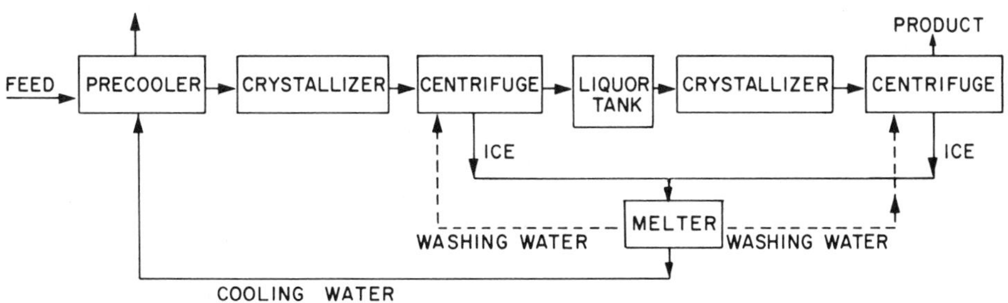

FIG. 9-20. Two-stage freeze-concentration system.

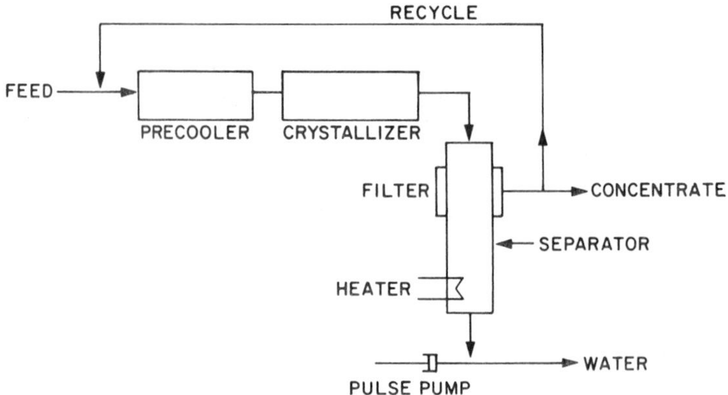

FIG. 9-21. Freeze-concentration system using a filter bed.

E. Normal Freezing

The freeze-concentration methods referred to previously involved the pumping of a liquid-ice crystal mixture in the form of a slush from a crystallizer to a separator where the crystals are removed. The separation often presents difficulties and in small-scale applications where batch operation is feasible these difficulties can be overcome by the use of normal freezing. In this method, a solid ice front is produced at the wall of a crystallizing vessel; this wall progresses slowly, thus directing the flow of concentrated liquid away from the wall toward a location where it is drawn off. Three possible arrangements are shown in Figure 9-22. In Figure 9-22a is shown a tube with stirred liquid which is being immersed gradually into a freezing bath. The ice layer grows from the bottom toward the top of the tube. Concentrated solution may be withdrawn from the top. In Figure 9-22b, a "cold finger" condenser is inserted into a stirred bath containing the solution which is to be concentrated. The ice layer progresses from the outside wall of the condenser. In Figure 9-22c, a stirred bath of liquid is surrounded by a cooling jacket, with the ice growing from the walls toward the center of the bath; concentrate is eventually removed from the center.

Applications of normal freezing are quite limited. For reviews of this method the reader is referred to Kepner et al. [13] and Gouw [11].

F. Problems Caused by Precipitation of Solids Other than Ice

In some food liquids, including tea infusions, the solids are only sparsely soluble at freezing temperatures. As a consequence, some soluble compounds, including polymers, precipitate in the crystallizer and are lost with the ice phase. This precipitate can also interfere with efficient operation of crystallizers and separators. To overcome this problem in freeze concentration of tea, a design shown in Figure 9-23 has been patented by Struthers Scientific and International Corporation [26]. In this design, the tea extract is cooled and then passed through a settling tank, where some of the solutes precipitate. The extract continues on to the crystallizer

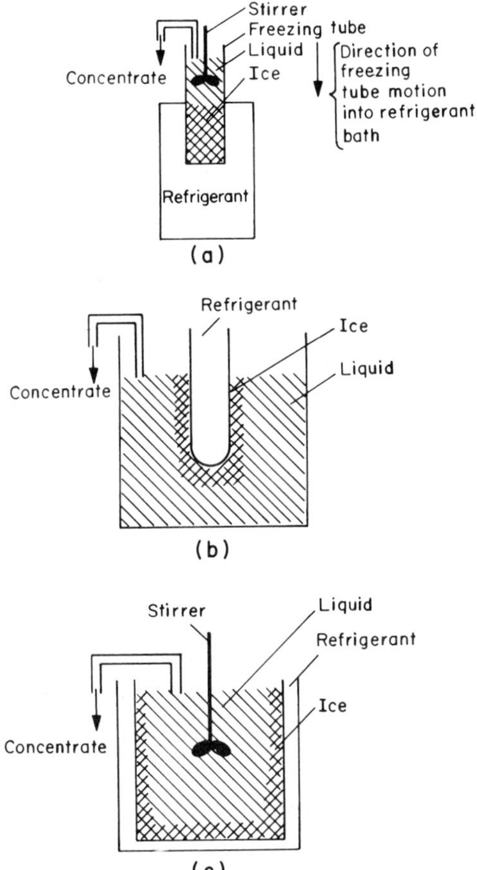

FIG. 9-22. Schematic representation of systems for normal freezing, discussed in Section IV. E.

and separator, where some of the water is removed as ice. The precipitate is redissolved in a small quantity of water and is then returned to the concentrate. The particular patent referred to here pertains to preconcentration of tea for subsequent freeze drying but can be applied to coffee as well.

A patent issued to Ganiaris [10] of the same company refers to problems with precipitation of coffee solids. The patented design overcomes the problem of solute precipitation by minimizing residence time in the crystallizer.

G. Concentration by Gas Hydrate Formation

Gas hydrates are icelike compounds, consisting of small "guest" molecules which are physically entrapped in hydrogen-bonded cages of water molecules (the "host"). Many of the fluids of consequence in food applications are capable of hydrate

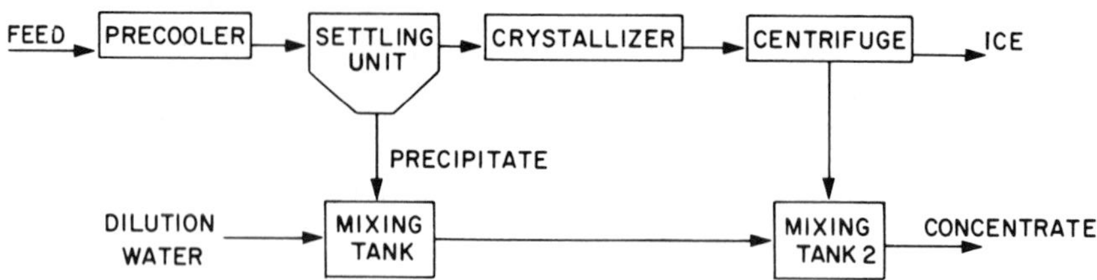

FIG. 9-23. System for precipitation of a portion of suspended solids prior to freeze concentration.

formation in the presence of various gases, such as halogenated, short-chain hydrocarbons. Many such hydrates are capable of existing at temperatures well above 0° C provided pressures are maintained at an elevated level.

Application of the above principle to freeze concentration (or more appropriately concentration by gas hydrate formation) was reported by Huang et al. [12]. They showed that fruit and vegetable juices could be concentrated by formation of CH_3Br and CCl_3F hydrates.

Werezak [30] removed water from various substances, including coffee extract and solutions of sucrose and sodium chloride, by forming hydrates of fluorocarbons, ethylene oxide, methyl chloride, and sulfur dioxide. The author described a process for concentrating coffee using ethylene oxide as the hydrate former. Much more research is needed to determine whether this process can be made mechanically and economically feasible.

H. Food Applications

Food applications of freeze concentration are still very limited. In a recent review article, Muller [22] reported the following applications:

1. Concentration of wine, in order to bring alcohol content to the legal limit, is practiced in France. (Addition of pure alcohol, which would seem the less difficult solution, is forbidden by law.)
2. A process has been developed for concentration of beer which is reported to be commercially utilizable.
3. Pilot scale and mobile units have been tested for use in concentration of apple juice, coffee, skimmed milk, and orange juice.
4. An operation for concentration of orange juice was in operation in Florida several years ago. It is no longer in existence.
5. Several commercial installations are reported in operation for concentrating vinegar, the largest having an ice removal capacity of 2000 lb/hr.

Recent interest has been centered in particular on concentration of coffee [14].

IV. MEMBRANE PROCESSES FOR CONCENTRATION

A. Introduction

Evaporation is the most economical and most common concentration process. Its limitations are due to undesirable effects of high temperature or to losses of volatile components when operating temperatures are lowered through use of vacuum. Freeze concentration is conducted at low temperature and is thus more likely to result in high-quality products. The mechanical separation of the ice, however, is a difficult and expensive process if losses of solids are to be avoided.

Another approach to production of high-quality products is the use of membrane processes. The principle involved is the interposition of a membrane between the feed stream and a "waste" or transfer stream, and the establishment of conditions providing a driving force for transport of water across the membrane from the "feed" to the "transfer" stream. The rate of transport, of course, is dependent on the driving force as well as on resistance, which depends on membrane properties and on conditions in the fluid streams on each side of the membrane. The possibility of obtaining high-quality products depends on the use of semipermeable membranes, which means membranes with much less resistance to water transport than to transport of components which are to be retained.

B. Driving Forces for Membrane Processes

When a membrane separates two regions, the following driving forces can be used to cause flow of a molecular or ionic species:

1. concentration difference
2. pressure difference
3. difference in electric potential
4. temperature difference

The different membrane processes resulting from the application of these driving forces are listed in Table 9-3. Figure 9-24 shows schematically the principles involved in dialysis, osmosis, thermoosmosis, and reverse osmosis. Osmosis and dialysis are based on concentration differences. In thermoosmosis high temperature is used to increase the partial pressure of solvent in the feed stream to create a driving force for transport. In reverse osmosis, hydraulic pressure is applied to the feed stream to counteract and overcome the osmotic driving force, hence "reverse" osmosis.

Ionized species, such as salts, can be used in separations utilizing electric current. The processes utilize membranes permeable only to ions which are positively charged (cation membranes) or only to ions negatively charged (anion membranes). When electric potential exists across the membrane, ionic flow occurs. This process can be used to reduce the ion content of solutions and is called "electrodialysis." In concentration, use is made of the fact that ions transported across membranes carry water of hydration. The process may therefore be used for removal of water, that is, for concentration. It is then called "electroosmosis."

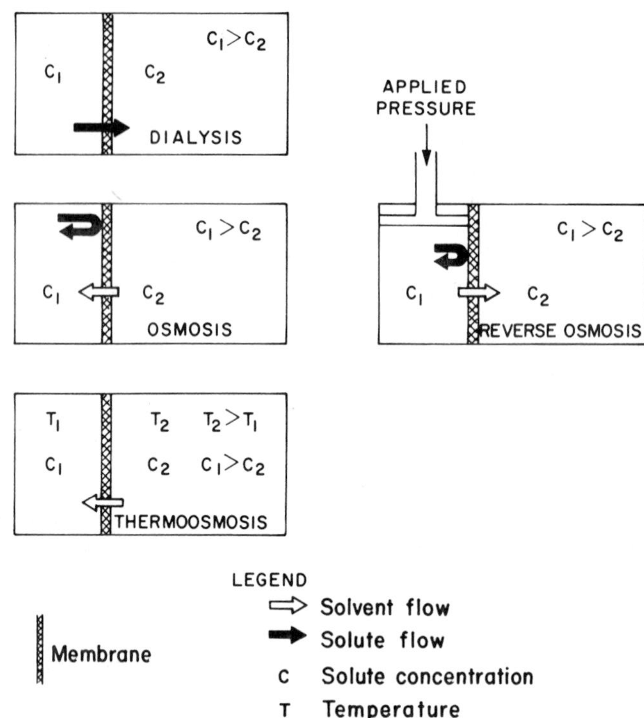

FIG. 9-24. Schematic representation of some membrane processes.

In Figure 9-25 are shown two arrangements of membranes for electroosmotic concentration of fluid foods containing electrolytes. In arrangement (a), a stack of membranes alternating with flow channels is located between the anode and the cathode of the completed electric circuit. Each flow channel containing the feed stream is sandwiched between two transfer stream channels from which it is separated by membranes, a cation membrane on the cathode side and an anion membrane on the anode side. Ions and associated water can thus flow out of the feed stream channels via the membranes, but flow of ions into the feed channel is prevented because the electric gradient causes the ions to move toward a membrane through which they cannot pass. In order to provide an ionic flow sufficient to remove the desired amount of water, salt can be added to the feed stream.

In arrangement (b), only cation membranes are used but there is an alternation between highly permeable (loose) membranes, and relatively impermeable (tight) membranes. The flux of ions and water into the feed stream is less than the flow out of it, resulting in net water removal.

C. Membranes

Membranes used for the separation processes described above are composed of natural or synthetic polymers. They may be classified in a variety of ways but

TABLE 9-3

Membrane Processes in Separation of Food Components

Driving force	Component separated	Name of process
Concentration difference	Solvent	Permeation, osmosis
	Solute	Dialysis
Temperature difference	Solvent	Thermoosmosis, pervaporation
Pressure difference	Solvent	Reverse osmosis, ultrafiltration
Electrical potential	Solute	Electrodialysis
	Solvent	Electroosmosis

the following categories are particularly descriptive of their functional transport process properties.

1. Porous Membranes

When pores of the membranes are very much larger than the molecular size of the permeating substances, the membranes are referred to as "porous." Other essentially synonymous terms are "microporous," "millipore membrane," and "ultrafiltration membranes." In this type of membrane, pore size ranges from 10^{-2} to 10^{2} μm. Hydraulic flow of solvent and low molecular weight solutes occurs through these membranes, whereas solutes of high molecular weight cannot pass.

2. "Solution" Membranes

When the "holes" within the membrane are small, that is, of the order of magnitude of the diffusing molecules, it is usually accepted that transport occurs by "solution" or "activated" diffusion. This type of diffusion is best described by a model in which the permeating molecules are assumed to condense or dissolve at one face of the membrane, are transported in the membrane as a dissolved species, and are then reevaporated or reextracted at the other face. Transport rates and separation factors are determined by solubility and diffusivity of the permeants in the membrane material.

3. Membranes Intermediate Between Porous and Solution Membranes

It is often difficult to establish a sharp distinction between microporous membranes with very narrow pores and solution membranes of high permeability, both of which may be used in ultrafiltration.

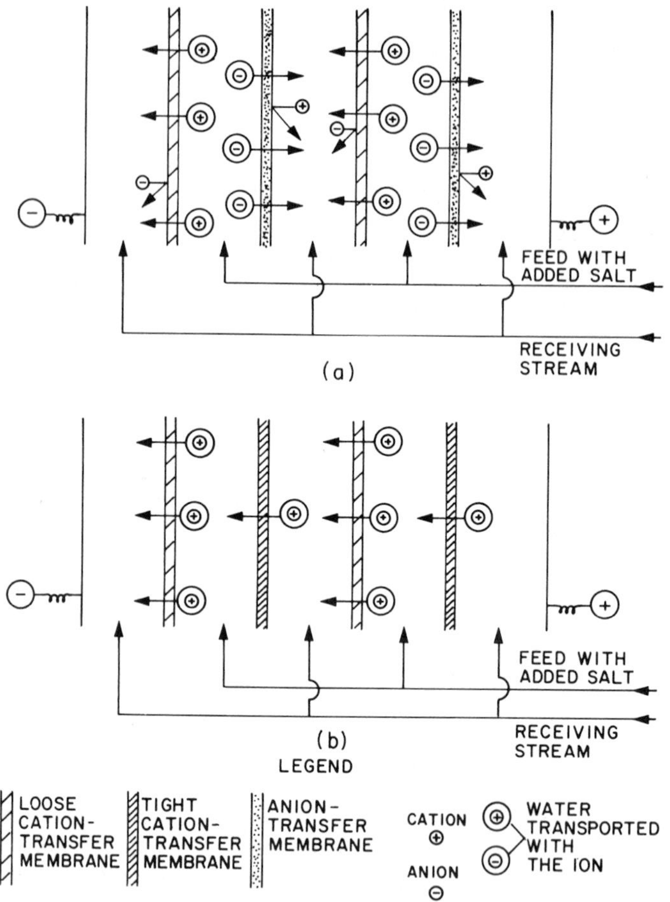

FIG. 9-25. Schematic representation of two types of electroosmosis systems: (a) System using cation and anion membranes; (b) system using only cation membranes.

An intermediate type is the heterogeneous membrane, such as the "Loeb-type" cellulose acetate membrane, which consists of a thin layer of dense, "solution" membrane upon a much thicker substrate of a loose, porous membrane. A representation of resistance to flow with this type of membrane is shown in Figure 9-26.

The cellulose acetate membrane of the Loeb type is prepared by converting a solution membrane into a porous membrane, except for a thin dense "skin." This is achieved by chemical treatment. Other means of preparation of heterogeneous membranes, including coating and graft polymerization, are possible.

It can be noted in Figure 9-26 that in addition to resistance due to the membrane itself, flow is also impeded by resistance of stagnant fluid films on each side of

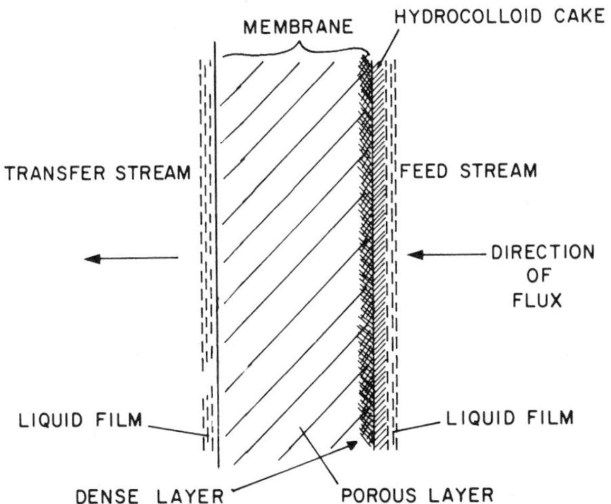

FIG. 9-26. Schematic representation of resistances to solvent flow across a heterogeneous cellulose acetate membrane separating two liquid streams.

the membrane, and especially by the deposition at the membrane-fluid interface of a film of condensed hydrocolloidal solutes, usually polymers or proteins. This "fouling" of the membrane is similar to the fouling which occurs in evaporators and is sometimes called "concentration polarization." The resistance of this fouling layer often critically decreases the flux across the membrane.

Another consideration of importance in the properties of solution membranes is the mobility of the polymeric chains of which the membranes are composed. In the case of noncrystalline (amorphous) membranes, the mobility is relatively high at high temperatures, especially in the presence of solvents having an affinity for the membrane. Consequently, the material is considered to be in the rubbery state as opposed to the glassy state, which occurs at lower temperatures and in which mobility is greatly restricted. The permeability of a membrane increases and its selectivity decreases as the transition from glassy to rubbery state occurs due to a rise in temperature. In partial crystalline (microcrystalline) polymers, the permeability is further reduced by the presence of the crystalline regions, which are not permeable and which may, in addition, restrict polymeric chain mobility in adjacent amorphous regions.

4. Ion-exchange Membranes

Ion-exchange membranes contain fixed charges in the matrix of the membranes. As a consequence, they tend to attract oppositely charged ions (counter-ions) and allow their transport. Co-ions (ions of the same sign as the fixed charges) are excluded from the membrane.

Cation membranes usually contain fixed negatively charged sulfonate ions; anion membranes usually contain fixed quartenary ammonium ions. The degree to which these membranes are hydrated is dependent on pH.

D. Reverse Osmosis

The flux of water across a membrane is given by Eq. (9-20):

$$J_w = \frac{\overline{D}_w \overline{C}_w}{RT} \frac{(\Delta\mu)_w}{\Delta X} \tag{9-20}$$

where

J is flux (mole/cm^2 sec)
$\bar{D}$ is average diffusion coefficient in membrane (cm^2/sec)
R is the gas constant (erg/mole °K)
T is absolute temperature (°K)
$\bar{C}_w$ is average concentration of water in membrane (moles/cm^3)
ΔX is membrane thickness (cm)
$\Delta\mu$ is the chemical potential difference across membrane (ergs/mole)
the subscript w refers to water

For the case of reverse osmosis, shown in Figure 9-24, the difference in chemical potential may be related to applied pressure and to osmotic pressure by Eq. (9-21):

$$\Delta\mu_w = \overline{V}_w (\Delta P - \Delta\Pi) \tag{9-21}$$

where

$\overline{V}_w$ is molar volume of water (cm^3/mole)
ΔP is applied pressure (dynes/cm^2)
Δ is osmotic pressure (dynes/cm^2)

In practical units, therefore, it is possible to relate flux of water to applied and to osmotic pressure as shown in Eq. (9-22):

$$J_w = K_w (\Delta P - \Delta\Pi) \tag{9-22}$$

where

J_w is water flux in convenient units
K_w is a permeance constant incorporating membrane properties and various constants shown in Eq. (9-21)
ΔP and $\Delta\Pi$ are pressures expressed in units consistent with the units of J_w

When the transfer stream is relatively dilute, the calculations may be simplified and approximations to membrane performance obtained by assuming that osmotic pressure is proportional to solute concentration in the feed stream.

The flux of solute across the membrane is then given by Eq. (9-23):

$$J_s = K_s C_{s(feed)} \tag{9-23}$$

where

J_s is flux of solute
K_s is a permeance constant for solute
$C_{s(feed)}$ is solute concentration in the feed stream

If a rejection factor is defined, as given in Eq. (9-24):

$$\text{Rejection} \equiv 1 - \frac{C_{s(transfer)}}{C_{s(feed)}} \tag{9-24}$$

where $C_{s(transfer)}$ is solute concentration in transfer stream, then the rejection factor can be related to operating variables by Eq. (9-25):

$$\text{Rejection} \equiv 1 - \frac{K_s \rho}{K_w (\Delta P - \Delta \Pi)} \tag{9-25}$$

where ρ = density. (Units of ρ must be consistent with those of K_w and K_s. If, for instance, K_s is given in ft/hr and K_w is in lb hr^{-1} ft^{-2} atm^{-1}, then ρ is in lb ft^{-3}.)

It is evident from Eqs. (9-22) and (9-25) that the flux of water increases with increasing applied pressure (ΔP), with increasing K_w, and with decreasing solute concentration in the feed stream. The selectivity of the process, that is, its ability to remove water while retaining solute, improves with increasing ratio K_w/K_s but decreases with increasing ΔP.

At the same operating pressure and given the same type of membrane the selectivity of the process decreases with increasing membrane permeability (increasing porosity). For instance, Merson and Morgan [17], working with the Loeb membrane at a pressure of 1000 psi, found that when the flux of water was 30 gal/ft^2 day there was almost no transport of dextrose (the solute in the feed stream). At fluxes higher than 50 gal/ft^2 day, however, substantial loss of the ester was observed.

E. Food Applications of Reverse Osmosis

The "Loeb" cellulose acetate membrane has been applied in several different types of pilot plant equipment to evaluate various proposed food applications. Among the uses proposed by USDA researchers are [21] (1) concentration of various fluid foods, including juices, egg white, coffee, and syrups; and (2) water recovery from waste streams.

Major food applications tested on a pilot plant scale indicate some promise but problems remain. Processes of food importance are summarized in Table 9-4 [7, 16, 17, 29].

TABLE 9-4

Some Reverse Osmosis Applications Evaluated in Pilot Plant Equipment

Application	Typical operating pressure (psi)	Typical water flux (gal/ft^2 day)	Major product advantage	Major problems
Concentration of maple syrup to 65° Brix	500	4	Full flavor retention	Sanitation of membrane, concentration polarization
Concentration of egg white	400	5	Good functional properties (no thermal denaturation)	Concentration polarization, sanitation problems
Concentration of fruit juices	1500	8	No thermal damage	Some of the polar flavor constituents lost with the water; pectin and other hydrocolloids cause concentration polarization
Concentration of whey	500	10	Economical fractionation into derived components	Membrane fouling

F. Ultrafiltration

Ultrafiltration can be differentiated from reverse osmosis by the relative porosity of the membranes. In the case of ultrafiltration only large molecules are retained and most of the small solute molecules, including salts, sugars, and most flavor compounds, are transported with the water. Since osmotic pressure depends on the number of molecules in solution and since the number of large molecules is small, the osmotic pressure across the membrane is negligible.

The water flux is caused by hydraulic flow and is proportional to applied pressure.

$$J_W = K'_W \Delta P \tag{9-26}$$

where K'_W is a constant.

Rejection factors depend almost entirely on the size of the solute molecule, whereas in reverse osmosis membrane polarity and affinity for membrane material are equally important.

According to published data, ultrafiltration is highly suitable for purification of biological materials, as well as for dewatering and desalting of protein solutions. Typical operating pressures are in the range of 10-100 psi, and flux rates of 10-50 gal/ft^2 day are reportedly attainable with membranes retaining essentially all of the protein.

Major problems remaining to be solved include equipment design for efficient sanitation and the need to counteract membrane fouling by precipitation of a protein "cake" [8, 5].

G. Electroosmosis

As discussed above, electrically driven processes can be used either for desalting, for ion exchange, or for concentration. The process of water removal, that is, concentration, is called "electroosmosis" and is represented schematically in Figure 9-25.

The flux of ions of a given sign across a membrane is given by Eq. (9-27):

$$J_i = K_i \frac{\Delta C_i}{\Delta X} + K'_i \frac{\Delta \psi}{\Delta X} \tag{9-27}$$

where

ΔC_i is the difference in concentration of ions
K_i is the permeance of the membrane caused by the concentration gradient
K'_i is a constant relating ion flux to electric potential gradient
ΔX is membrane thickness
J_i is ion flux (mole equivalent per unit area and time)
$\Delta\psi$ is electric potential difference

The associated flux of water depends on the type and concentration of ions, type of membrane, and other factors. Typical values are: 100-1000 cm^3 of water per mole equivalent of ions.

The rate of removal which may be attained depends on current density. For example, if the arrangement shown in Figure 9-25a is assumed and if a current density of 1 farad/ft^2 hr is assumed, then 1 mole equiv of cations per square foot of cation membranes and 1 mole equiv of anions per square foot of anion membranes are transported. If we assume that each mole equivalent transports 0.5 liters of water, then we obtain a flux of 1 liter/ft^2 of membrane pair each hour, or about 6 gal/ft^2 of membrane pair each day. It should be noted that in order to accomplish the separation, salt must be added to the feed stream, since the natural electrolyte content is usually much lower than the amount needed for electroosmotic transport.

Electroosmotic concentration has been proposed for concentration of whey, of "stickwater" (residue of manufacture of fishmeal), and of various biological preparations. Problems in operation of electroosmosis equipment are similar to those in reverse osmosis: membrane fouling, sanitation problems, and cost.

H. Thermoosmosis and Pervaporation

In a system proposed by Thijssen and Paardekooper [28], an evaporator and a condenser are separated by a membrane permeable to water but impermeable to flavor compounds, and the liquid is evaporated at a temperature above that in the condenser. The major advantage of such an arrangement is the retention of valuable organic compounds that otherwise would be steam distilled with the water. The major difficulty lies in the lack of membranes which allow a high flux of water and retention of food flavors.

I. Desalination by Membrane Processes

Much of the work on removal of water by membrane processes has been conducted in the field of water purification and desalination. Excellent reviews of this subject are given by Rickles [24] and Merten [18].

V. COST OF FOOD CONCENTRATION

The cost of food concentration is obviously a major consideration in process selection. In spite of its overall importance, however, it is difficult to obtain authoritative data on the subject. Among the difficulties involved is the dependence of cost on a large number of variables: feed concentration; product concentration and maximum tolerable temperature; and the specific requirements of each process for sanitation, quality, and inspection. Evaporation processes, for instance, are relatively inexpensive when conducted at high temperatures and when flavor recovery is not required.

Thijssen [27] compared the cost of concentration of a liquid food from 10 to 35% solids for several processes. He arrived at the following cost per pound of water removed, using a typical set of assumptions.

	Cost (cents)
Two-stage evaporator with flavor recovery tower	0.3
Thin-film evaporator with flavor recovery tower	0.4
Reverse osmosis	0.2
Freeze concentration	0.2

Fallick [8] estimated the cost of concentration by ultrafiltration of fluid foods requiring no flavor or salt retention. He estimated the cost at 0.02-0.08 cent/lb of water removed.

Dunkley [7] estimated the cost of whey dewatering by ultrafiltration at 0.1-0.2 cent/lb of water removed. Muller [22] estimated freeze-concentration cost at 0.4-1 cent/lb of water removed.

More reliable figures for concentration costs require additional plant experience with the newer concentration processes.

SYMBOLS

A	area (ft^2)
C	concentration ($moles/cm^3$ or lb/ft^3)
D	diffusion coefficient (cm^2/sec)
"Dh"	decomposition hazard index
G	vapor flow rate (lb/hr)
J	flux through a membrane (mole cm^2 sec or lb/ft^2 hr or gal/ft^2 day)
K_{corr}	constant in Eq. (9-14)
K_i, K'_i	ionic permeance constants defined in Eq. (9-27)
K_s	permeance constant defined in Eq. (9-23)
K_w	permeance constant defined in Eq. (9-22)
K'_w	permeance constant defined in Eq. (9-26)
L, L'	liquid flow rate (lb/hr)
M.W.	molecular weight
P	pressure
R	gas constant
Sp	volumetric pumping speed
T	temperature
U	overall heat transfer coefficient (Btu hr^{-1} ft^{-2} $^\circ F^{-1}$)
V	volume
V_s	volume of a closed system to be evacuated
$\overline{V}_w$	molar volume of water (cm^3/mole)
X	distance
ΔX	membrane thickness (cm)
X_{wall}	wall thickness (ft)
c_p	specific heat (Btu/lb $^\circ$F)

h	film coefficient of heat transfer (Btu hr^{-1} ft^{-2} $^\circ F^{-1}$)
k_{wall}	thermal conductivity of wall material (Btu ft^{-1} hr^{-1} $^\circ F^{-1}$)
ln	natural logarithm
n	number of moles
g	heat transfer rate (Btu/hr)
r_θ	total resistance to heat flow (hr ft^2 $^\circ F$ Btu^{-1})
r_{wall}	wall resistance to heat flow (hr ft^2 $^\circ F$ Btu^{-1})
t	time
Δt	time interval
w	solids concentration (lb solids/lb total)
λ	latent heat of vaporization (Btu/lb)
$\Delta\Pi$	osmotic pressure
$\Delta\mu$	chemical potential difference (ergs/mole)
ρ	density
$\Delta\psi$	electric potential difference

Subscripts

0,1,2,3 ...	subscripts referring to specific flow streams or regions as shown in pertinent figures
A, B, C, D ...	referring to specific components of a phase as defined in text
feed	refers to feed stream
I	ice
i	ions
transfer	refers to transfer stream
e	component removed from the system
s	solute
w	water

REFERENCES

1. G. D. Armerding, Advan. Food Res., 15, 305 (1966).

2. R. A. Carlson, J. M. Randall, R. P. Graham, and A. I. Morgan, Jr., Food Technol., 21, 194 (1967).

3. A. L. Carter and R. R. Kraybill, Chem. Eng. Progr., 62(2), 99 (1966).

4. D. J. Casimir and J. F. Kefford, CSIRO Food Preserv. Quart., No. 1-2, 1968, p. 20.

5. E. S. Chian and J. T. Selldorf, Process Biochem., 4(9), 47 (1969).

6. J. M. Coulson and J. F. Richardson, Chemical Engineering, Vol. 2, Pergamon Press, Oxford, 1960.

7. W. L. Dunkley, in Proceedings of a Symposium on Reverse Osmosis in Food Processing, U. S. Department of Agriculture, Report ARS 74-51, August, 1969, pp. 19-28.

8. G. F. Fallick, Process Biochem., 4(9), 29 (1969).

9. A. W. Farrall, Engineering for Dairy and Food Products, John Wiley and Sons, New York, 1963.

10. N. Ganiaris, Canadian Pat. 811,389 (1969).

11. T. H. Gouw, in Progress in Separation and Purification (E. S. Perry, ed.), Vol. 1, Wiley-Interscience, New York, 1968, pp. 187-246.

12. C. P. Huang, O. Fennema, and W. D. Powrie, Cryobiology, 2, 240 (1966).

13. R. E. Kepner, S. van Straten, and C. Weurman, Agr. Food Chem., 17, 1123 (1969).

14. F. K. Lawler, Food Eng., 12(4), 73 (1969).

15. M. Loncin, Die Grundlagen der Verfahrenstechnik in der Lebensmittelindustrie, Verlag Sauerländer, Aarau, Switzerland, 1969.

16. E. Lowe, E. L. Durkee, R. L. Merson, K. Ijichi, and S. L. Cimino, Food Technol., 23, 757 (1969).

17. R. L. Merson and A. I. Morgan, Jr., Food Technol., 22, 631 (1968).

18. U. Merten, Desalination by Reverse Osmosis, MIT Press, Cambridge, Mass., 1966.

19. S. Mizrahi, Sc.D. Thesis, Technion (Israeli Institute of Technology), Haifa, Israel, 1968.

20. A. I. Morgan, Jr., Chimia, 21, 280 (1967).

21. A. I. Morgan, Jr., E. Lowe, R. L. Merson, and E. L. Durkee, Food Technol., 19, 1970 (1965).

22. J. G. Muller, Food Technol., 21, 51 (1967).

23. M. S. Peters, Elementary Chemical Engineering, McGraw-Hill, New York, 1954.

24. R. N. Rickles, Membranes, Technology, and Economics, Noyes Development Corp., Park Ridge, N. J., 1967.

25. F. H. Slade, Food Processing Plant, Vol. 1, CRC Press, Cleveland, Ohio, 1967.

26. Struthers Scientific and International Corporation, British Pat. 1,133,629, 1968.

27. H. A. C. Thijssen, J. Food Technol., 5, 211 (1970).

28. H. A. C. Thijssen and E. J. C. Paardekooper, U. S. Pat. 3,367,787 (1964).

29. J. C. Underwood and C. O. Willits, Food Technol., 23, 787 (1969).

30. G. N. Werezak, in Unusual Methods of Separation (J. A. Gerster, ed.), Vol. 91, AIChE Symposium Series, Washington, D. C., 1969, p. 6.

Chapter 10

DEHYDRATION OF FOODS

Marcus Karel

CONTENTS

I. INTRODUCTION

Dehydration is the operation in which water activity of a food is lowered by removal of nearly all the water normally present through vaporization or sublimation. It differs from the concentration processes described in Chapter 9 in the generally lower final water content. Dehydration as defined above does not include removal of water by extraction with suitable solvents. Furthermore, production of foods of low water activity by incorporation of such osmotic agents as sugar or salts is not included in this definition. We shall discuss such foods, however, since they are similar in properties and uses to dehydrated foods.

The principles governing the binding of water to food and the relations of water content to water activity were discussed in Chapter 8. In Chapter 7 the general principles governing mass transfer in foods were discussed, and in Chapter 2 the principles of heat transfer in foods were introduced. Dehydration of foods is a combined heat- and mass-transfer operation and its understanding requires knowledge of the above principles.

Dehydration is a process of considerable economic importance both in the United States and abroad. In the early 1960's the four products dried in the United States in the largest quantities were sugar, about 10^7 tons; coffee, 1.4×10^6 tons; cornstarch, 10^6 tons; and flour mixes, 10^6 tons. Other important dehydrated products included breakfast food, pet foods, pasta products, and dried fruits [18]. The situation in recent years has remained essentially unchanged. Breakfast foods, snacks, and prepared mixes have increased in volume produced and a large portion of the total coffee produced (about 2×10^8 lb) is either spray-dried or freeze-dried instant coffee. Table 10-1 shows statistics for some of the important dehydrated foods and Table 10-2 shows details for dried milk products. In other countries, particularly Europe, dehydration of vegetables is a more important segment of the food industry than it is in the United States.

In general, however, the major food products dehydrated on a large scale continue to include sugar, starch, coffee, milk products, flour mixes and breakfast foods, snacks (including potato chips), macaroni and pasta products, fruits, and vegetables for soup mixes and as bulk ingredients [20].

Dehydration can be carried out in various kinds of driers. Since dehydration is a combined heat- and mass-transfer operation, the driers can be classified according to the methods by which these transfers are accomplished. For more detailed information on specific drying processes the reader is referred to more specialized reviews [26, 29, 31, 34, 36, 42, 45]. The heat and mass transfer are affected by the total pressure at which the drier operates. In this chapter we shall not consider vacuum driers, since Chapter 11 is devoted to one operation which in most of its forms is a type of vacuum drying, namely, freeze drying. Driers operating at atmospheric pressure can be batch or continuous, and the heat transfer can be by convection, conduction, radiation, or dielectric heating, as shown in Table 10-3.

The most widely used dehydration methods involve exposure of foods to heated air. In these driers the primary mode of heat transfer is by convection. We shall endeavor to illustrate the process of dehydration by examples chosen from air-drying methods.

II. MASS AND ENERGY BALANCE IN AN AIR DRIER

Consider a continuous air drier operating in a steady state (Fig. 10-1). Heat transfer is either by convection alone or by convection combined with one of the other modes of heat transfer. The symbol q_D in Figure 10-1 represents heat flow supplied to the drier in addition to the heat associated with the air entering the drier. q_D might, for instance, be supplied by an electric heater. q_{loss} represents

TABLE 10-1

Some Dehydrated Products Produced in the United States

Product	Amount produced (lb)	Year
Milk products	2.2×10^9	1971
Breakfast foods and flour mixes	4×10^9	1970
Sugar	20×10^9	1968
Cornstarch	3×10^9	1968
Dried fruit	9×10^8	1970
Dried potatoes	1.3×10^8	1969

heat loss through the walls of the drier. Using an hour of operation as a basis for our balance, we have:

$$s_1 m_1 + G_2 H_2 = s_1 m_3 + G_2 H_4 \qquad (10\text{-}1)$$

Water in Water out

where

s_1 is the weight of food solids entering drier in 1 hr (lb)
m is the moisture content (lb H_2O/lb solids)
G_2 is the weight of dry air entering drier in 1 hr (lb)
H_2 is the humidity of the air (lb H_2O/lb dry air)

Subscripts 1, 2, 3, and 4 refer to specific streams as indicated in Figure 10-1. The operation can be counter-current or co-current, as indicated in Figure 10-1.

The mass balance assumes that no water accumulates in the drier (steady state) but that each hour an amount of water equal to $s_1(m_1 - m_3)$ is removed from the food and transferred to the air.

TABLE 10-2

Dry Milk Production (Thousands of Metric Tons)

Product	United States 1964	United States 1968	Canada 1968	Netherlands 1968	France 1968
Whole milk	40	43	4	43	30
Skim milk	998	728	169	104	677
Whey	169	226	24	34	42

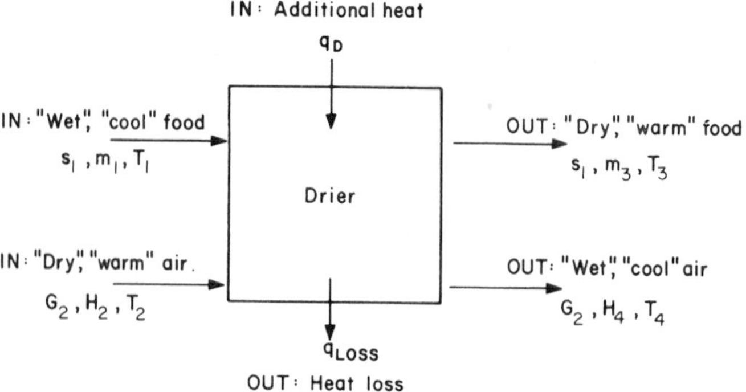

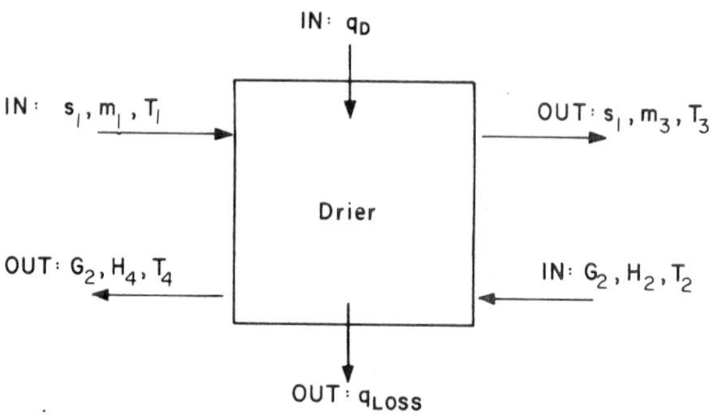

FIG. 10-1. Mass and energy balance in an air drier operating in steady state.

$$s_1(m_1 - m_3) = G_2(H_4 - H_2) \tag{10-2}$$

The energy balance must take into consideration sensible heat changes (heating of the food and cooling of the air); the latent heat of vaporization of water, ΔH_v; and additional heats of adsorption for bound water, ΔH_a (see Chapter 8). For the present balance, assume that m_3 is well above the BET monolayer value and that none of the vaporized water has a latent heat exceeding ΔH_v.

Assume that vaporization occurs at the exit temperature of the food, T_3. It then follows that:

1. Heat to vaporize water = $s_1(m_1 - m_3)\,\Delta H_v$
2. Heat to bring solids to exit temperature = $(s_1)\,(c_{p_s})\,(T_3 - T_1)$
3. Heat to bring moisture to vaporization temperature = $(s_1)\,(c_{p_L})\,(m_1)\,(T_3 - T_1)$

TABLE 10-3

A Classification of Some Atmospheric Driers

Mode of heat transfer	Driers: Batch operation	Driers: Continuous operation
Convection	Kiln drier Cabinet drier	Tunnel drier Conveyor or belt drier Spray drier Fluidized-bed drier
Conduction	Heated-shelf drier Agitated pan drier	Drum drier
Radiation	Infrared heated-shelf drier	Infrared heated belt drier
Internal generation of heat	Microwave oven	Dielectric continuous oven Microwave tunnel
Mixed	Shelf drier	Rotary drier

4. Heat to bring vaporized water to exit air temperature = $(s_1)(m_1 - m_3)(c_{p_v})(T_4 - T_3)$
5. Heat supplied by cooling the dry air = $G_2(c_{p_a})(T_2 - T_4)$
6. Heat supplied by cooling the moisture associated with the incoming air = $(G_2)(H_2)(c_{p_v})(T_2 - T_4)$

The balance is given by Eq. (10-3):

$$q_{loss} + s_1(m_1 - m_3)\Delta H_v + s_1(m_1 - m_3)c_{p_v}(T_4 - T_3) + s_1(c_{p_s})(T_3 - T_1) + s_1(m_1)(c_{p_L})(T_3 - T_1) = q_D + G_2(c_{p_a})(T_2 - T_4) + G_2(H_2)(c_{p_v})(T_2 - T_4) \tag{10-3}$$

where

c_{p_s} is specific heat of solids
c_{p_L} is specific heat of water in food
c_{p_v} is specific heat of water vapor
c_{p_a} is specific heat of dry air

Consideration of the above mass and heat balances indicates that properties of air-water mixtures are important to drying, and some methods of charting these properties are therefore considered in the next section.

III. PROPERTIES OF AIR-WATER MIXTURES

For many purposes it is convenient to tabulate the properties of moist air using as the basis a unit amount, usually 1 lb of dry air. The properties of mixtures of ideal gases have been discussed in Chapter 7, and under normal conditions of temperature and pressure, a mixture of water and dry air (for all practical purposes 78% N_2, 21% oxygen, and 1% argon, with a "molecular weight" of 29) behaves ideally. The following relationships can therefore be written:

$$H = \frac{p_a}{P_T - P_a} \cdot \frac{18}{29} \tag{10-4}$$

where

p_a is partial pressure of water in air
H is lbs of water per lb of dry air
P_T is total pressure

We shall further define, or restate the following definitions:

P° is the vapor pressure of water at a given temperature
H° is H corresponding to P°
$(H/H^{\circ}) \times 100$ is percent absolute humidity
$RH = p_a/P^{\circ}$ is percent relative humidity

We should note that H and RH are related by:

$$H = (H^{\circ})(RH)(0.01)\left[\frac{P_T - P^{\circ}}{P_T - P_a}\right] \tag{10-5}$$

However, when P° is relatively low (at low temperatures), the quantity $(P_T - P^{\circ})/(P_T - P_a)$ is approximately equal to 1 and the percent absolute humidity approximately equals percent relative humidity.

In considering energy balances on moist air it is convenient to introduce the concept of enthalpy associated with 1 lb of dry air. This quantity is defined as the total energy required to bring the air from some reference point (usually 32°F and H = 0) to its condition of interest. The enthalpy includes both sensible and latent heats:

$$i = c_{p_a}(T - 32) + H(\Delta H_v) + c_{p_v} H(T - 32) \tag{10-6}$$

where i is enthalpy (Btu/lb dry air).

Note that under the definitions used here ΔH_V is the latent heat of vaporization of water at 32° F. Another frequently used set of properties is "humid heat" (c_H) and "humid volume" (v_H):

c_H is the heat to raise moist air, associated with 1 lb of dry air, 1° F
v_H is the volume of 1 lb of dry air and associated moisture

Note that c_H and v_H are not constant but vary with T and with H.

Finally we shall define two characteristic temperatures: the dewpoint and the wet bulb temperature (T_w). Dewpoint is the temperature at which the air is saturated with water ($p_a = P^o$). T_w is the temperature attained by a surface of water when there is evaporation from it to a stream of air with a given temperature T (also called "dry bulb" temperature) and H. Underlying assumptions are that all of the heat of vaporization is supplied by the air and that the air flow is so large that this heat and mass transfer produces negligible changes in the properties of the air.

The above properties are usually represented graphically in one of several types of psychrometric charts. A psychrometric chart containing H, T, T_w, RH, and v_H for air temperatures from 32° to 500° F is shown in Figure 10-2. The enthalpy of air at any given T and H is readily calculated using Eq. (10-6) and assuming the values of 1075 Btu/lb for ΔH_V; 0.24 Btu/lb ° F for c_{p_a} and 0.45 for c_{p_v}. c_H is readily calculated from:

$$c_H = 0.24 + Hc_{p_v} \tag{10-7}$$

It is therefore possible to obtain from the psychrometric chart, or through simple calculations using data read off that chart, the key parameters required for energy and mass balance. Figure 10-3 shows how the various properties are located on the psychrometric chart.

In addition to the location of properties of an air-water mixture at given T and H, it is possible to represent on the psychrometric chart processes of change in air properties. Figure 10-4 shows how (1) heating at constant H (line A-B), (2) isothermal humidification (line A-C), and (3) adiabatic humidification (line A-D) can be represented on this chart. The first two are self-explanatory; the adiabatic line represents a situation in which all heat required to vaporize water into air is obtained by cooling the air.

IV. TRANSPORT OF WATER IN FOOD

Dehydration is a combined heat- and mass-transfer operation and involves the transport of heat to and within the food and the transport of water in the food and then away from it. In Chapter 7 we briefly reviewed the fundamentals of mass transfer and, in particular, developed the concept of k_g, the film transfer coefficient in the gas phase. We also mentioned transport of water in porous solids but deferred a detailed discussion to the present chapter. Water may be transported in the food either as bulk liquid or as vapor. In the liquid state, transport may occur due to concentration gradients that develop as water evaporates from the surface, thereby decreasing its concentration. (A more conventional way of looking at it is to say that the concentration of solids increases.)

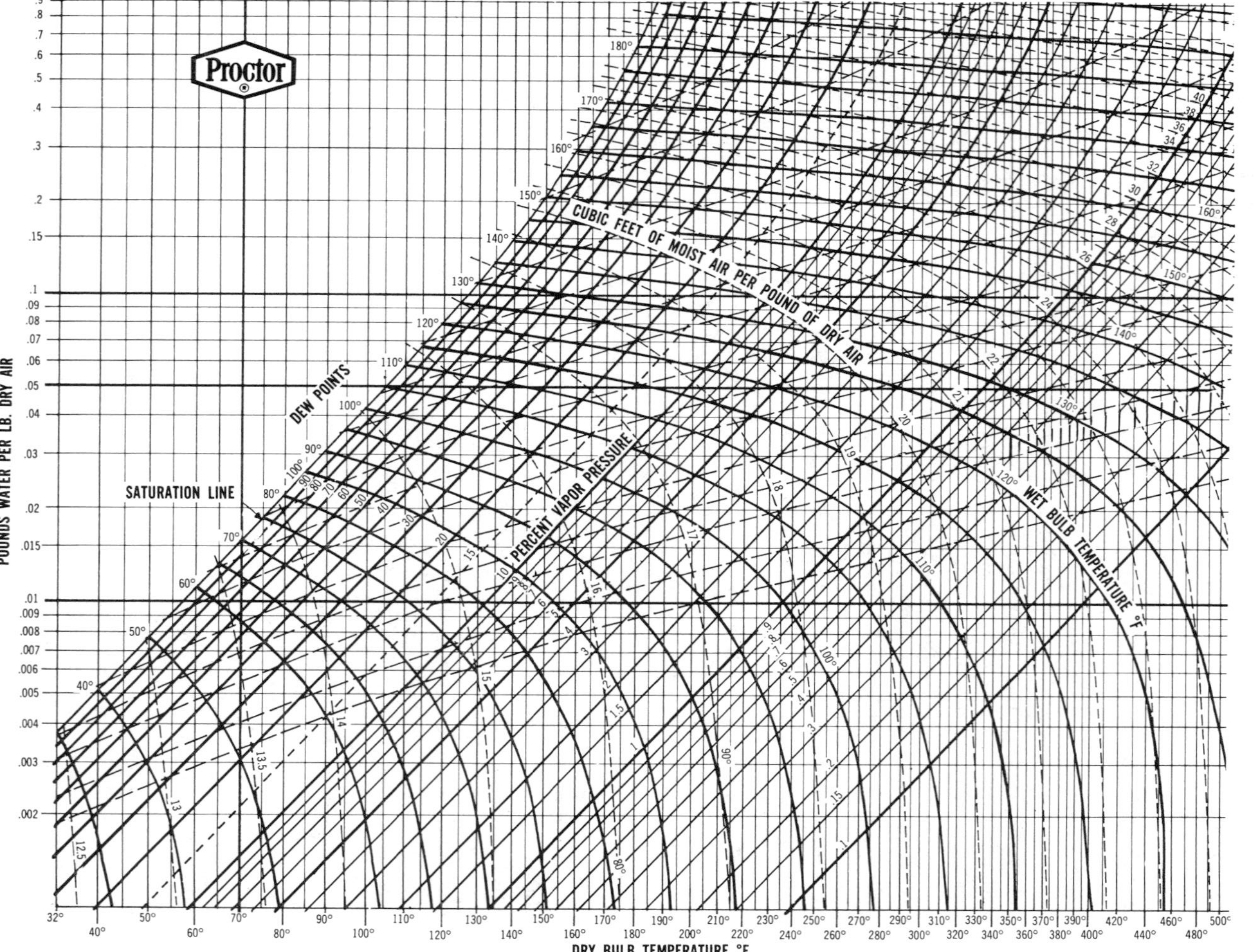

FIG. 10-2. Psychrometric chart (courtesy of Proctor and Schwartz, Inc.).

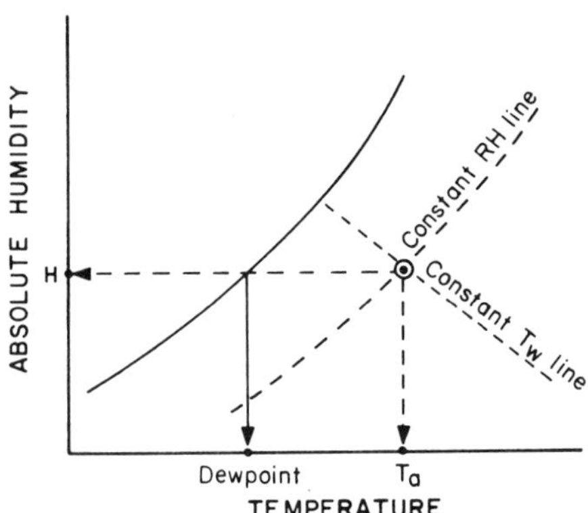

FIG. 10-3. Location of air properties on a psychrometric chart.

This type of flow (using unidirectional flow, as in the case of a slab) is governed by the following equation.

$$-\frac{dw}{dt} = -D_c A \frac{dC_w}{dX} = -D_c A \rho_s \frac{dm}{dX} \tag{10-8}$$

where

$-\frac{dw}{dt}$ is water lost by a drying slab of food (lb/hr)
D_c is the diffusivity of liquid water (ft^2/hr)
ρ_s is the bulk density of solids (lb/ft^3)
C_w is the concentration of water (lb/ft^3)
X is distance in direction of flow
A is area (ft^2)

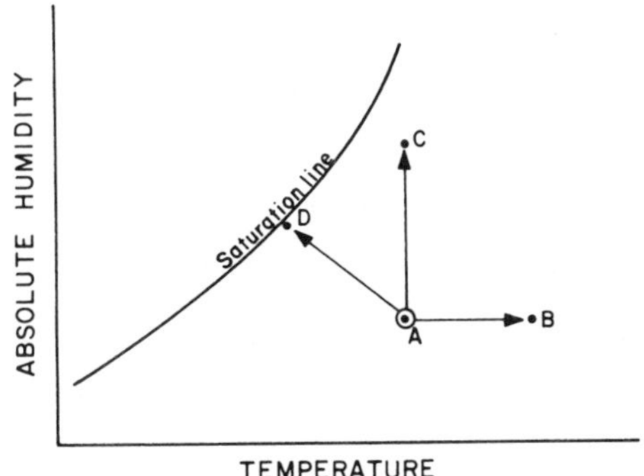

FIG. 10-4. Representation of processes on the psychrometric chart. Line AB represents heating at constant absolute humidity; AC represents isothermal humidification; AD represents adiabatic humidification.

Another important way in which internal water transport occurs is by capillary forces. Recall that for a liquid, such as water, which wets the walls of capillaries typically present in foods, there exists a pressure difference between liquid in a capillary and a connected bulk liquid. In a vertical capillary represented in Figure 10-5 the pressure difference causes a capillary rise given by Eq. (10-9)

$$\ell = \frac{2\gamma}{(r)\,(g)\,(\rho_L)} \qquad (10\text{-}9)$$

where

ℓ is capillary rise (cm)
γ is free surface energy (ergs/cm^2)
r is capillary radius (cm)
g is acceleration due to gravity (cm/sec^2)
ρ_L is density of water (g/cm^3)

The flow due to capillarity in a complex system of capillaries, such as exists in foods, is difficult to calculate and it is also difficult to differentiate between flow due to capillarity and that due to concentration gradients. For our purposes, therefore, it is more convenient to define an overall "liquid conductivity," D_L, which is defined in Eq. (10-10).

$$-\frac{dw}{dt} = -AD_L\rho_s\frac{dm}{dX} \qquad (10\text{-}10)$$

The liquid conductivity varies very strongly with water content, as indicated schematically in Figure 10-6.

As the water content of food decreases, the transport through capillaries and pores occurs primarily in the vapor phase. There are several mechanisms possible, some of which were mentioned in Chapter 7. We shall again, however, define a single overall coefficient, b defined by Eq. (10-11), which indicates the permeability of the food to transport of water vapor.

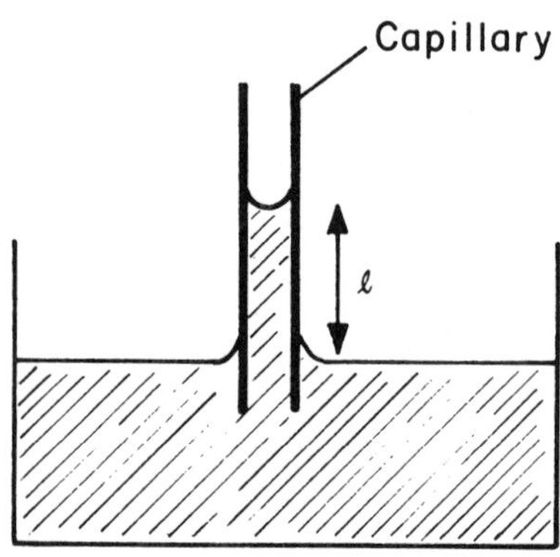

FIG. 10-5. Capillary rise.

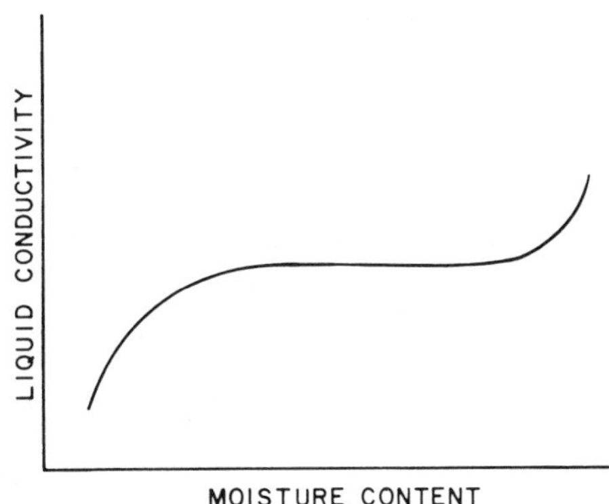

FIG. 10-6. Effect of moisture on liquid conductivity (D_L).

$$-\frac{dw}{dt} = -Ab\frac{dp}{dX} \tag{10-11}$$

where

p is partial pressure of water in the food
b is permeability (lb ft^{-1} $torr^{-1}$ hr^{-1})

V. DEHYDRATION UNDER CONSTANT EXTERNAL CONDITIONS

A. Periods of Drying for a Hygroscopic and a Nonhygroscopic Slab

Consider dehydration of an ideal nonshrinking slab of food in a stream of air at constant T and H. (In practice, the properties of the air can remain relatively constant if there is a very large air flow compared to the weight of water in the slab.) The behavior of the drying food depends on the manner in which the moisture is "bound" (see Chapter 8).

In air drying under constant environmental conditions the process of drying may be divided into a "constant rate" period and one or more "falling rate" periods. In this respect we may divide materials to be dehydrated into two categories, nonhygroscopic and hygroscopic. Our definition of a nonhygroscopic material is one in which the partial pressure of water in the material is equal to the vapor pressure of water. In hygroscopic materials, however, the partial pressure of water becomes less than the vapor pressure of water at some critical level of moisture which is referred to as m_h. We can therefore define (m_i = initial moisture content) for nonhygroscopic foods:

$$p = P^o \qquad \text{for } 0 < m \leq m_i$$

For hygroscopic foods the relationships are:

$p = P^\circ \qquad \text{for } m_h < m \le m_i$

$0 < p < P^\circ \qquad \text{for } 0 < m \le m_h$

We can describe the "constant rate" period as one in which the surface of the slab is maintained at a moisture level assuring that $p = P^\circ$ at the wet bulb temperature T_w. This is true as long as the moisture content at the surface is greater than zero for nonhygroscopic materials or greater than m_h for hygroscopic ones.

Under these conditions, the resistance to heat and mass transfer is located solely in the air stream, and since in an idealized system the conditions of the air are constant, this resistance does not change with time. The driving force for mass transfer $(p - p_a)$ and for heat transfer $(T_a - T_w)$ also remains constant; hence the constant rate period.

$$-\frac{dw}{dt} = Ah(T_a - T_w)(\Delta H_v)^{-1} \qquad (10\text{-}12)$$

where

h is a film coefficient of heat transfer (Btu hr^{-1} ft^{-1} $^\circ F^{-1}$)
T_a is air temperature ($^\circ$ F)
T_w is wet bulb temperature of air ($^\circ$ F)

$$-\frac{dw}{dt} = Ak_g(P^\circ - p_a) \qquad (10\text{-}13)$$

where

k_g is a gas film coefficient of mass transfer (lb hr^{-1} $torr^{-1}$ ft^{-2})
p_a is partial pressure of water in air (torr)

The constant rate period continues as long as the supply of water to the surface suffices to maintain saturation of the surface to give a constant surface temperature and a level of water pressure equal to P° at the wet bulb temperature. The constant rate period ends when this condition no longer applies. One explanation for the end to the constant rate period is that "dry" patches eventually appear at the surface, and thus the area for the evaporative process decreases [13]. We shall assume, however, that in our slab geometry the conditions are uniform at any given slab depth, including depth zero, the surface level.

In our idealized system we can divide the drying of hygroscopic food into three periods and that of nonhygroscopic into two periods (Figs. 10-7 and 10-8). In the first period water removed from the surface is constantly resupplied by capillary flow to the surface, as shown in Eq. (10-14):

$$-\left(\frac{dw}{dt}\right)_I = k_g A(P^\circ - p_a) = -D_L \rho_s A \frac{dm}{dX} \qquad (10\text{-}14)$$

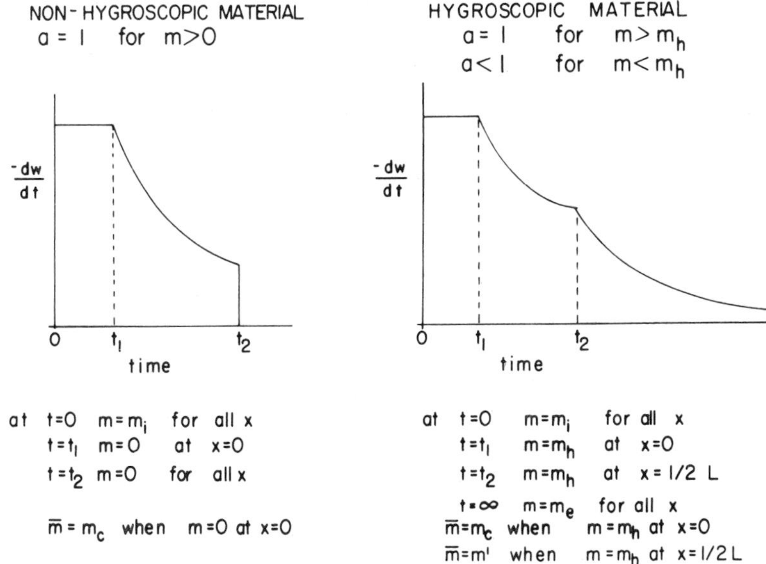

FIG. 10-7. Representation of periods of drying for hygroscopic and nonhygroscopic slabs dried under constant conditions of temperature and humidity (drying rate is plotted against time).

where

$(dw/dt)_I$ is constant drying rate
D_L is overall internal liquid diffusivity in slab (ft^2/hr)
ρ_s is bulk density of dry solids in slab (lb ft^{-3})
m is moisture content (lb H_2O/lb solids)
X is distance (ft)

The end of the constant rate period occurs when the internal water supply is inadequate to saturate the surface. Krischer and Kröll [28] and Görling [11] present the end of "constant rate" period data in terms of "break point" curves. Under idealized conditions [28] the average moisture content (critical moisture) of the slab at which the constant rate period ends is given by:

$$m_c = \frac{1}{3} \frac{k_g(P^\circ - p_a) L}{D_L \rho_s} \qquad (10\text{-}15)$$

where

L is slab thickness (ft)
m_c is critical moisture content (lb/lb)

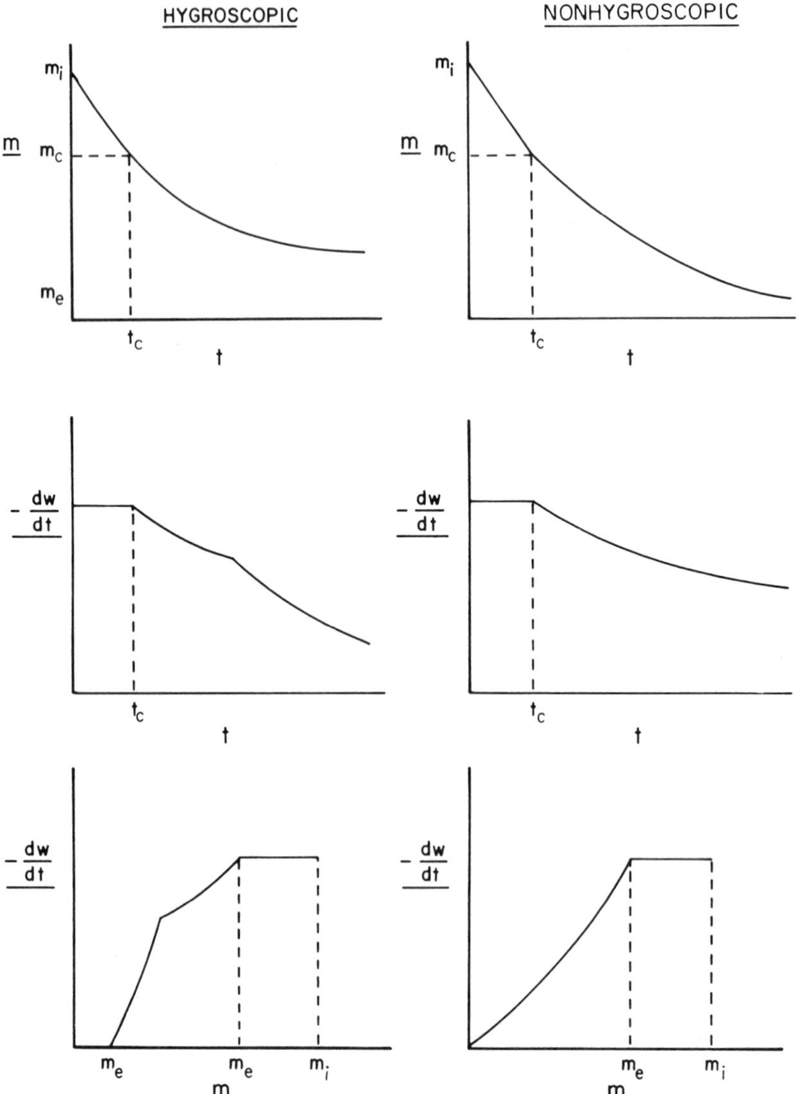

FIG. 10-8. Schematic representation of changes in moisture content as function of time, and of drying rate as function of moisture content and of time, for the hygroscopic and nonhygroscopic slabs defined in Fig. 10-7.

The relationship shown in Eq. (10-15) is an idealized one derived on the basis of often nonapplicable assumptions concerning the type of moisture profile existing in food. Equation (10-15) or its exact equivalent, Eq. (10-16), is often used in a graphical representation of the expected ideal behavior and of the actual empirically determined behavior (Fig. 10-9).

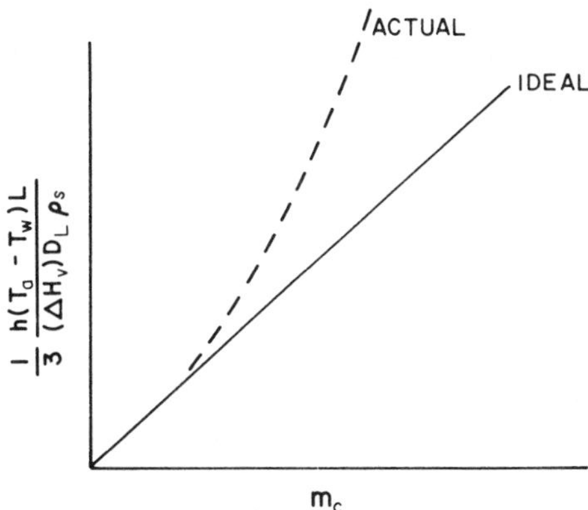

FIG. 10-9. Idealized and actual relations between critical moisture content and drying parameters.

$$m_c = \frac{1}{3} \frac{h(T_a - T_w)L}{(\Delta H_v)D_L \rho_s} \tag{10-16}$$

As can be seen in Figure 10-9, the relation in Eq. (10-16) allows a way of correlating air properties (T_a, T_w, h) and product characteristics (ρ_s, L, D_L) with the moisture content at which the constant rate period ends (m_c). Note that in ideal as well as in actual situations m_c is not a property of the food since it depends on food thickness as well as on air properties.

The reasons for the deviation from linearity of the so-called "breakpoint curves" shown in Figure 10-9 are primarily the nonconformance of real materials with the following idealized assumptions on which Eq. (10-15) and Eq. (10-16) are based:

1. $m_{X=L} = 0$ when $t = t_c$ ($m_{X=L}$ = m at surface of slab; t_c = t when constant rate period ends).
2. D_L is assumed to be independent of m.
3. ρ_s and L are assumed to be constant (no shrinkage).
4. D_L is assumed not to change with time (in fact chemical reactions may cause such change, as for instance in the case of formation of a cross-linked skinlike surface layer).

Following the constant rate period comes the first falling rate period. During this period it is assumed that transport from a level in the slab at which the saturation conditions still prevail ($p = P^\circ$) to the surface occurs by transport in the gas phase, and that transport from the surface to the bulk of the air occurs through the gas film in the air. Hence the drying rate is given by:

$$-\frac{dw}{dt} = \frac{A}{\frac{z}{b} + \frac{1}{k_g}} \cdot (P^\circ - p_a) \tag{10-17}$$

where z = distance from surface to the layer at which ($p = P^o$). The rate is assumed to fall during this period due to the increasing internal resistance, the cause of which is the increase in distance z from the slab surface to the saturation surface within the slab.

Finally, the end of the second period is reached when the center of the slab no longer has enough water to maintain a partial pressure equal to P^o. In nonhygroscopic foods, this corresponds to the end of drying; in hygroscopic food this point corresponds to the center of slab attaining a local moisture content m_h.

With hygroscopic foods a third period is encountered during which the partial pressure of water throughout the food is everywhere below the saturation level P^o. In this situation drying occurs by desorption throughout the food and the usual idealized assumption is to consider that at the beginning of the third period the moisture content is uniform throughout the slab. Of course, this assumption is patently untrue, but it aids in evaluation of the differential Eq. (10-18), which governs desorption in this third period.

$$\frac{dm}{dt} = b/\rho_s \frac{d^2p}{dX^2} \tag{10-18}$$

If an assumption is made that the relationship between p and m is linear (another patently untrue but convenient oversimplification), Eq. (10-19) can be written:

$$\frac{dm}{dt} = b\sigma/\rho_s \frac{d^2m}{dX^2} = D\frac{d^2m}{dX^2} \tag{10-19}$$

where

σ is a linear isotherm constant relating p to m
D is effective diffusion coefficient for water vapor in food during the third drying period

The desorption process which occurs during the third period of drying has been treated thoroughly by a number of authors, including King [23]. Assuming that mass transfer occurs only in the vapor phase, that there is no shrinkage, and that total latent heat of vaporization is constant, but taking into consideration the dependence of D on process variables, King [23] has suggested the relation:

$$\frac{\delta m}{\delta t} = \frac{\delta}{\delta X}\left(D\frac{\delta m}{\delta X}\right) \tag{10-20}$$

This is of course equivalent to Eq. (10-19) when D is constant. King [23], however, has noted that D depends on process variables and has presented a series of equations representing different cases of desorption, including (a) control by internal resistance, (b) control by external resistance, and (c) when both internal and external resistances are important. The usual practice has been to assume that the effective diffusion coefficient D is constant over the entire falling rate period, or at least over significant portions of this period, so that the drying process can be represented by one or more straight lines when the quantity

$\ln (m - m_e)$ is plotted against time [17] (m_e being the equilibrium moisture content). These straight lines result from the assumed validity of Eq. (10-21), which represents the solution of Eq. (10-20) using assumptions including control by internal resistance, isotropic medium, and D constant over each segment of the drying period.

$$\left(\frac{m - m_e}{m' - m_e}\right) = \frac{8}{\pi^2} \exp\left(-\frac{\pi^2}{4}\frac{D}{L^2}t\right) \qquad (10\text{-}21)$$

where

m is average moisture content
m' is moisture content at the beginning of the drying time segment

The problem of applying Eq. (10-21) lies in the lack of reliable data on effective diffusion coefficients for foods. Typical values of D observed in the literature range from 10^{-8} cm^2/sec to 10^{-5} cm^2/sec, often for very similar materials. It should be noted, however, that the values of D were often obtained from drying data in which the assumptions underlying Eq. (10-21) were violated.

B. Simplified Approach to Analysis of Drying of Hygroscopic Products Under Constant External Conditions

In Figure 10-8 we have observed that hygroscopic products have a complex drying behavior. It is possible, however, to simplify the analysis of drying by representing the drying curves as shown in Figure 10-10. Two periods of drying are shown here; namely a constant rate period during which $-(dw/dt)_I$ is constant and given equally well by Eqs. (10-12) or (10-13). The falling rate period can be represented by a straight line (Fig. 10-10) and the drying rate $-(dw/dt)_F$ is given by:

$$\left(-\frac{dw}{dt}\right)_F = (\text{const})\ (m - m_e) \qquad (10\text{-}22)$$

where m_e is m in equilibrium with p_a. Since

$$m = w/s \qquad (10\text{-}23)$$

and

$$s = AL\rho_s \qquad (10\text{-}24)$$

we can write by combining Eq. (10-24) with Eq. (10-13)

$$\left(-\frac{dm}{dt}\right)_I = \frac{k_g (P^\circ - p_a)}{\rho_s L} \qquad (10\text{-}25)$$

and since the right-hand side of Eq. (10-25) represents a constant, we can readily integrate to give drying time for the constant rate period:

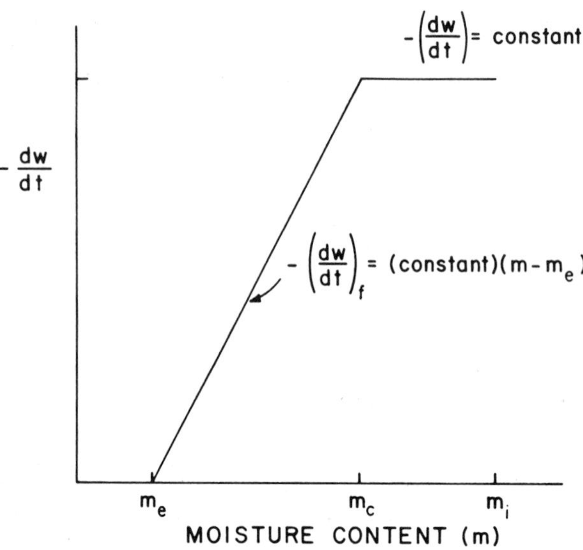

FIG. 10-10. Simplified representation of the drying rate of a slab as a function of moisture content.

$$\int_{m_c}^{m_i} dm = \int_{m_i}^{m_c} -dm = \frac{k_g (P^\circ - p_a)}{\rho_s L} \int_0^{t_c} dt \tag{10-26}$$

and

$$t_c = \frac{(m_i - m_c)(\rho_s L)}{(k_g)(P^\circ - p_a)} \tag{10-27}$$

By inspection of the geometry of the drying rate curves in Figure 10-10 it becomes immediately apparent that the constant in Eq. (10-22) represents the "rise over run" of the falling rate portion of the curve and that Eq. (10-22) can be rewritten

$$\left(\frac{dw}{dt}\right)_F = \frac{(dw/dt)_I}{m_c - m_e} (m - m_e) \tag{10-28}$$

By combining Eqs. (10-28), (10-24), and (10-12) and rearranging,

$$\frac{-dm}{m - m_e} = \left[\frac{k_g (P^\circ - p_a)}{(m_c - m_e)\rho_s L}\right] dt \tag{10-29}$$

Note that the bracketed term on the right hand side of the equation represents constant quantities. We can now integrate to give the duration of the falling rate period defined as the time required to bring m from m_c to some desired final moisture (m_f):

$$t_F = \frac{(m_c - m_e)\,\rho_s L}{k_g (P^\circ - p_a)} \cdot \ln \left(\frac{m_c - m_e}{m_f - m_e} \right) \tag{10-30}$$

From this equation a plot of the logarithm of $(m - m_e)$ should be linear with time in the falling period, as shown in Figure 10-11. Note that m_f must be above m_e, since to reach m_e $t = \infty$ is required. The above equation is sometimes called the "proportional to thickness" law, because t_F appears to be proportional to L. This law is often contrasted with the "diffusion law" derived on the basis of Eq. (10-19), which, on integration using idealized boundary conditions, gives:

$$t_F = \frac{4L^2}{\pi^2 D} \ln \left(\frac{m_c - m_e}{m - m_e} \right) \tag{10-31}$$

Some texts list materials for which each of these integrations is most suitable. Actually the difference in power of L to which t_f is allegedly proportional is more apparent than real. It has been seen that m_c is ideally proportional to L [Eq. (10-16)], and since in the so-called "proportional to thickness law," Eq. (10-30), t_f is proportional to the product of $L \cdot (m_c - m_e)$ it is in reality approximately proportional to L^2, just as in the "diffusion law," Eq. (10-31). There is no reason why either of the two equations cannot be applied to so-called "diffusion-controlled" materials. Equation (10-31) is sometimes applied to the entire drying process (including the period during which a constant rate is expected and the total drying curve is divided into segments each with a different D [17]. Since real foods exhibit significant deviations from the ideal behavior approximation by several segments, utilizing Eq. (10-31) is often as close to reality as the assumption of a constant period followed by falling periods.

The above derivations are based on the assumption that drying is performed on an infinitely long and wide slab with a finite thickness L. Actually, the approach is quite suitable to less extensive slabs and to beds of food particles and may even be applied to correlate empirical results obtained with individual particles of irregular shape, provided a characteristic distance L can be calculated for such particles.

C. Temperature Gradients in Drying Materials

Dehydration in the falling rate periods is an unsteady-state process not only with respect to moisture gradients but also with respect to temperature. The temperature at the beginning of the falling rate period is equal to T_w and rises throughout the falling rate period toward the dry bulb temperature of air T_a. The relative effects of temperature and moisture gradients inside the food may as a first approximation be related to the Lewis number, $L_e = (\alpha/D)$, where α is thermal

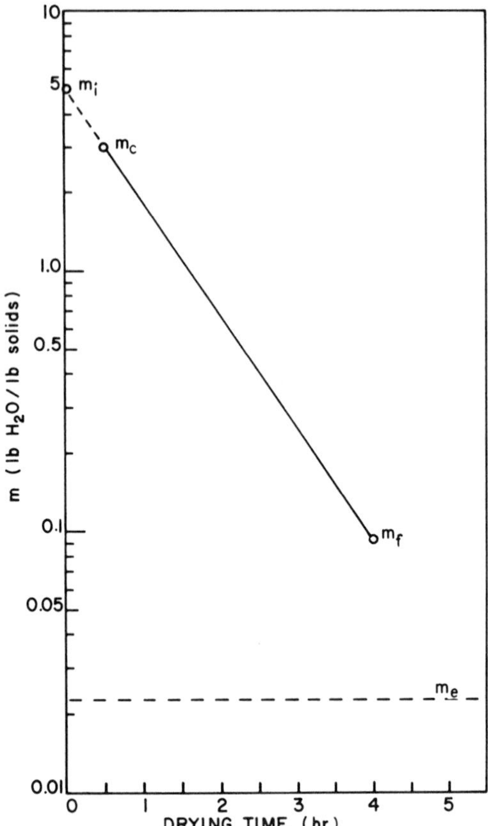

FIG. 10-11. Semilogarithmic plot relating moisture content to time.

diffusivity in units consistent with those of D. Young [48] has noted that when L_e is greater than 60 the thermal gradients can be neglected. In fact, a number of investigators have measured temperature distributions in foods during drying and have found little or no gradients [5, 17]. Harmathy [13] has studied theoretical distribution of moisture and temperature during drying of a porous slab and has found negligible temperature gradients. It is therefore feasible in many cases to treat dehydration of a slab in air having constant wet and dry bulb temperature as if the temperature of the slab varies with time but not with position within the slab.

D. Shrinkage During Drying

A serious and significant deviation from the assumed idealized situation lies in the changes of geometry and structure of drying food materials.

In dehydration processes other than freeze drying, structural changes in the food materials occur readily since there is the possibility for liquid flow, solute redistribution, and shrinkage. As a result, the mass transport properties change drastically during the dehydration process. Fish [10] has reported that in scalded potatoes the diffusion coefficient for water is 10^{-8} to 10^{-7} cm^2/sec when the moisture is still at a level of 15-20% (dry basis) but that it drops to less than 10^{-10} when the moisture is less than 1%. Similar data are reported for other materials, but in some foods diffusion coefficients do not decrease as sharply with decreasing water content as is found in the case of potatoes [10]. Thus, Jason [17] has observed that the dehydration behavior of fish muscle can be described by using two diffusion coefficients, the first, D' valid down to about 5% free water and the second, D'', down to almost complete dehydration. He has found that D' is about $2\text{-}4 \times 10^{-6}$ cm^2/sec for various species of fish, and D'' $0.1\text{-}1.0 \times 10^{-6}$ cm^2/sec [17].

The quality of dehydrated foods is often limited by changes in their texture and their rehydration capability. Tough, "woody" texture; slow and incomplete rehydration; and loss of the juiciness typical of fresh food are the most common quality defects of dehydrated foods. The physicochemical basis for these changes is as yet not fully understood. In the case of plant materials loss of cellular integrity and crystallization of polysaccharides are thought to play a key role in these changes. It is known that crystallization of such polysaccharides as starch and cellulose is, in fact, promoted by removal of water. In the case of dehydrated animal tissues tenderness losses are due to aggregation of muscle proteins.

Freezing and thawing prior to dehydration are sometimes used to improve rehydration characteristics of dehydrated vegetables. In this case, the effect is attributable to increased internal porosity caused by the cavities left by large ice crystals. Absorption of water during rehydration is facilitated by this technique but the water-binding capability of the food material is not necessarily enhanced. Thus, foods which have been made porous by slow freezing (large crystals) may imbibe m ore water during rehydration but this water is often loosely held.

A related method, designed to prevent shrinkage of the food tissues during dehydration, has been patented by Haas [12]. He proposes freezing foods that have been previously pressurized with 500-1500 psig of methane, nitrogen, CO, air, freon, or ethane (CO_2 is not effective). The pressurized, frozen substances are then air dried and they retain their shapes. Other treatments to increase porosity include pressurizing of vegetable pieces in high-pressure chambers followed by explosive decompression (puffing).

E. Nutrient Destruction and Other Deteriorative Reactions Occurring During Drying

Reactions occurring during drying can result in quality losses, particularly nutrient losses, and in other deleterious changes caused by nonenzymatic browning. Attempts have been made to analyze these changes in a quantitative manner and to relate them to process conditions. Such quantitative analysis facilitates the determination of optimal drying conditions with respect to nutrient retention and organoleptic quality.

Kluge and Heiss [25] have measured rates of nonenzymatic browning in a model food system (glucose, glycine, and cellulose) and have used these rates to develop relations for the calculation of browning during drying. Labuza [30] has suggested the use of computers for prediction of the extent of deterioration occurring during storage. The principle is as follows. Given the rates of deterioration as functions of moisture content and temperature and the knowledge of moisture and temperature distributions within the food throughout the dehydration process, a suitable integration procedure with respect to time and space can be carried out to obtain the total deterioration occurring in the process.

Actually, two limiting cases can be recognized which make the analysis less cumbersome than implied by the general statement.

1. In air dehydration it is usually possible to assume that the temperature gradient within the food is negligible. Thus it is possible to calculate the moisture-temperature profiles by assuming that moisture content varies with time and location and that temperature varies with time but not with location in the food.
2. In freeze dehydration, within the "ice-free" layer (deterioration in the frozen layer during drying often can be neglected), it can be assumed that the temperature varies with time and location but that the moisture gradients are small. In fact, in freeze drying it is often permissible to simplify the situation even further and to assume that as long as any ice is present, the temperature difference across the dry layer remains unchanged.

As an example, consider the extent of nonenzymatic browning during dehydration of potato slabs. We shall use data on kinetics of nonenzymatic browning in potato published by Hendel et al. [14] and some simplifying assumptions concerning moisture and temperature distribution in the potato slab. The rate of nonenzymatic browning depends, of course, on both moisture content and temperature.

Figure 8-14 in Chapter 8 presented typical data derived from the results of Hendel et al. [14]. These data were used to calculate browning rates in potato as follows.

The moisture-temperature distribution for any drying condition was derived using idealized assumptions including the following: (1) slab geometry; (2) no shrinkage; (3) constant conditions of air temperature, humidity, and velocity; (4) drying occurring in the falling rate period, with moisture content of 3.5 g/g at the beginning of that period; (5) the moisture distribution in potato slabs and the average moisture derived on the basis of idealized assumptions; (6) the diffusion coefficient independent of concentration but depending on temperature, as shown in Eq. (10-32); (7) equilibrium moisture contents (m_e) for each air temperature condition based on isotherms for the potato-water system published by Görling [11]; and (8) temperature assumed not to vary with position but to vary with time according to Eq. (10-33). The dependence of diffusivity on temperature is given by:

$$D = D_0 \ e^{-E_D/RT} \qquad (10\text{-}32)$$

where

D is diffusivity (cm^2/min)
D_0 is a constant, assumed to be 66 cm^2/min
E_D is the activation energy for diffusion, assumed equal to 7500 cal/mole

The temperature depends on time:

$$\ln \left(\frac{T_a - T_w}{T_a - T} \right) = \mu t \tag{10-33}$$

where μ = slope of the line ln (m/m_i) vs time. Figure 10-12 shows the moisture distribution in the slab, and Figure 10-13 shows the instantaneous browning rates within the potato slab as a function of position and degree of drying.

Figure 10-14 shows the browning that occurs at various locations in a drying slab when drying is conducted at an air temperature of 71°C.

Finally Figure 10-15 shows the predicted amount of browning for different final moisture contents for the slab dried at 71°C. It is evident that there is a difference between the total browning calculated by summing up browning at each location (B_Σ) compared with browning computed by using an average value of moisture for each drying time ($\bar{B}$).

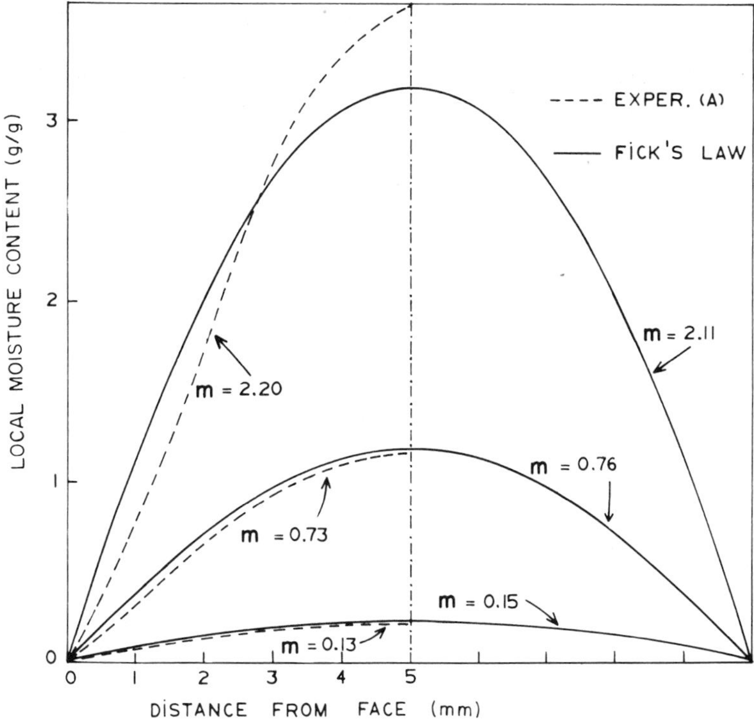

FIG. 10-12. Local moisture content distribution in an idealized slab during drying in air as a function of average moisture content of the slab (m).

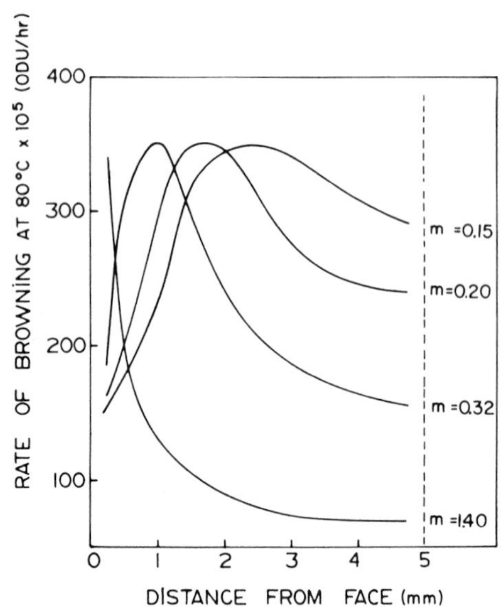

FIG. 10-13. Instantaneous browning rates at specified locations in a drying potato slab at specified values of average moisture content (m) (m in lb water/lb solids).

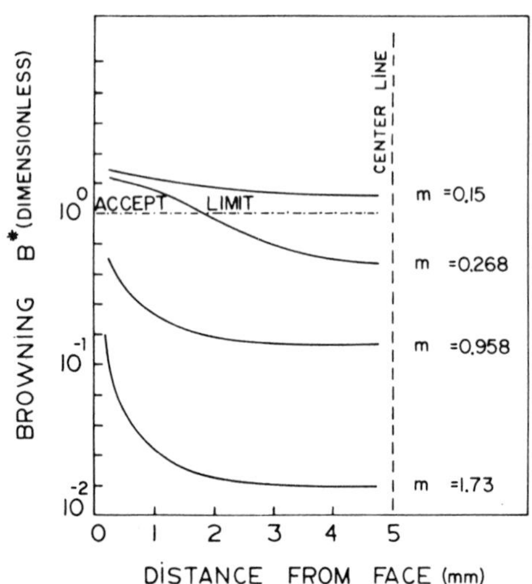

FIG. 10-14. Browning gradients in a potato slab dried in air at 71° C to specified values of average moisture content (m).

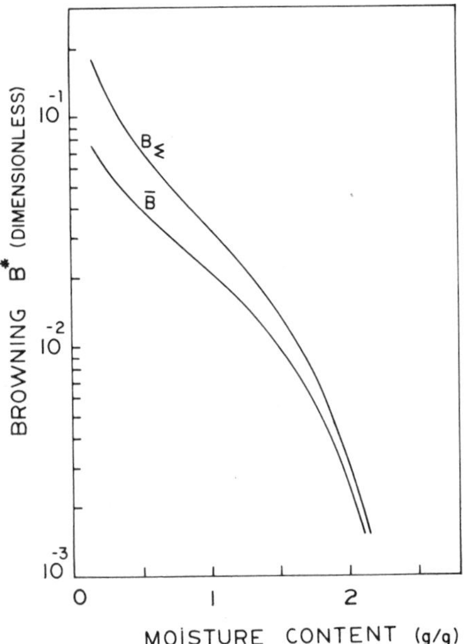

FIG. 10-15. Comparison of browning in a potato slab by using average moisture contents, $\bar{B}$, with browning calculated with consideration of moisture gradients existing during drying at 71°C (Σ B).

VI. INDUSTRIAL DRYING OF FOODS

A. Atmospheric Batch Driers

Batch driers are used when different types of material are to be dried and where the operations are small or seasonal. The kiln drier is a batch drier that is still used extensively for drying apple rings and slices in the United States. Elsewhere, various agricultural commodities are dried in kiln driers. A simple kiln drier is shown schematically in Figure 10-16. It consists of a two-story building with a slotted floor separating the drying rooms in the upper story from the rooms on the lower floor. The rooms on the lower floor contain burners heating air which, along with the combustion gases, supplies heat to a layer of fruit (up to 8 in. or more deep) resting on the slotted floor. The loading, unloading, and periodic raking of the drying layer require substantial labor, causing this aspect to be among the major costs in kiln drying. The drying times are quite long, even though drying of apples is stopped at a moisture content above 20%, thus avoiding the excessively long second falling rate period. At this moisture content fruit is very susceptible to browning and is therefore treated with sulfite prior to drying. A traditional feature of kiln drying has been the incorporation of sulfur burning, thus generating sulfur dioxide in the combustion gases. Modern operations tend to rely more extensively on sulfites delivered by dipping treatments. Following drying the apple slices are allowed to equilibrate in bulk containers, because the apples show substantial variations in moisture content from slice to slice as well as within

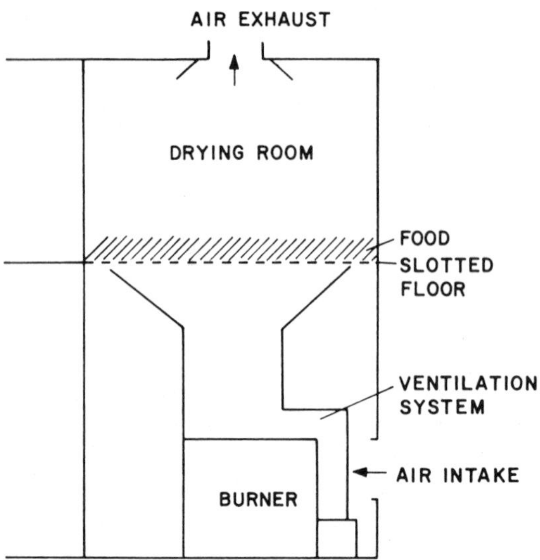

FIG. 10-16. Schematic representation of a kiln drier.

each slice. Low-moisture apples are prepared from these kiln-dried ("evaporated apples" is the term used for them in the industry) apple slices by reducing their size to flakes, dice, or nuggets and drying them in a continuous hot air drier. Low-moisture apple slices or wedges also can be prepared by vacuum drying of the "evaporated apples" and this method produces a superior but more costly product.

Cabinet or tray driers represent another type of atmospheric batch-drying equipment that is useful for small-scale operation as well as for pilot plant scale operation. Figure 10-17 is a schematic representation of a tray drier consisting simply of an insulated cabinet, trays containing layers of material to be dried, and a means for circulating heated air. The equations for drying of slabs approximate in general the drying situation in kiln or cabinet driers. When heat is supplied entirely by the air and drying is in the "constant rate" period, the humidification of air (as the food dries the air becomes humidified) progresses in accord with a wet bulb line on the psychrometric chart (e.g., A-D in Fig. 10-4). However, it is often more convenient to incorporate auxiliary heaters in the cabinet, as shown in Figure 10-18. Under these circumstances the drying process is represented by a line lying somewhere between lines A-D and A-C in Figure 10-4. The effect of additional heating within the cabinet is that more water can be absorbed per pound of air. At the same time, however, the maximum surface temperature of the food is likely to increase above T_W, so additional temperature control may be required when overheating is dangerous to food quality.

The effective drying area, A , and the characteristic thickness, L, are difficult to define in the batch driers described above. When hot air is flowing through a bed of particles at high velocity, the area approaches as a limit the total surface area of individual particles (for instance, in the case of n cubical particles each with a side of d, the total area = $6nd^2$) and the characteristic thickness is the

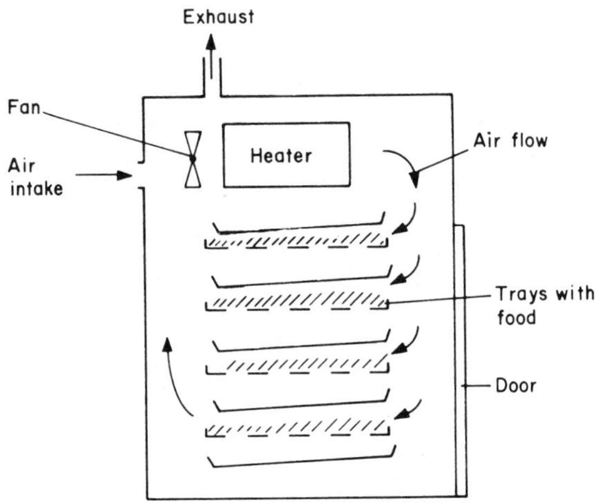

FIG. 10-17. Schematic representation of a simple tray drier.

TABLE 10-4

Expected Effects of Several Process Parameters on Drying Rates of Food Pieces Dried in Layers (Basis: Unit Surface Area of Tray)

	Effect with increasing value of each parameter		
Parameter	Drying rate in constant rate period (w'_I)	Duration of constant rate period (t_c)	Drying rate in falling rate period (w'_{II})
Air velocity (G)	$w'_I \propto G^{0.8}$	Decreases	Small effect
Air temperature (T)	$w'_I \propto (T-T_w)$	Decreases	Increases
Air humidity (H)	Decreases	Increases	Decreases
Piece size	In flow over trays, minimal effect; in crossflow, decreases	Decreases	$w'_{II} \propto L^{-2}$
Bed depth[a]	In flow over trays, minimal effect; In crossflow, increases	Depends on degree of packing in the layers	

[a] Note that the basis is unit area of tray and not unit weight of food. The effects are quite different for the two bases.

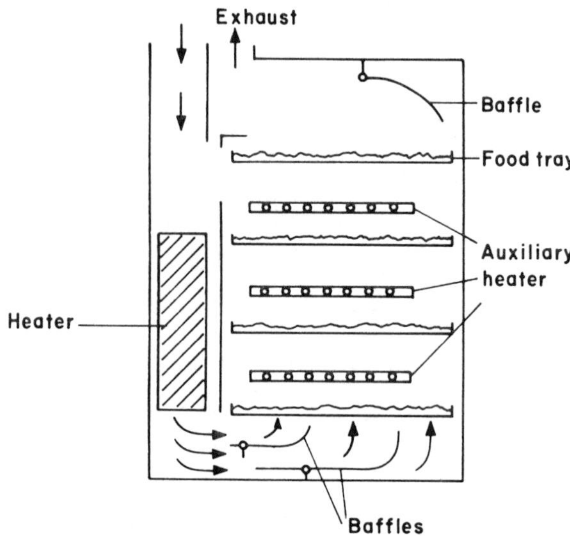

FIG. 10-18. Schematic representation of a tray drier with auxiliary heaters.

mean radius of the particle. In the other extreme, when air flows over beds of densely packed particles, the area reduces to the surface area of the trays and L is the bed thickness.

It is difficult to quantitatively analyze the course of drying in batch driers. However, it is possible to predict the effects of various operational parameters on rate of drying and these are summarized in Table 10-4. A computer program has been used to calculate the drying rate of fish fillets in a kiln drier and the predictions have been found in good agreement with actual performance [7].

B. Continuous Hot-air Driers

In many food-dehydration operations drying is conducted continuously in tunnel driers, or belt driers. Figure 10-19 compares schematically the operations of tunnel driers with co-current and counter-current flow. Cross-current flow and a great variety of mixed systems are also possible. Figure 10-20 shows a mixed cross- and counter-current operation in a tunnel, and Figure 10-21 a mixed co- and counter-current operation of a belt drier. A two-stage operation for preparing a snack product in a belt drier is shown in Figure 10-22. The drier operates in counter-current flow in the first stage, which is designed to reduce the moisture content to a level where cutting of the sheet into the desired shape is easiest to perform. The final drying is carried out in a stage with crossflow. Note that in the stage with crossflow the average piece dimension is important since drying is limited by internal resistance to diffusion of water. Cutting of the product effectively reduces the internal diffusional resistance.

The temperature distribution in a drier varies with the type of heating and the arrangement of air flow. In a convective drier operating in the constant rate

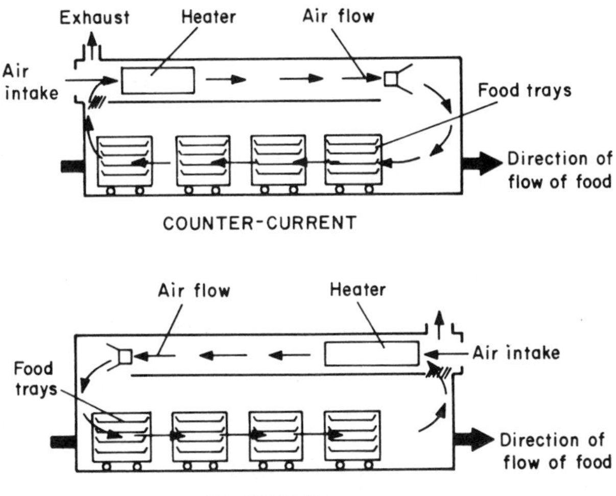

FIG. 10-19. Schematic comparison of co-current and counter-current tunnel driers.

period only the temperature distribution is shown in Figure 10-23. If, as is usually the case, the drying takes place primarily in the falling rate period, the temperature distributions are approximated by Figure 10-24. Tunnel as well as band driers may also contain internal heaters, in which case the air temperature can be maintained at any desired level above the wet bulb temperature.

Flow arrangements are chosen depending on the properties of the product. Counter-current operation provides the highest temperatures at the point where the moisture content is lowest and therefore counter-balances the increasing resistance to drying. However, this arrangement increases the danger of "scorching" the product. Often "scorching" implies damage due to nonenzymatic browning, and as we have seen in Chapter 8, this reaction has a maximum rate at intermedi-

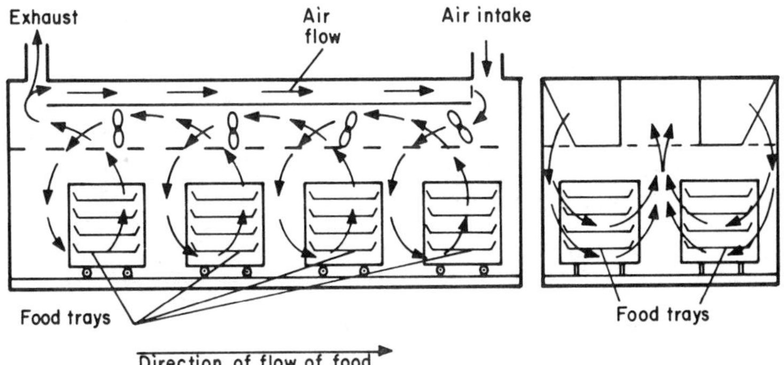

FIG. 10-20. Schematic representation of air flow in a tunnel drier with mixed (crossflow and counter-flow) flow.

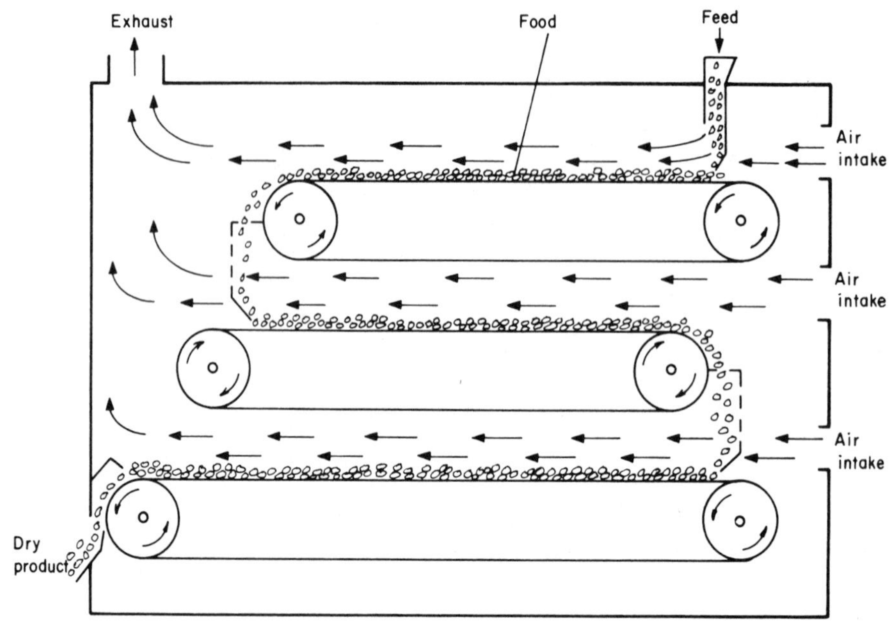

FIG. 10-21. Schematic representation of a belt drier with mixed co-current and counter-current air flow.

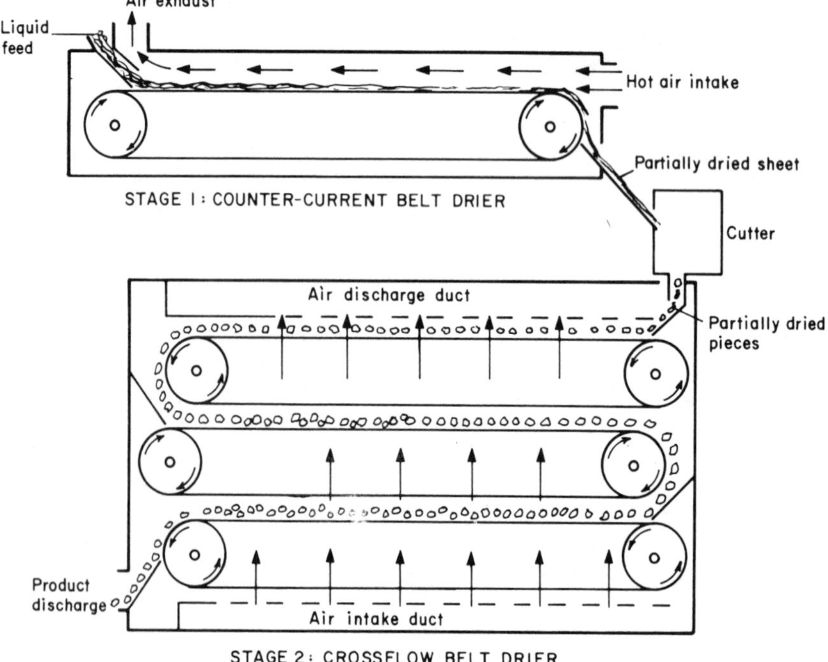

FIG. 10-22. Schematic representation of a two-stage belt drier for drying of a snack product.

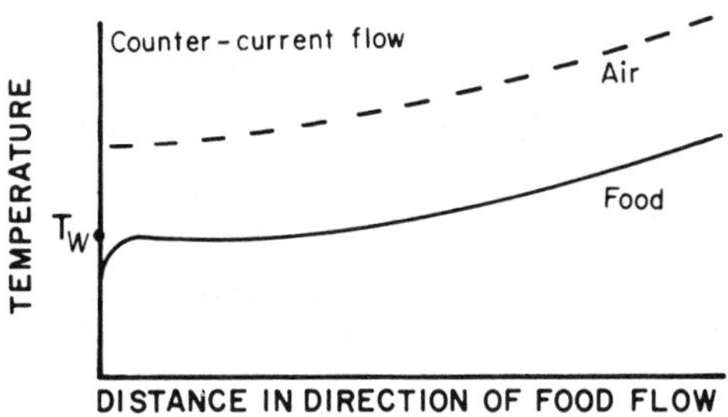

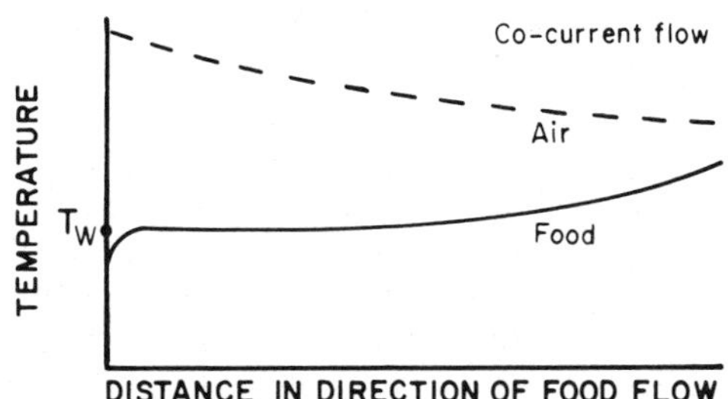

FIG. 10-23. Schematic comparison of temperature patterns in co-current and counter-current tunnel driers operating primarily in the falling rate period.

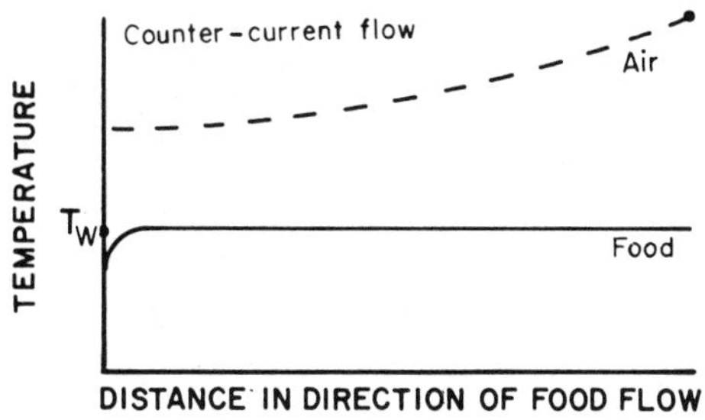

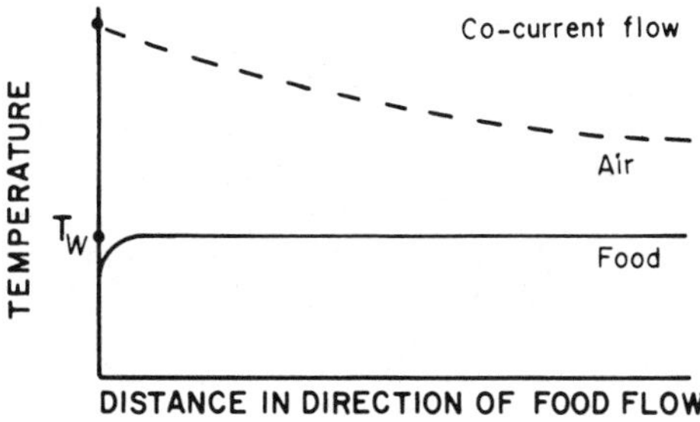

FIG. 10-24. Schematic comparison of temperature patterns in co-current and counter-current tunnel driers with operation in the constant rate period only.

ate moisture contents, the exact position of the maximum varying from food to food. Co-current operation allows the use of high entering air temperatures and therefore results in rapid initial drying, which is often desirable but in a few cases is deleterious to the texture of products.

Another important type of continuous drier, suitable for large-scale production of "flowable" products, is the rotary drier. Granulated sugar is dried in such driers. For additional information the reader should consult one of the several excellent books on dehydration which are listed in the references.

C. Atmospheric Conductive Driers — Drum Driers

Heat is transferred to the drying substances by conduction in drum driers. Drum driers consist of hollow drums constructed of carefully machined iron or stainless steel. The drums are heated internally by steam. The food to be dried must be in liquid or slurry form and is applied to the outer surface by one of the several methods indicated in Figure 10-25, namely by dipping, splashing, or spraying or by the use of spreading devices or feeding rollers. The drier may consist of single, "twin," or "double" drums as indicated in Figure 10-26. With double drums the thickness of the film on the drum depends on the clearance between the drums, which can be adjusted. When liquid is applied by dipping, the thickness is dependent on physical properties of the liquid and on the peripheral velocity of the drum. The most important physical parameter of the liquid is its surface tension (γ) and its contact angle (θ), with the film thickness considered proportional to $(\gamma \cos\theta)^{3/2}$. The film therefore can be decreased in thickness by lowering surface tension through addition of surface-active agents. The contact angle for aqueous solutions on metals is close to zero (and hence $\cos\theta = 1$), but the presence of fat or of greasy films on the drum can adversely affect the uniformity of the applied liquid film and can have ruinous effects on both the drying rate (thick films dry slowly; part of the total drum area is not utilized) and the quality of product (overheating of some portions and underdrying of other portions).

In an efficient operation, the rate of drying depends only on heat transfer from the interior of the drum to the product and is given by

$$\frac{dw}{dt} = UA\Delta T\,(\Delta H_V^{-1}) \qquad (10\text{-}34)$$

where U is the overall heat-transfer coefficient (Btu hr^{-1} ft^{-2} $°F^{-1}$). Since the thickness of the drum wall is usually small compared to the diameter of the drum, the area, A, can be regarded simply as the outer surface area of the drum. When the interior is heated by steam under pressure, the temperature of the drying liquid film usually remains close to its boiling point until the water content decreases substantially, at which point the temperature rises rapidly. Hence, ΔT is a mean value based on the difference between the constant internal temperature and the mean of the rising surface temperature. The overall heat-transfer coefficient U depends on h_V, k_W, ΔX_W, and h_L:

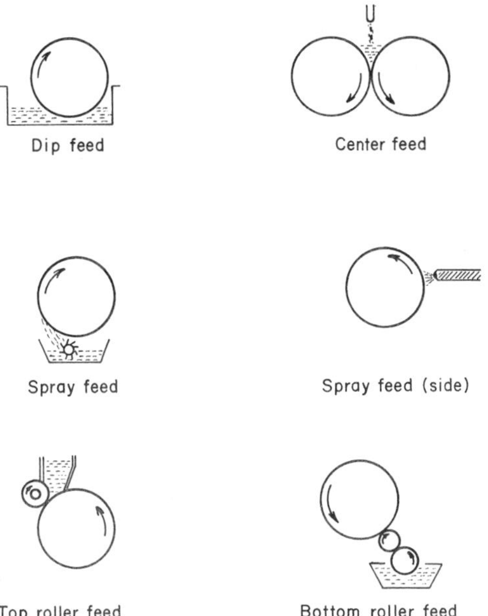

FIG. 10-25. Methods of applying feed to drum driers.

$$U = \frac{1}{\frac{1}{h_v} + \frac{\Delta X_w}{k_w} + \frac{1}{h_L}} \qquad (10\text{-}35)$$

where

h_v is the film coefficient of heat transfer for steam (Btu ft^{-2} °F^{-1} hr^{-1})
ΔX_w is drum wall thickness (ft)
k_w is drum wall conductivity (Btu ft^{-1} °F^{-1} °F^{-1} hr^{-1})
h_L is film coefficient of heat transfer for the drying film (Btu ft^{-2} °F^{-1} hr^{-1})

The factor having the greatest effect on U is the condition of the liquid film. Drying rates for drum driers can be extremely high when thin, uniform films of low viscosity are evaporated, and it is thus permissible to use high temperatures. Some boiling films, for instance, can be evaporated at rates as high as 50 lbs/hr ft^2, but food materials are more likely (because of high viscosity and of heat sensitivity) to dry at rates of 1-5 lbs of water removed per hour per square foot.

In addition to assuring an adequate heat transfer, the drying system must provide for removal of water vapor. At atmospheric pressure, with air entering at 120°F and 50% saturation and discharging at 90% saturation, the removal of 3 lbs of water per hour would require a minimum dry air flow of about 2000 cu ft/hr.

The removal of product may present special requirements for materials which tend to be plastic at the temperature of discharge. In this instance the product is often cooled just prior to the discharge point by a directed stream of cold, dry air.

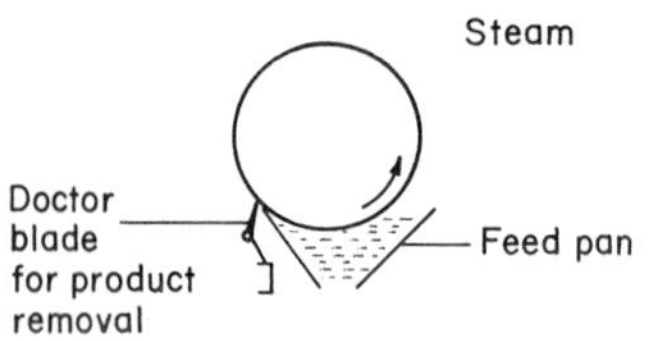

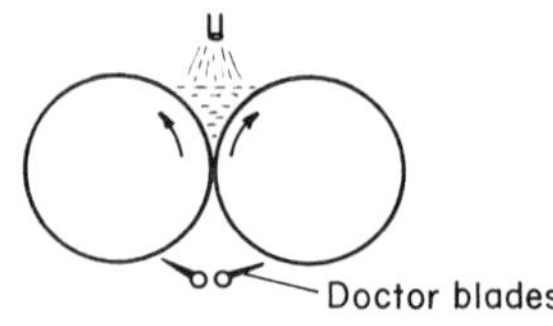

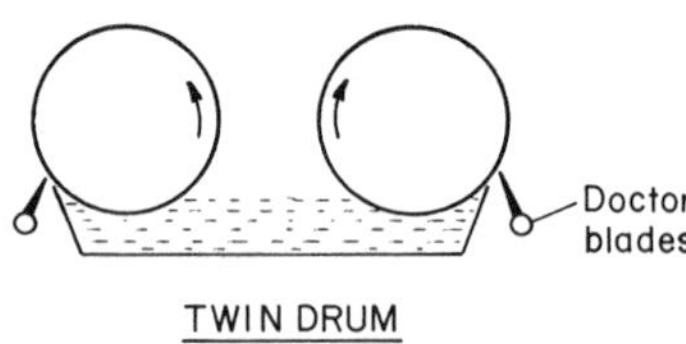

FIG. 10-26. Schematic representation of types of drum driers.

Additional modifications may include controlled-temperature zones along the periphery of the drums achieved by air flow at different temperatures.

The present applications of drum drying in the food industry are limited. Potato flakes and " instant" cereal products are the major products drum dried in the United States. In Europe potato products are drum dried on a large scale. The major advantages of the process are: (1) relatively low cost (Spadaro et al. [43] have estimated it at 0.8 cents/lb), and (2) continuous operation at high throughput rates.

D. Spray Drying

1. Principle of Operation

Spray drying is the most important method for dehydrating liquid food products. Among the widely recognized applications are drying of milk products, coffee, and eggs. A wide variety of other products have been successfully spray dried, either in pilot plants or in commercial installations, and a partial list is shown in Table 10-5. Historically, the development of spray drying in the food industry has been closely associated with dehydration of dairy products.

TABLE 10-5

Some Food Substances that Have Been Successfully Spray Dried

Bananas	Potatoes
Blood	Proteins from various plant sources
Cake mixes	Soy isolates and hydrolyzates
Casein and caseinates	Starch and derivatives
Cheese	Tea
Citrus juices	Tomato puree
Coffee	Yeast
Corn hydrolyzates	Cream
Egg	Icecream
Fish concentrates	Milk replaces
Milk	Yogurt

In principle, spray drying consists of the following steps. A liquid or a paste is atomized into a chamber, where it is put in contact with a stream of hot air and rapidly dried. The dry particles, suspended in the air stream, flow into separation equipment where they are removed from the air, collected, and packaged or subjected to further treatment, such as instantizing (Fig. 10-27). Each of the steps, so simply described, actually constitutes a complicated and delicately balanced engineering operation [22, 32]. Thus the properties of feed, including viscosity, surface tension, and chemical composition, must be controlled; the atomization procedure is extremely important and often imperfectly understood; the flow and heat- and mass-transfer events in the chamber are complicated; and the separation of the dry material from the air stream is often difficult to achieve efficiently. The characteristics of the product may place additional restraints on the operation. For instance, drying of some types of materials, such as protein hydrolyzates, may require that the spray drier walls be cooled to prevent powder from sticking to them. The complicated nature of the flow and heat-transfer patterns in the chamber makes this operation difficult to describe in simple engineering equations; furthermore, and perhaps more significantly, they make pilot plant experience difficult to apply to subsequent commercial scale-up.

2. Atomization

Since atomization determines the size distribution of droplets, it is the most important feature of a spray drier. Atomization can be achieved by means of high-pressure nozzles in which a fluid acquires a high-velocity tangential motion while being forced through the nozzle orifice. The fluid swirls out in a cone-shaped

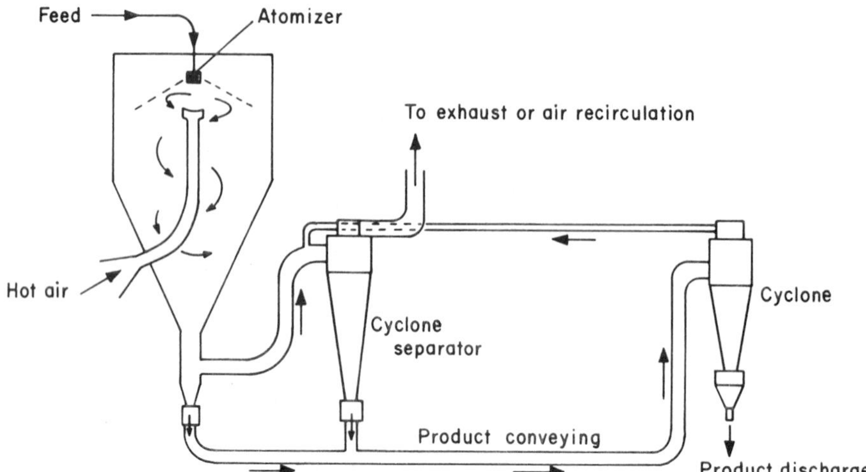

FIG. 10-27. Schematic representation of a typical spray drying operation.

sheet, which breaks up into droplets. Two-fluid nozzles use a second fluid, such as air or steam, to break up the stream of the feed into a mist of fine droplets. Both of the above types of atomizers require homogeneous feed containing no particles capable of clogging the nozzle. For nonhomogeneous slurries and suspensions, use is made of centrifugal disk atomizers. The feed is delivered to a disk spinning at high velocities (2000-20,000 rpm) and is thrown off the disk in the form of ligaments, sheets, or elongated ellipsoids, all ultimately forming spherical droplets. The size distribution depends on disk speed and on viscosity and surface tension of the fluid. Recently, attention has been given to sonic and ultrasonic means of droplet production, but no applications to foods have appeared.

3. Heat and Mass Transfer in the Drying Chamber

Few details are known about actual heat- and mass-transfer processes in the drying chamber, primarily because the pertinent variables, such as temperature, humidity, and droplet distribution, are difficult to measure and to estimate [21]. Theoretical analysis of drying processes under conditions approximating the events in a single droplet in a spray drier, however, can be studied readily. Van der Lijn et al. [46] have studied droplet heat and mass transfer under spray-drying conditions and have concluded that (1) a history of drying in a single droplet can be constructed; (2) in the initial period, temperature increases to the wet bulb temperature; (3) in the second period, a concentration gradient builds up in the drop and water activity at the surface decreases, thus causing the surface temperature to rise above that of the wet bulb temperature; (4) in the third period, internal diffusion is entirely limiting; and (5) there is a critical moisture content below which the surface becomes impermeable to aroma compounds, thus preventing losses of flavor. (High inlet temperatures and perfect co-current patterns tend to minimize flavor losses.)

The drying time for a single droplet can be approximated by

$$t = \frac{r^2 \rho_L \Delta H_v}{3h(\Delta T)} \cdot \frac{m_i - m_f}{1 + m_i} \tag{10-36}$$

4. Separation of Dry Particles

Separation is carried out partly within the drying chamber itself and partly in secondary separation equipment. In Chapter 9 (Fig. 9-17) we discuss entrainment separators operating on the principle of "momentum separation." This principle is used extensively in spray driers. In addition, it is often necessary to have other types of separators to remove very fine particles from exhaust gases. Cyclone separators (which operate on the "momentum separation" principle) allow, under some conditions, very extensive powder losses, thus necessitating the addition of expensive filters, scrubbers, or electrostatic precipitation equipment. Masters [33], for instance, has estimated that filters are four to six times more expensive than cyclones, both in capital and in operating costs.

In general it is relatively easy to remove 95-98% of the solids, but removal of the remainder becomes progressively more difficult and expensive. In spite of the increased cost it is often necessary to have very efficient air cleaning, not only because of the value of the product but also because of the necessity to avoid air pollution.

5. Structure of Spray-dried Materials

The highly porous nature of hollow spray-dried particles is readily apparent from examination of the dry particles. The scanning electron microscope has been particularly valuable. Buma [2] has used scanning electron microscopy and other methods to study the structure of spray-dried milk particles and the distribution of free fat in them. He showed that spray drying results in hollow particles, the shells of which have a glassy structure, primarily amorphous lactose, that entraps a part of the total fat. Other portions of the fat are "free" in the sense of either being located directly on the surface or being accessible to solvents used for fat extraction. Scanning electron micrographs of spray-dried caseinate and spray-dried lactose show that lactose gives spherical particles with no dents or folds, whereas caseinate, as does spray-dried skim milk, gives folded particles [3]. Buma and Henstra [3] have attributed the folds in dried skim milk to the presence of casein, which, unlike amorphous lactose, shrinks unevenly during drying.

Phase transformations of carbohydrates affect their water-binding characteristics. In spray-dried milk the matrix of the shells of hollow particles is based on amorphous lactose. In many other spray-dried products, such as juices and extracts, various amorphous soluble substances compose the spray-dried particle matrix. Recrystallization in storage or partial crystallization in the drying process itself can cause caking or important changes in water absorption, color, texture, and ability to entrap flavors.

6. Selected Developments in Industrial Practice of Spray Drying of Foods

The major items spray dried are eggs, instant coffee, and milk products, including nonfat milk, whey, and casein. One of the largest egg-drying installations is located in Riverside, California, and it removes 4000 lbs H_2O an hour in a drying chamber 14 ft in diameter [40]. Some milk-drying installations operate at throughputs in excess of 50,000 lbs/hr; very large driers are also used for coffee [27].

Spray drying of fruit and of tomato products is difficult. They contain a high percentage of sugars and other soluble solids, yield amorphous thermoplastic flasses on dehydration, and tend to adhere to the walls of the drying chamber. Various remedies have been suggested. An example is a 250 ft tower used for drying tomato paste. This tower uses air at a temperature below 125° F that has been dehumidified by passage through silica gel. The capacity of this Swiss plant is several thousand pounds of powder a day.

Fruit juices can be spray dried when substances are added which reduce the tendency of the powder to stick. The "sticky point," or temperature at which the powder tends to stick when tested under standardized conditions, increases (and this increase is desirable) when polymers or proteins are added. Liquid glucose is also a suitable additive for extremely sensitive substances, such as orange juice. The "sticky point" also increases rapidly with decreasing moisture, so that once the powder is reduced to a moisture content of 2-3% it can be safely handled at room temperature, provided humidity of the room is low enough.

E. Instantized Powders

It is desired to have powders that disperse rapidly and completely in water. To fulfill this requirement powders must have the following properties:

1. a large wettable surface
2. sinkability (must not float on surface)
3. dispersibility of solubility
4. resistance to sedimentation

The wettability depends on the total surface area of the powder and on surface properties of the powder particles [37, 41]. Figure 10-28 is a schematic representation of forces acting on a drop of liquid resting on the surface of a solid. The balance of forces results in the establishment of the contact angle shown in Figure 10-28. Equation (10-37) gives the ideal dependence of this angle on surface properties.

$$\cos \theta = \frac{\gamma_s - \gamma_{sL}}{\gamma_L} \qquad (10\text{-}37)$$

where

θ is the contact angle
γ_s is the free surface energy of the solid (in contact with air) (ergs/cm^2)

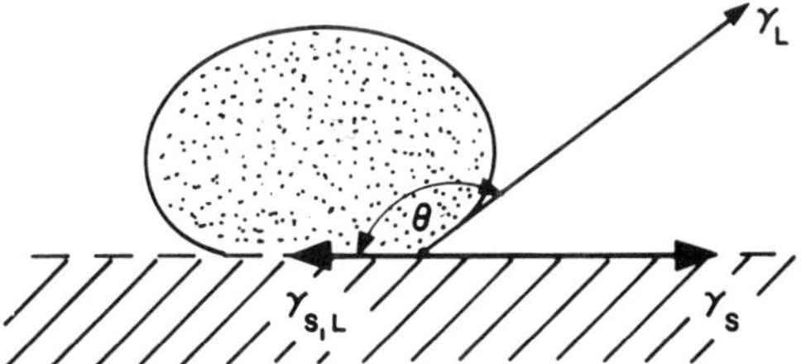

FIG. 10-28. Contact angle of a liquid drop on a solid surface.

γ_L is the free surface energy of the liquid (in contact with air)
γ_{sL} is the interfacial free surface energy between solid and liquid

Wetting of a solid by a liquid, as shown in Figure 10-29, occurs when the spreading coefficient S is positive:

$$S = \gamma_s - \gamma_{sL} - \gamma_L \tag{10-38}$$

Spreading coefficient S is related to the contact angle by

$$S = \gamma_L (\cos\theta - 1) \tag{10-39}$$

and is positive only when $\theta = 0^\circ$. Hence, one of the requirements for wetting is a zero contact angle. Inspection of Eq. (10-38) shows that the spreading coefficient can be increased by increasing γ_s, which in practical terms means increasing the polarity of the particle surface.

In real food materials, inadequate surface polarity normally results from free fat located on the surface. Counteracting measures involve the addition of such surfactants as lecithin. The other possibility is to lower the surface tension of the rehydration liquid. Addition of wetting agents to the powder can accomplish this aim if they rapidly dissolve in the rehydration medium and do not otherwise impair the powder properties.

Assuming that the polarity of the powder surface is as high as possible, the other obvious course is to increase the internal surface area by making the

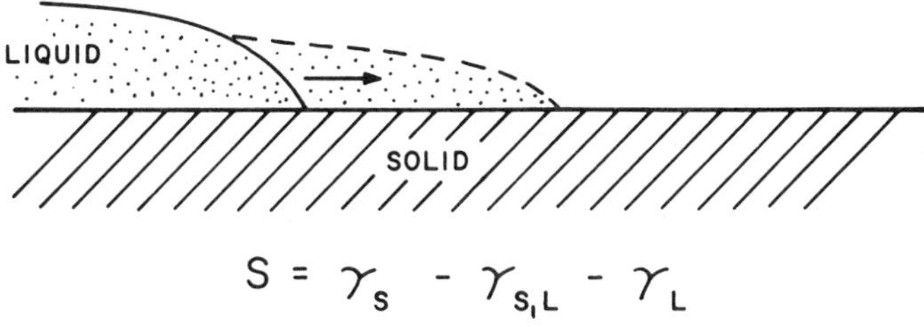

FIG. 10-29. Spreading of a liquid on a solid surface.

particles small. However, this approach fails. When the particles become very small, wetting is greatly impaired and the particles tend to clump together. Sticky clumps, wet outside and dry inside, are created by this system of incompletely wetted powder and entrapped air. A product with good sinkability readily penetrates the liquid and stays there.

In practice, the industry has achieved good dispersibility by a combination of surface treatments mentioned above (e.g., addition of lecithin) and agglomeration.

In agglomerated products, the individual, very small particles are fused into much larger aggregates in which the points of contact between the particles are so few that practically all of their surface area is available for wetting. These aggregates are stable enough, however, to assure against clumping on exposure to the rehydration liquid and they break apart only after each particle is well surrounded by the rehydration medium.

Figure 10-30 shows schematically the appearance of agglomerated powders. In most cases the desirable size of aggregates is 100-250 μm.

Instant powders are usually produced by processes in which spray-dried powder is rewetted. The degree of this rewetting is carefully controlled; the particles are brought into contact and allowed to grow into aggregates; then the aggregates are redried, cooled, and separated by size [38, 39].

Processes for instantizing milk powders have been developed in the United States by several different companies. The processes differ primarily in the type of equipment used in the various steps. For instance, the humidification necessary for surface rewetting can be conducted by (1) exposing free-falling particles to low-pressure steam; (2) air conveying particles into a chamber and exposing them to a controlled amount of steam, or slightly superheated steam; or (3) passing an air-stream mixture through a fluidized bed of particles. Subsequent drying and the period of time the particles are in contact with each other in the turbulent air stream also can vary from process to process [35, 41, 44]. Most of the processes developed for milk also can be applied to other spray-dried particles. A schematic representation of an agglomeration process is shown in Figure 10-31.

F. Osmotic Drying

Dehydration of foods by immersion in liquids with a water activity lower than that of the food forms the essence of osmotic drying. Solutions of sugars or of salt are usually employed, giving rise to two simultaneous counter-current flows:

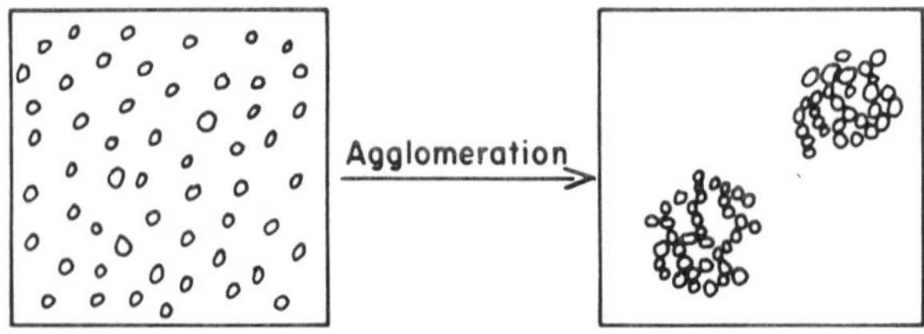

FIG. 10-30. Schematic representation of particle agglomeration.

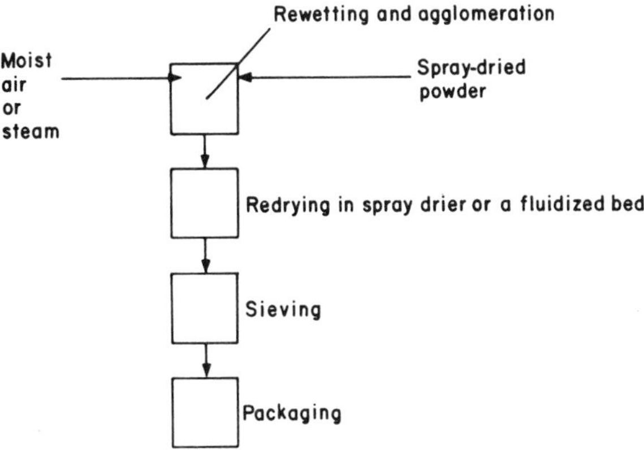

FIG. 10-31. Steps in agglomeration processes.

solute diffusion from solution into the food and water diffusion out of the food into the solution. Diffusivity of sugars is much lower than that of water, making it possible to design processes which result in substantial water removal with only marginal sugar pickup. Slow processes, on the other hand, approach equilibrium for both sugar and water, resulting in "candied" sugar-rich fruits. Disruption of structural barriers within the fruit pieces can also be expected to increase diffusivities of both water and sugar, thereby allowing faster approach to equilibrium and favoring "candying" as opposed to dehydration.

Figure 10-32 shows schematically how the water content and sugar concentration change with time in fruit pieces exposed to concentrated sugar solutions. It is evident that if the process of equilibration is interrupted early it is possible to remove a great deal of water without substantial sugar infusion. Furthermore, since the diffusion of sugar into tissues is slow, most of the sugar "picked up" by the fruit is on or near the surface and can be removed by a quick rinse with water. The rinsing does increase the moisture content, but in processes in which osmotic drying is a pretreatment for vacuum or freeze-drying this amount is not significant. In other cases the increased sugar content of fruit or the increased salt content of salted fish (which is an osmotically dried product) may be acceptable without rinsing.

Farkas and Lazar [8] have found osmotically dried apples to have a soluble solids content of about 40% of total solids, compared to 35% in air-dehydrated apples; thus, the sugar pickup is about 25% on a total solids basis.

Osmotic dehydration is best achieved in a stirred bath at elevated temperature. For example, golden delicious apples cut into wedges and half rings can be osmotically reduced in weight by 50% by treatment in a 70° Brix sucrose at 122°F [8]. In my laboratory similar results have been obtained with apples, melons, and peaches and the Hawaii Experimental Station has reported satisfactory production of osmotically treated bananas. Coltart [6] has patented a process for producing osmotically dehydrated fruits by immersing them in a rotating drum containing a 40-90% sugar solution at a temperature of 70°-180°F.

Camirand et al. [4] have osmotically dried shrimp, meat balls, and fruit to which they have previously applied a pectate coating that is permeable to water but not to sugars. Because of the coating, however, the drying times are very long, ranging up to 144 hr.

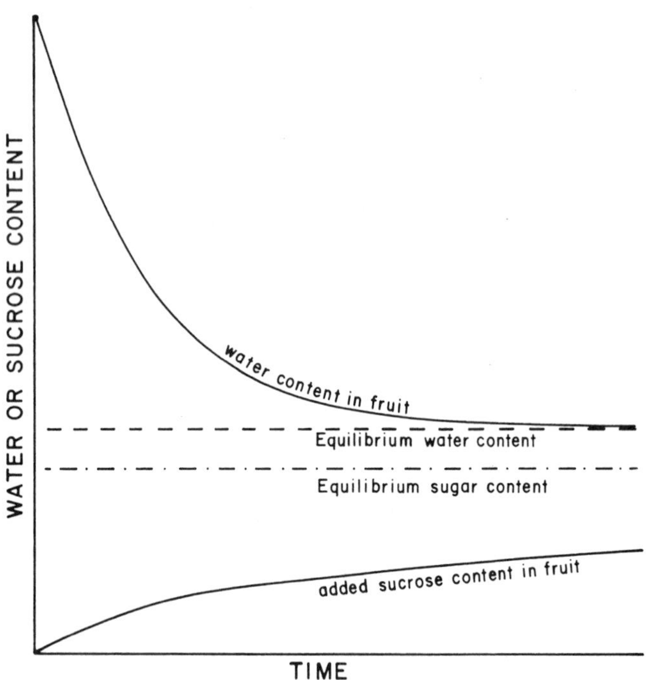

FIG. 10-32. Approach to equilibrium water and sugar contents during osmotic drying.

G. Intermediate-moisture Foods (IMF)

Intermediate-moisture foods are characterized by a water activity low enough to prevent the growth of bacteria and by conditions minimizing the potential for growth of other microorganisms. Definitions of such foods are therefore not rigorous. "Dried fruits," also known in the industry as "evaporated fruits," such as prunes and apricots, certainly belong in this category, as do jams, some types of sausage, and some pie fillings.

Water activities and associated moisture contents vary widely among these well-known foods, ranging generally from 0.7 to 0.9 water activity and from 20 to 50% water by weight. Growth of organisms resistant to low water activities usually is prevented by chemical additives (e.g., sulfite in fruit; nitrite in meat products; preservatives of various kinds in jams). Pasteurization and/or low-temperature storage also can be used to extend the lives of products with water activities near the critical level. These approaches are sometimes used with semistable sausage products that have relatively high water activities of 0.83-0.97.

IMF have received increased attention since the development of new products based on the following technological principles: lowering of water activity by adding a solute, such as glycerol, sucrose, glucose, or salt and retarding microbial growth of adding antimicrobial agents (primarily antimycotic), such as propylene glycol and/or sorbic acid. Water activity can be further depressed by partial dehydration and enzyme activity can be prevented by heating.

The technological renaissance of IMF made its initial impact in the area of pet foods. The pet IMF, known in the trade as "soft-moist," are based on a combination of meat byproducts with soyflakes and sugar (glucose or sucrose) to give a product with a moisture content of about 25% and a water activity of about 0.83. Propylene glycol (about 2%) and potassium sorbate (0.3%) provide antimycotic activity, and the glycol also serves as a plasticizer. Emulsifiers, salt, nutritive supplements, and other functional components are also added.

Development of IMF for human consumption has been encouraged by the United States military and NASA because of logistic advantages of such food. The major problem in developing such foods has been the inadequate number of safe, effective, and palatable osmotic agents and/or preservatives [1, 19, 20].

IMF for human consumption generally have glycerol contents of 10-50%, water contents of 15-40%, and water activities of 0.80-0.85. Two major methods have been used to achieve this composition; solid food pieces can be freeze dried and then soaked or infused with a solution containing the osmotic agents, or solid food pieces can be cooked in an appropriate solution to bring the final water content to the desired level.

The commercial application of IMF is still in its infancy but is likely to develop further in the future. A number of patents have been issued for various types of intermediate-moisture foods. Intermediate-moisture items also have been used in the NASA Apollo and Skylab programs, and the composition of an intermediate-moisture fruit cake used on Apollo flights is shown in Table 10-6.

H. Other Methods of Drying

The discussion above has been, by necessity, limited to only a few methods of drying. Some additional methods merit at least a brief mention.

1. Fluidized-bed Drying

We have seen that it is possible to conduct drying by passing hot air through beds of food particles, as in a crossflow tunnel or belt drier. As the velocity of the air passing through a bed of particles is increased, a situation is eventually reached where the pressure drop across the bed balances the weight of the bed. At greater air velocities the bed expands, particles become suspended in air, and the bed is said to be fluidized. The bed remains uniform and behaves as a uniform fluid when the so-called Froude number is below unity.

$$\text{Froude number} = u^2/2gr$$

where

u is air velocity
g is the gravitational constant
r is the average particle radius

TABLE 10-6

Ingredients of the Apollo 17 Fruit Cake [a]

Ingredient	Percent by weight
Flour, wheat, soft	7.3
Flour, soy	7.3
Sugar	19.0
Shortening	7.8
Eggs, whole, fresh	6.96
Salt	0.4
Baking powder	0.4
Water	2.2
Cherries, candied	10.4
Pineapple, candied	8.6
Pecans, shelled	13.8
Raisins, bleached	15.6
Clove powder	0.06
Nutmeg	0.06
Cinnamon	0.12
	100.00

[a] After Heidelbaugh and Karel [49].

As the velocity increases further the bed is disturbed and assumes the appearance of a "boiling fluid." At still higher air velocities individual particles are carried away (conveyed) by the air.

Fluidizing is a very effective way of maximizing the surface area of drying within a relatively small total space. The characteristic dimension limiting diffusion to the surface of the particle is the radius of the particle, and for small particles drying rates can be very high. The major limitation of fluidized drying is the limited range of particle sizes which can be effectively fluidized.

Fluidized-bed drying can be carried out as a batch or continuous drying process, and a variety of modifications, including operation at gravitational forces greater than 1 x g (in a centrifugal drier), is possible. There are several recent reviews of fluidized drying, including a book by Vanecek et al. [47] and articles by Holdsworth [16] and Farkas et al. [9].

2. Pneumatic Driers

In these driers particles of food are carried along by a stream of air as they are simultaneously exchanging mass and heat with the air. In principle, the drying is similar to spray drying, except it is applied to solid particles. It has the advantage of high drying velocities, similar to those of spray and fluidized-bed drying, and is limited primarily by the type and size of particle that can be conveyed effectively without damage by an air stream.

3. Foam Mat Drying

The desire to increase rates of drying for fluid and semisolid foods has led to the development of this method. A viscous foam with a large internal surface area is formed and deposited in a relatively thin layer upon a perforated belt. The layer is then dried, usually in a convective drier with crossflow of air. Practical applications of this method are still limited [15, 24].

SYMBOLS

(The specific units for each symbol represent examples only; other consistent English or metric units could also be used.)

A	area (ft^2)
$\overline{B}$	degree of browning calculated using average conditions
B_Σ	degree of browning calculated by integration of point conditions
C	concentration (lb/ft^3)
D	diffusivity (ft^2/hr)
D_0	constant
D_c, D_L	diffusivities for specific conditions defined in text
E_D	energy of activation (kcal/mole)
G	gas flow rate (lb/hr)
H	absolute humidity (lb H_2O/lb dry air)
ΔH_v	latent heat of vaporization (Btu/lb)
L	thickness of slab (ft)
L_e	Lewis number
P°	vapor pressure (torr)
P_T	total pressure (torr)
R	gas constant
RH	relative humidity (%)

T	temperature (°F)
T_w	wet bulb temperature (°F)
U	overall heat-transfer coefficient (Btu ft^{-2} hr^{-1} $°F^{-1}$)
X	distance (ft)
b	permeability (lbs hr^{-1} ft^{-1} $torr^{-1}$)
c_p	specific heat (Btu/lb °F)
c_H	humid heat (Btu/lb °F)
g	acceleration due to gravity (ft/sec^2)
h	film coefficient for heat transfer (Btu ft^{-2} $°F^{-1}$ hr^{-1})
i	enthalpy (Btu/lb)
k	thermal conductivity (Btu ft^{-1} hr^{-1} $°F^{-1}$)
k_g	film coefficient of mass transfer (lbs hr^{-1} ft^{-2} $torr^{-1}$)
ℓ	height
m	moisture content (lb H_2O/lb dry food)
m_c, m_e, m', m_h	specified moisture contents
p	partial pressure of water (torr)
q_D	heat supplied to drier by means other than hot air
q_{loss}	heat loss from drier
r	radius of particle or droplet
s	weight of solids (lbs)
t	time (hr)
u	air velocity (ft/sec)
v_H	specific volume of moist air (ft^3/lb)
w	weight (lb)
z	thickness of "dry" layer, in air drying
α	thermal diffusivity (ft^2/hr)
γ	surface tension (dyne/cm)
θ	contact angle
μ	slope of moisture content gradient with time
ρ	density (lb/ft^3)
σ	constant

Subscripts

F	refers to falling rate period
I	refers to constant rate period
L	refers to properties of liquid
a	refers to air properties
f	refers to final conditions
i	refers to initial conditions
s	refers to properties of solid
v	refers to properties of water vapor

REFERENCES

1. M. C. Brockmann, Food Technol., 24, 896 (1970).
2. T. J. Buma, Free Fat and Physical Structure of Spray-Dried Whole Milk, Doctor of Natural Sciences Thesis, Landbouwhogeschool, Wageningen, The Netherlands, 1971.
3. T. J. Buma and S. Henstra, Netherlands Milk Dairy J., 25, 278 (1971).
4. W. M. Camirand, R. R. Forrey, K. Popper, F. P. Boyle, and W. L. Stanley, J. Sci. Food Agr., 19, 472 (1968).
5. J. Chirife, J. Food Sci., 36, 327 (1971).
6. M. L. Coltart, Process and Apparatus for Producing Fruit Products by Osmotic Dehydration, Canadian Pat. 878-427, 1971.
7. P. E. Doe, J. Food Technol., 4, 319 (1969).
8. D. F. Farkas and M. E. Lazar, Food Technol., 23, 688 (1969).
9. D. F. Farkas, M. E. Lazar, and T. A. Butterworth, Food Technol., 23(11), 125 (1969).
10. B. P. Fish, in Fundamental Aspects of the Dehydration of Foodstuffs, Society of Chemical Industry, London, 1958, p. 143.
11. P. Görling, in Fundamental Aspects of the Dehydration of Foodstuffs, Society of Chemical Industry, London, 1958, p. 42.
12. G. J. Haas, Food Eng., 43(11), 58 (1971).
13. T. Z. Harmathy, I & EC Fundamentals, 8, 92 (1969).

14. C. E. Hendel, V. G. Silveira, and W. O. Harrington, Food Technol., 19, 433 (1955).

15. M. S. Hertzendorf and R. J. Moshy, CRC Crit. Revs. Food Technol., 1(1), 25 (1970).

16. S. D. Holdsworth, J. Food Technol., 6, 371 (1971).

17. A. C. Jason, in Fundamental Aspects of the Dehydration of Foodstuffs, Society of Chemical Industry, London, 1958, p. 103.

18. M. A. Joslyn, in Fundamentals of Food Processing Operations (J. L. Heid and M. A. Joslyn, eds.), AVI Publ. Co., Westport, Conn., 1967, p. 501.

19. M. Kaplow, Food Technol., 24, 889 (1970).

20. M. Karel, CRC Crit. Revs. Food Technol., 3, 329 (1973).

21. R. B. Keey, Proceedings, International Symposium on Heat and Mass Transfer Problems in Food Engineering, Oct. 24-27, 1972, Vol. 2, Univ. of Wageningen, Wageningen, The Netherlands, 1972, p. F4-1.

22. R. B. Keey, Drying, Principles and Practice, Pergamon Press, Oxford, 1972.

23. C. J. King, Food Technol., 22, 509 (1968).

24. C. J. King, Proceedings, 3rd International Congress of Food Science and Technology, Washington, D.C., Institute of Food Technologists, Chicago, Ill., 1970, p. 565.

25. G. Kluge and R. Heiss, Verfahrenstechnik, 6, 251 (1967).

26. F. Kneule, Das Trocknen, Verlag Sauerlander, Aarau, Switzerland, 1967.

27. M. E. Knipschildt, J. Soc. Dairy Technol., 22, 201 (1969).

28. O. Krischer and K. Kröll, Trocknungstechnik. I. Die Wissenschaftlichen Grundlagen der Trocknungstechnik, 2nd ed., Springer Verlag, Berlin, 1963.

29. O. Krischer and K. Kröll, Trocknungstechnik. II. Trockner und Trocknungsverfahren, Springer Verlag, Berlin, 1959.

30. T. P. Labuza, CRC Crit. Revs. Food Technol., 3, 217 (1972).

31. M. Loncin, Die Grundlagen der Verfahrenstechnik in der Lebensmittelindustrie, Verlag Sauerlander, Aarau, Switzerland, 1969.

32. C. W. Lyne, Brit. Chem. Eng., 16, 370 (1971).

33. K. Masters, Spray Drying, CRC Press, Cleveland, Ohio, 1973.

34. P. Y. McCormick, Ind. Eng. Chem., 62(12), 84 (1970).

35. R. E. Meade, Food Eng., 43(7), 88 (1971).

36. G. Nonhebel and A. A. H. Moss, Drying of Solids in the Chemical Industry, CRC Press, Cleveland, Ohio, 1971.

37. R. Noyes, Dehydration Processes for Convenience Foods, Noyes Development Corporation, Park Ridge, N. J., 1969.

38. N. Pintauro, Agglomeration Processes in Food Manufacture, Noyes Development Corporation, Park Ridge, N. J., 1972.

39. J. Pisecky and V. Westergaard, Dairy Ind., 37, 144 (1972).

40. W. Plummer and J. P. Geddes, Food Eng., 40(12), 63 (1968).

41. E. Samhammer, Ernährungs-Umschau, 5, 163 (1972).

42. F. H. Slade, Food Processing Plant, Vol. 1, CRC Press, Cleveland, Ohio, 1967.

43. J. J. Spadaro, J. I. Wadsworth and H. L. E. Vix, ASHRAE J., 8(9), 55 (1966).

44. J. Ulrich, Zucker, 24, 753 (1971).

45. W. B. Van Arsdel, M. J. Copley, and A. I. Morgan, Food Dehydration, 2 vols., AVI Publ. Co., Westport, Conn., 1973.

46. J. Van der Lijn, W. H. Rulkens, and P. Kerkhof, Proceedings, International Symposium on Heat and Mass Transfer Problems in Food Engineering, Oct. 24-27, 1972, p. F3-1, Wageningen, The Netherlands, Vol. 2, 1972.

47. V. Vanecek, M. Markvart, and R. Drbohlav, Fluidized Bed Drying, Leonard Hill, London, 1966.

48. J. H. Young, Trans. ASAE, 12, 720 (1969).

49. N. D. Heidelbaugh and M. Karel, in Freeze Drying and Advanced Food Technology (S. A. Goldblith, L. Rey, and W. W. Rothmayr, eds.), Academic Press, London, 1975, pp. 619-641.

Chapter 11

FREEZE DEHYDRATION OF FOODS

Marcus Karel

I. INTRODUCTION

Freeze dehydration is an operation in which water is removed from foods by transfer from the solid state (ice) to the gaseous state (water vapor) as illustrated in the schematic diagram in Figure 11-1. As can be seen in this diagram, this operation (sublimation) can only be accomplished when the vapor pressure and temperature of the ice surface at which the sublimation takes place are below those at the triple point (4.58 torr and a temperature of approximately 32°F). The relationship between the vapor pressure of ice and its temperature below the triple point is shown in Figure 11-2. Partial pressure of all water in foods containing ice crystals must, of course, be equal to the vapor pressure of this ice, if an equilibrium is in existence. Accordingly, the diagram in Figure 11-2 also

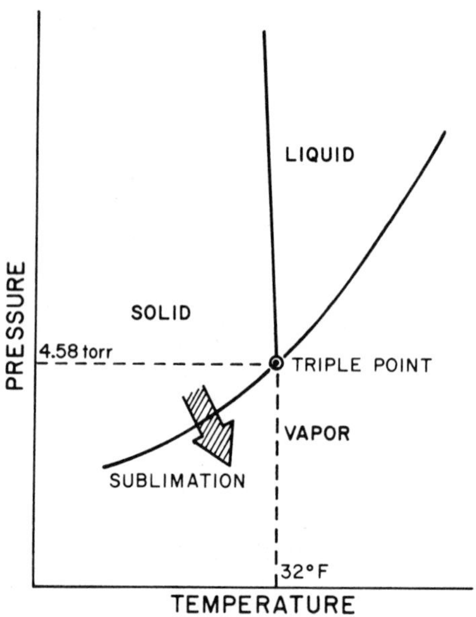

FIG. 11-1. Phase diagram for water and representation of the sublimation process.

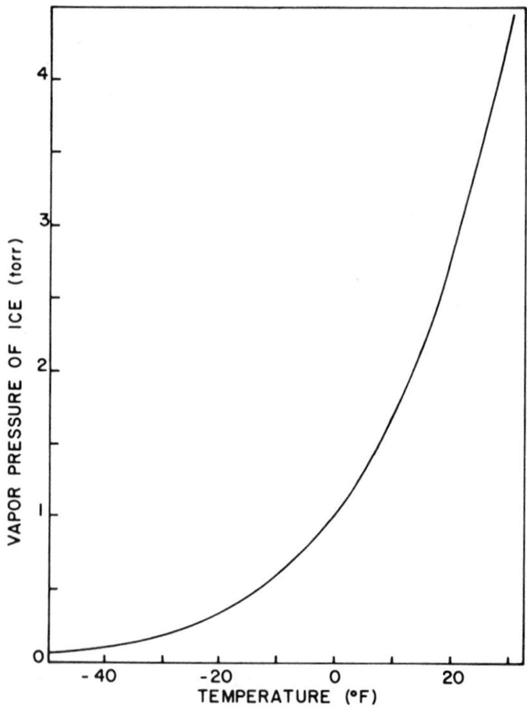

FIG. 11-2. Vapor pressure of ice.

represents the vapor pressure of water in frozen foods as a function of temperature, given the existence of an ice phase. Actual measurements of vapor pressure over frozen foods generally agree with this rule [45], although a deviation has been reported [6].

Food structure is affected differently by sublimation than by other methods of dehydration because of an absence of a liquid phase during freeze dehydration. Ideally, the distribution of moisture during sublimation is that shown in Figure 11-3a. There is an interface at which the moisture content drops from the initial level (m_0) in the frozen layer to the final moisture content of the dry layer (m_f), which in turn is determined by equilibrium with partial pressure of water (p_s) in the space surrounding the dry layer. Actually, this ideal representation is not true and the process is represented more accurately by Figure 11-3b, which shows the existence of some gradients in the "dry layer." The figure shows that there exists a "transition region" in which there is no longer any ice but in which the

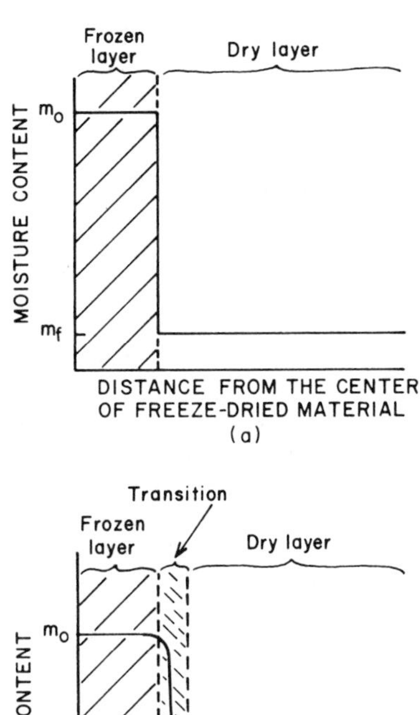

FIG. 11-3. Schematic representation of the moisture gradient in freeze-drying material. (a) Idealized gradient, (b) probable gradients occurring in freeze drying.

moisture content is still substantially higher than the final content of the dry layer (m_f). Recent detailed studies on the nature of this transition layer show it to be relatively narrow, and the assumption of a uniformly receding interface of zero thickness does not lead to large errors in the engineering analysis of sublimation [21, 25]. Furthermore, chemical, physical, and microscopic analyses all confirm that flow, redistribution of components, and disruption of structure in the transition region are virtually prevented if the temperature of the frozen layer is sufficiently low to avoid "eutectic" melting [37].

Structural changes can occur even in the completely dry layer if the temperature is high enough (collapse temperature) [24, 28], but major structural and compositional changes due to the mobility of an aqueous liquid phase are prevented (shrinkage, concentration of water-soluble components). The potential for production of high-quality food products through sublimation is primarily due to this absence of structural changes because there is no liquid phase. The low temperature at which freeze drying is conducted is a secondary factor contributing to quality.

II. MASS AND HEAT TRANSFER IN FREEZE DRYING

A. General Considerations

Every pound of water being sublimed at the ice surface must be transported away from this surface and this requires the supply of energy corresponding to its latent heat of sublimation (ΔH_s). Hence freeze drying is an operation involving both mass transfer and heat transfer, and the rate of drying depends on the magnitude of resistances to these transfers. It may be instructive to start the consideration of the freeze-drying operation with some highly idealized and conceptually simple situations.

Consider (Fig. 11-4a) a slab of ice maintained at constant temperature T_i with a corresponding vapor pressure p_i connected to a sink containing a condenser at a temperature T_c so low that the corresponding vapor pressure p_c is negligible. We stipulate further that the resistance of the space between the ice and the condenser is also negligible. Under these highly ideal and unrealistic conditions, sublimation achieves its maximum rate G_{max} given by:

$$G_{max} = \frac{K_1 A p_i}{\sqrt{T_1}} \tag{11-1}$$

where

G_{max} is the maximum rate of evaporation (lb/hr)
A is sublimation area (ft^2)
T_i is the absolute temperature of ice ($^\circ$R)
p_i is vapor pressure of ice (torr)
K_1 is a constant depending on molecular weight of subliming substance

For the specific case of ice, G_{max} is given by:

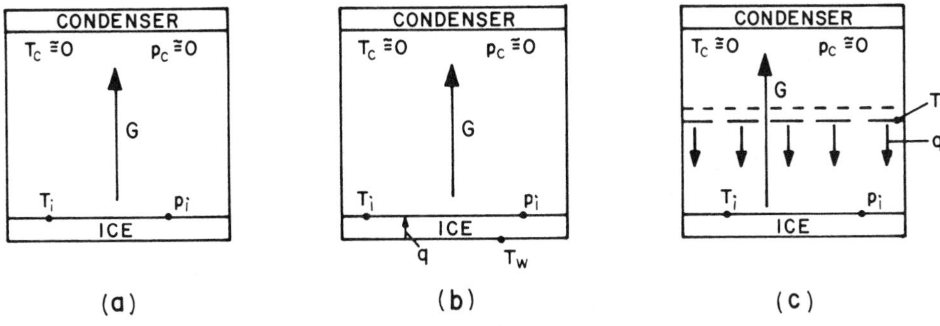

FIG. 11-4. Schematic representation of processes in sublimation of a block of ice. (a) Problem of heat transfer ignored; (b) heat transfer by conduction through the ice; (c) heat transfer by radiation to the surface.

$$G_{max} = \frac{2447Ap_i}{\sqrt{T_i}} \qquad (11\text{-}2)$$

In the ideal case considered above, we have ignored the problem of supplying the latent heat of sublimation. Let us now consider a block of ice at 15°F. G_{max} for this case corresponding to a p_i of approximately 2 torr and referring to an area of 1 ft^2 is:

$$G_{max} = \frac{(2447)(1)(2)}{\sqrt{470}} = 222 \text{ lb/hr}$$

This sublimation rate requires the supply of latent heat of sublimation equal to:

(222 lb/hr) x (1,200 Btu/lb) = 266,000 Btu/hr

How can this heat be supplied? Consider the possibility of supplying it through the block of ice, as shown in Figure 11-4b. In this case, the backface of the ice block is maintained at T_w (which cannot exceed 32°F since melting must be avoided), and heat is transferred by conduction. It has been shown in Chapter 6, Sec. III. D that ice is a reasonably good heat conductor, and it can be assumed here to have a thermal conductivity (k_i) of 1.35 Btu hr^{-1} ft^{-1} °F^{-1}. The heat flux q is equal to:

$$q = A(T_w - T_i)k_i/x_i \qquad (11\text{-}3)$$

where

q is heat flux (Btu/hr)
x_i is thickness of ice block (ft)

We shall now determine the maximum x_i permitting the sublimation at the maximum rate G_{max}:

$$x_i = \frac{A(T_w - T_i)}{q} = \frac{(1)(32 - 15)}{266,000} = 8.35 \times 10^{-5} \text{ ft}$$

This calculation shows that, in spite of the good conductivity of ice, the heat supply demand of sublimation is so high and the available temperature gradient (T_w - T_i) so low that only a very thin ice layer can be utilized if G_{max} is to be attained. Of course, the other alternative is to accept a lesser heat supply, resulting in a lower ice temperature T_i, which in turn lowers the sublimation rate.

We shall also consider the possibility of supplying the heat of sublimation by radiation. In this case, the heat flow shown in Figure 11-4c depends on temperature of a radiator T_r as follows (assuming greatly idealized radiant transfer conditions):

$$q = (A)(0.173)[(T_r/100)^4 - (T_i/100)^4] \tag{11-4}$$

When we calculate the T_r for the above conditions, we find $T_r = 3500°R$ or over 3000°F. Thus, heat supply is a major limitation of sublimation.

In the preceding example, we chose an unrealistically simplified situation by ignoring any resistance to mass transport and by choosing an unrealistically low condenser temperature.

In reality, freeze drying, like dehydration, is a coupled mass-transfer and heat-transfer process. Figure 11-5 shows a schematic representation of the process and of the resistances to mass and heat transfer. Thus, the rate of sublimation is given by:

$$G = \frac{A(p_i - p_c)}{R_d + R_s + K_1^{-1}} \tag{11-5}$$

where

G is the rate of sublimation
R_d is the resistance of the "dry" layer in the food
R_s is the resistance of space between food and condenser
K_1 is a constant defined in Eq. (11-1)

At the same time, the heat of sublimation ΔH_s must be supplied and therefore

$$G = q/\Delta H_s \tag{11-6}$$

Freeze drying, therefore, represents coupled heat and mass transfer and both must be considered simultaneously in an analysis of this operation. Essential components for freeze drying include vacuum chambers, condensers, and vacuum pumps, and these have been discussed in Chapter 9, Sec. II. H in connection with vacuum evaporation.

We shall consider three cases which represent the three basic types of possibilities in vacuum freeze drying (Fig. 11-6):

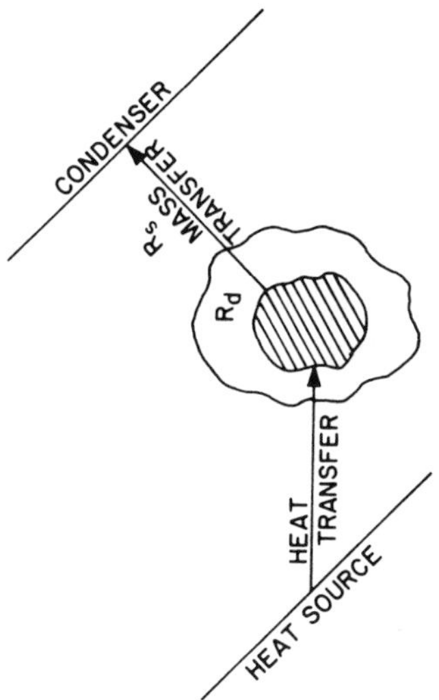

FIG. 11-5. Schematic representation of heat- and mass-transfer resistances in freeze drying.

1. Heat transfer and mass transfer pass through the same path (dry layer) but in opposite directions.
2. Heat transfer occurs through the frozen layer and mass transfer through the dry layer.
3. Heat generation occurs within the ice (by microwaves) and mass transfer through the dry layer.

An additional possibility closely related to the first case above is freeze drying at atmospheric pressure rather than in a vacuum.

B. Heat and Mass Transfer Through the Dry Layer

Consider the following case, which is simplified but is nevertheless typical of most vacuum freeze-drying operations. The material to be dried is heated by applying radiation to the dry surface, and its internal frozen layer temperature is determined by the balance between heat and mass transfer. For simplicity's sake, we shall consider a slab geometry with negligible end effects, as shown in Figure 11-7. We shall assume that the maximum allowable surface temperature T_s is reached instantaneously and that the heat output of the external heat supply is adjusted in such a manner as to maintain T_s constant throughout the drying cycle. We also assume that the partial pressure of water in the drying chamber, p_s, is constant and that all of the heat is used for sublimation of water vapor.

Under these conditions, the heat transfer at any instant is given by:

$$q = Ak_d(T_s - T_i)x_d^{-1} \tag{11-7}$$

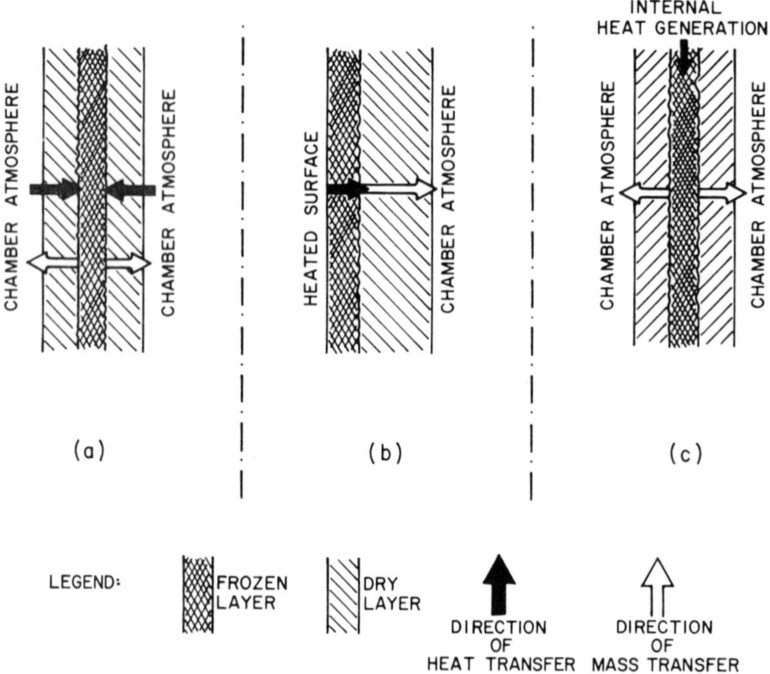

FIG. 11-6. Types of heat transfer defining different types of freeze drying. (a) heat transfer by conduction through dry layer; (b) heat tr ansfer through frozen layer; (c) internal heat generation by microwaves.

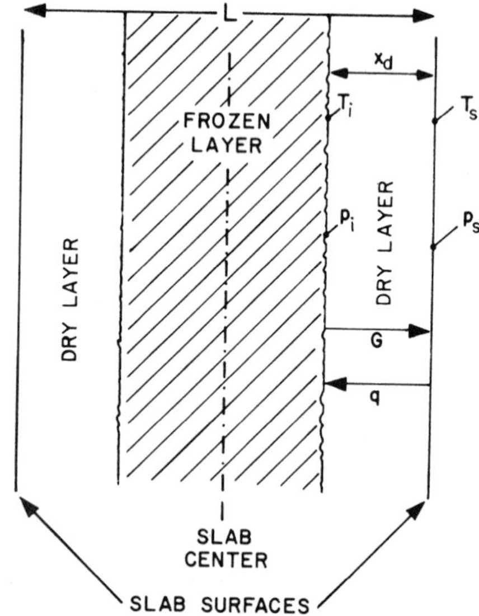

FIG. 11-7. Schematic representation of freeze drying of a slab with heat transfer through the dry layer.

where

x_d is the thickness of the dry layer (ft)
k_d is the thermal conductivity of dry layer (Btu ft^{-1} hr^{-1} $°F^{-1}$)

The sublimation rate is given by:

$$G = -\frac{dW}{dt} = Ab(p_i - p_s)x_d^{-1} \tag{11-8}$$

where

W is the weight of water in the slab (lb)
t is time (hr)
b is the permeability of the dry layer (lb ft^{-1} hr^{-1} $torr^{-1}$)
p_s is the partial pressure of water at dry layer surface (torr)

At the same time, given slab geometry and assuming that the moisture content at the ice-dry layer interface drops from an initial value m_0 to a final value m_f, we obtain a relation between loss of weight and the rate of recession of the interface.

$$-\frac{dW}{dt} = A\rho\,(m_0 - m_f)\frac{dx_d}{dt} \tag{11-9}$$

where

ρ is the bulk density of solids in the drying slab (lb/ft^3)
m_0 is initial moisture content (lb water/lb solids)
m_f is final moisture content (lb water/lb solids)

By combining Eqs. (11-8) and (11-9) we obtain Eq. (11-10):

$$x_d dx_d = \left(\frac{b}{\rho(m_0 - m_f)}\right)(p_i - p_s)\,dt \tag{11-10}$$

We can make the assumption that the rate of heat transfer is equal to the rate of sublimation multiplied by latent heat of sublimation, as assumed in Eq. (11-6):

$$Ak_d(T_s - T_i)x_d^{-1} = Ab(p_i - p_s)\,x_d^{-1}\Delta H_s \tag{11-11}$$

and after simplification we have a relation between pressure and temperature.

$$p_i = p_s + \frac{k_d}{b\Delta H_s}T_s - \frac{k_d}{b\Delta H_s}T_i \tag{11-12}$$

Since we assume that p_s, b, ΔH_s, k_d, and T_s are all constant, we have a linear equation relating p_i with T_i as shown in Figure 11-8. In that same figure, we

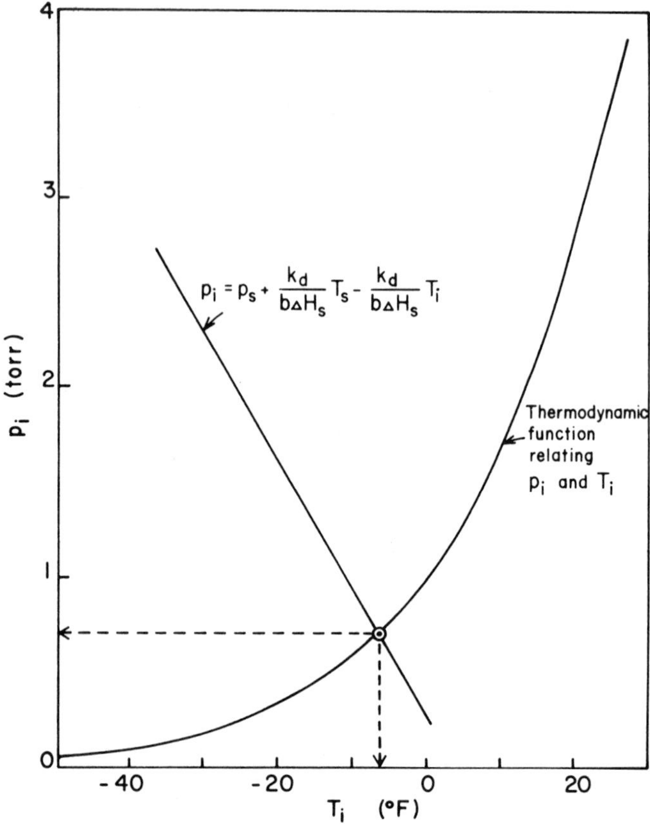

FIG. 11-8. Graphical determination of ice temperature and vapor pressure.

also show the thermodynamic relation between p_i and T_i previously shown in Figure 11-2. We see that there is only one point at which the curves representing the two equations meet. This means that if the assumptions inherent in our analysis are true, the frozen layer temperature T_i must remain constant throughout the drying cycle.

As a consequence, Eq. (11-10) can be readily integrated to calculate the freeze-drying time, because it contains only the two variables x_d and t.

$$\int_0^{L/2} x_d \, dx_d = \frac{b(p_i - p_s)}{\rho(m_0 - m_f)} \int_0^{t_d} dt \tag{11-13}$$

and the drying time is given by

$$t_d = \frac{L^2 \rho(m_0 - m_f)}{8b(p_i - p_s)} \tag{11-14}$$

where

L is thickness of the slab (ft)
t_d is drying time

Because of the equivalence of heat and mass transfer shown in Eq. (11-11) we can equally well integrate the heat-transfer equation, Eq. (11-7), and have a corresponding equation for drying time, Eq. (11-15):

$$t = \frac{L^2 \rho (m_0 - m_f) \Delta H_s}{8k_d(T_s - T_i)} \tag{11-15}$$

We see therefore that drying time depends on the following variables:

1. maximum permissible surface temperature (T_s)
2. initial and final moisture contents (m_0, m_f)
3. bulk density of solids (ρ)
4. latent heat of sublimation (ΔH_s)
5. thickness of the slab (L)
6. thermal conductivity of the dry layer (k_d)
7. permeability of the dry layer (b)

Frozen layer temperature depends neither on the overall thickness of the drying material nor on the thickness of the layer already dry.

Figure 11-9 shows the temperatures of the frozen layer, of the product center, and of the radiator during freeze drying, with heat transfer by radiation to the dry layer surface and by conduction through the dry layer. Ideally, the surface temperature is brought rapidly to the maximum permissible level and maintained at this level by changing the radiator temperature in accordance with a suitable program. In practice, the changes are based on an experimentally determined empirical program but the changes can be made by feedback devices. Calculations allowing the predetermination of the program on the basis of physical properties of foods have been published [31]. The maximum surface temperature is usually dictated by quality considerations, especially flavor and color changes.

The ice temperature is determined by interactions between surface temperature, chamber vapor pressure p_s, and the dry layer properties k_d and b. The frozen layer temperature must be maintained below a critical level, which depends on the nature of the product and on its thermal history [25]. Ideally, it should also be below the eutectic melting point, which may in some cases be 50°F or more below the melting point of ice. Typical ice temperatures existing during freeze drying of foods under conditions in which the total pressure is primarily due to water vapor and the heat transfer has taken place via the dry layer are shown in Table 11-1.

Work conducted at MIT on marine products [14] and on model systems, as well as work conducted by Bralsford [1, 2] on beef, confirms the validity of the relations shown in Figure 11-8, which determine the dependence of T_i on p_s and T_s.

The transport properties k_d and b in themselves depend on pressure and on the kind of gas filling the pores of the dry layer [40]. The dependence of thermal conductivity on pressure is shown in Table 11-2, which lists some selected values

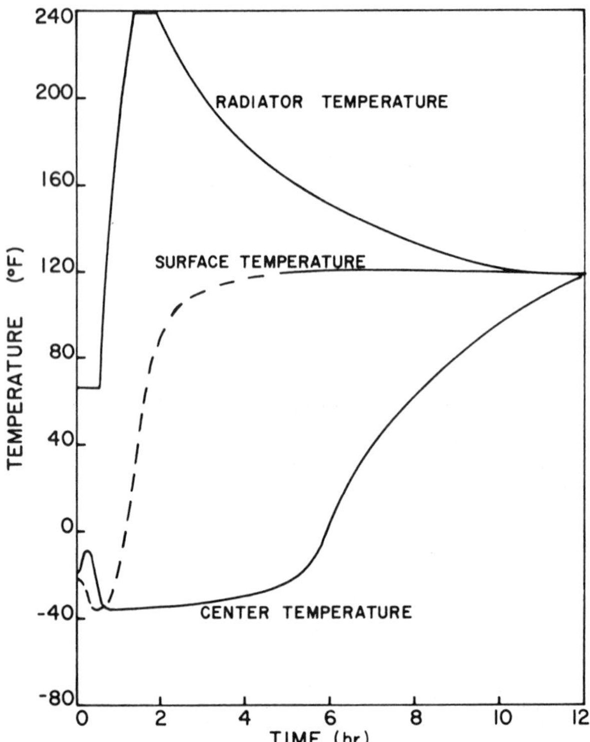

FIG. 11-9. Representation of temperature changes during dehydration.

of k_d. The permeability can be considered to be inversely proportional to total pressure, at least for pressures in excess of 1 torr [25]. Table 11-3 gives some typical values of permeabilities of foods.

It should be noted that the thermal conductivities of dry layers of foods are extremely low and compare with conductivities of such insulators as cork and styrofoam. As a consequence, the temperature drop across the dry layer is large, and with surface temperatures often limited to values below 150° F because of danger of discoloration and in some cases to values below 100° F because of danger of denaturation, the resultant ice temperature is usually well below 0°F. Except for foods with very low eutectic melting points, it is the surface temperature that limits drying rate.

As a consequence of the above limitation, drying rates attainable in practice are very much below the maximum rates attainable with ice. Thus, for materials loaded into the freeze drier at about 2-4 lb/ft^2 of tray surface, which corresponds to industrial practice reported in the literature [3], average drying rates are of the order of 0.30 lb of water removed per hour per square foot. The corresponding drying times are 6-10 hr.

A much more rapid rate, of course, can be achieved by decreasing the particle size and loading rates. This corresponds to reduction of the average thickness of the dry layer and thus of mass- and heat-transfer resistances. This approach,

TABLE 11-1

Frozen Layer and Maximum Dry Surface Temperatures in Typical Freeze-Drying Operations Conducted with Heat Input Through the Dry Layer

Food material	Chamber pressure (torr)	Maximum surface temperature (°F)	Frozen layer temperature (°F)	Reference
Chicken dice	0.95	140	-4	[41]
Strawberry slices	0.45	158	5	[41]
Orange juice	0.05-0.1	120	-45	[42]
Guava juice	0.05-0.1	110	-35	[42]
Shrimp	0.1	125	-20	[9]
Shrimp	0.1	175	0	[9]
Salmon steaks	0.1	175	-20	[9]
Beef, quick frozen	0.5	140	7	[11]
Beef, slow frozen	0.5	140	1	[11]

however, is limited only to selected products, since efficient operation requires specialized equipment, such as continuous freeze-driers.

In general, operating conditions during freeze drying of foods include maximum surface temperatures of 100° - 180°F and chamber pressures of 0.1-2 torr. Freeze drying of biological specimens, vaccines, and microorganisms is usually conducted with maximum surface temperatures of 70° - 90° F and chamber pressures below 0.1 torr.

It is possible to conduct freeze drying at atmospheric pressure, provided the gas in which the drying is conducted is very dry. All of the theoretical relations discussed above apply in this case, but in addition to mass- and heat-transfer resistances in the food, one must also consider those in the gas phase. The heat transfer improves but mass transfer deteriorates as the pressure increases, and it is mass transfer that becomes limiting in atmospheric freeze drying. As a consequence, for all but very small particles the drying rates are very slow.

C. Heat Transfer Through the Frozen Layer, with Mass Transfer Through the Dry Layer

Freeze dehydration of liquids and of solids capable of intimate contact with a heating surface can be conducted with heat transfer through the frozen layer as indicated in Figure 11-10. Practical systems in which we can observe situations approximating the idealized scheme of Figure 11-10 include, for instance, the

TABLE 11-2

Typical Thermal Conductivities of Freeze-dried Foods in the Presence of Various Gases

Food material	Pressure (torr)	Thermal conductivity (Btu $^{\circ}F^{-1}$ ft^{-1} hr^{-1}) Air or nitrogen	Helium	Reference
Turkey breast	0.1	0.02	0.02	[48]
	1.0	0.04	0.04	[48]
	10	0.045	0.07	[48]
	100	0.045	0.10	[48]
	760	0.045	0.11	[48]
Starch gel	1.0	0.0054	—	[43]
	760	0.023	0.086	[43]
Haddock	0.15	0.013	—	[27]
Beef	0.1	0.024	0.024	[18,20]
	10.0	0.035	0.066	[18,20]
	760.0	0.037	0.098	[18,20]
Apple	0.1	0.010	0.015	[18,20]
	10.0	0.022	0.049	[18,20]
	760.0	0.024	0.098	[18,20]

TABLE 11-3

Typical Water Vapor Permeabilities of Freeze-dried Foods, in the Presence of Various Gases

Food material	Pressure (torr)	Permeability (lb $torr^{-1}$ ft^{-1} hr^{-1}) x 10^3		Reference
		Air or nitrogen	Helium	
Beef	1.6	4.6	—	[22]
Turkey breast	0.1	5.1	4.9	[41]
	1.0	4.6	3.2	[41]
	10.0	1.7	0.7	[41]
	100.0	0.25	0.08	[41]
	760.0	0.03	0.01	[41]
Coffee (10% solids)[a]	0.15	0.02	—	[26]
Gelatin (2.6% solids)[a]	0.21	0.002	—	[26]
Coffee (20% solids)[a]	0.93	0.01	—	[36]

[a] Solids content before freeze drying.

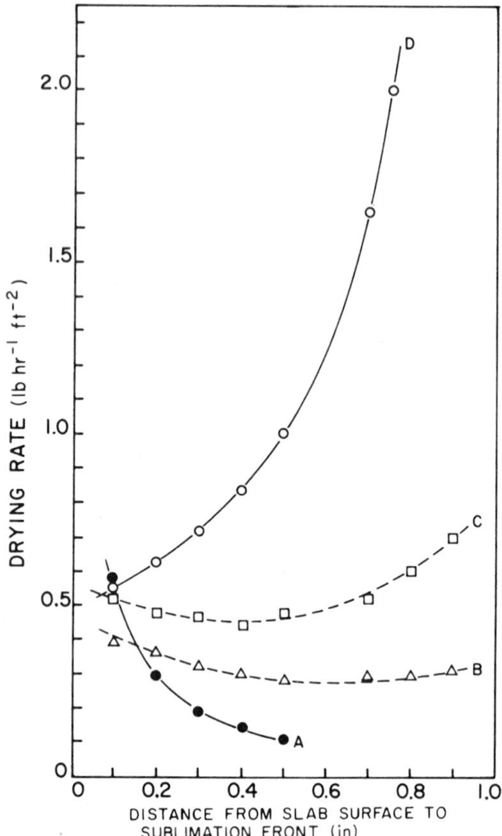

FIG. 11-10. Drying rate for hypothetical slab with properties specified in text: (A) Heat transfer by radiation to two faces of slab. (B) Heat transfer by conduction to "backface" of slab. Wall at 10°F. (C) Heat transfer by conduction to "backface" of slab. Wall at 28° F. (D) Heat transfer by conduction to "backface" of slab. Wall at 10° F, dry layer removed continuously.

tubular drier invented by Seffinga as well as several other patented procedures for freeze drying liquids [32]. Experimental results in a pilot scale drier were obtained by Lambert and Marshall [26]. The mass- and heat-transfer equations are given below for the case of a slab with negligible end effects.

The mass transport of water vapor is given by Eq. (11-8), just as in the previous case:

$$-\frac{dW}{dt} = Ab(p_i - p_s)x_d^{-1} \tag{11-8}$$

Similarly, because Eq. (11-9) is applicable, the rate of recession of the frozen layer and increase of the dry layer is given by Eq. (11-10).

$$x_d dx_d = \frac{b}{\rho(m_0 - m_f)} (p_i - p_s)\, dt \tag{11-10}$$

The heat transfer rate, however, is now given by:

$$q = Ak_i(T_w - T_i)x_i^{-1} \tag{11-16}$$

where

T_w is temperature of the wall in contact with the frozen layer (°F)
k_i is the thermal conductivity of the frozen layer (Btu ft^{-1} hr^{-1} $°F^{-1}$)
x_i is the thickness of the frozen layer (ft)

As a consequence, the relation between pressure and temperature of the ice interface becomes more complicated. In the case of heat transport through the dry layer, a linear equation relating p_i and T_i contains no additional variables [Eq. (11-12)]. In the present case, however, Eq. (11-17) contains an additional variable.

$$p_i = p_s + (k_i/b\Delta H_s)(x_d/x_i)(T_w - T_i) \tag{11-17}$$

It may appear that two additional variables are contained in Eq. (11-17), x_d and x_i, but these two variables are interrelated by Eq. (11-18) and one can be converted into a function of the other as shown in Eq. (11-19).

$$x_d = L - x_i \tag{11-18}$$

$$p_i = p_s + (k_i/b\Delta H_s)(x_d/L - x_d)(T_w - T_i) \tag{11-19}$$

The interface temperature and pressure are therefore no longer independent of time of drying, even if k_i, ΔH_s, x_d, b, p_s, and T_w remain constant. In fact, p_i must be evaluated as a function of x_d. This can be achieved readily by assuming an x_d and then determining the corresponding p_i, as shown in Figure 11-8. This is then repeated for another assumed value of x_d and, if enough points are taken, one obtains p_i as a function of x_d:

$$p_i = f(x_d) \tag{11-20}$$

Substituting this function into Eq. (11-13) allows its integration [Eq. (11-21)] by analytical, or if $f(x_d)$ is too complicated, by numerical methods.

$$\int_0^L x_d\, dx_d = \frac{b(p_i - p_s)}{\rho(m_0 - m_f)} \int_0^{t_d} dt \tag{11-13}$$

$$\int_0^L \frac{x_d}{f(x_d) - p_s}\, dx_d = \frac{b}{\rho(m_0 - m_f)} \int_0^{t_d} dt \tag{11-21}$$

The most convenient way to solve the above equations is through the use of computer, either digital or analog.

In the case described here, the balance between ease of heat transfer and ease of mass transfer changes continuously, with mass transfer becoming more difficult as drying progresses (longer path through the dry layer) and heat transfer becoming progressively easier (shorter path through the frozen layer). The resistance of the frozen layer to the transport of heat is in any case not too formidable, with the thermal conductivities often being as high as 1.0 Btu ft^{-1} hr^{-1} $°F^{-1}$ and therefore up to two orders of magnitude higher than the thermal conductivities of the dry layer. As a consequence, theoretically calculated averaged drying rates with heat input through the frozen layer often exceed those attained in drying with heat transfer through the dry layer. The drying rate, for materials dried with heat transfer through the ice layer can be increased very significantly by continuously removing most of the dry layer. This can be achieved, at least in theory, by a rotating knife or other scraping device, as suggested by Greaves [16]. The improvement theoretically achievable by this method of minimizing mass transfer resistance is evident from the calculated drying rates of a hypothetical slab of 1 in. thickness, with the following properties and drying conditions:

1. $b = 2 \times 10^{-2}$ lb hr^{-1} ft^{-1} $torr^{-1}$
2. p_s = 0.05 torr
3. $k_d = 5 \times 10^{-2}$ Btu hr^{-1} ft^{-1} $°F^{-1}$
4. k_i = 1.0 Btu hr^{-1} ft^{-1} $°F^{-1}$
5. ρ = 20 lb/ft^3
6. m_0 = 2.5 lb water/lb solids
7. m_f = 0.05 lb water/lb solids
8. ΔH_s = 1200 Btu/lb
9. maximum permissible surface temperature: 125°F

Data calculated for this slab are shown in Figure 11-10, which compares drying rates for heat transfer by conduction through the dry layer (from two faces of the slab) with three different cases for conduction through the frozen layer. It should be noted that drying is complete when the dry layer thickness reaches 0.5 in. for the case of conduction through the dry layer, but the dry layer thickness must equal 1 in. for heat transfer from the "backface". In the particular example chosen here, the drying times are as follows (referring to curves shown in Figure 11-10):

A = 8.75 hr
B = 13.5 hr
C = 7.2 hr
D = 4.0 hr

D. Freeze Drying with Heat Input by Microwaves

The limitations on heat transfer rates in conventionally conducted freeze-drying operations led very early to attempts to provide internal heat generation through the use of microwave power. Work on this problem began in the 1950's and the early attempts were reviewed by Burke and Decareau [3] and by Copson [5].

The generation of heat by microwaves depends on the presence of dipoles (in the case of foods, primarily water), which when placed in a rapidly changing electric field undergo changes in orientation which can result in "friction" and consequently generation of heat. Frequencies available for such industrial applications as freeze drying are limited by Federal authorities to two: 915 Mc/sec and 2450 Mc/sec. Equation (11-22) gives the amount of power generated in the material in the electric field [13]:

$$\text{Power (W/cm}^3) = E^2 \upsilon \varepsilon'' 55 \times 10^{-14} \tag{11-22}$$

where

E is electric field strength (V/cm)
υ is frequency (cps)
ε'' is loss factor

The loss factor ε'' is an intrinsic property of foods which depends strongly on temperature and composition. In particular, liquid water absorbs much more energy than ice or dry food components. This is evident from data shown in Table 11-4.

Theoretically, the use of microwaves should result in a greatly accelerated rate of drying, because the heat transfer does not require internal temperature gradients and the temperature of ice can be maintained close to the maximum permissible temperature for the frozen layer without the need for excessive surface temperatures. If, for instance, it is permissible to maintain the frozen layer at 10°F, then the drying time for an ideal process using microwaves for a 1 in. slab with properties identical to those described in the preceding section would be given by Eq. (11-14):

$$t_d = \frac{L^2 \rho (m_0 - m_f)}{8b(p_i - p_s)} \tag{11-14}$$

$$t_d = \frac{(1/12)^2 \; (20) \; (2.45)}{(8)(2 \times 10^{-2}) \; (1.55)} = 1.37 \text{ hr}$$

We should note that this drying time compares very favorably with the 8.75 hr required for the case of heat input through the dry layer, 13.5 hr for heat input through the frozen layer without dry layer removal, and even with the relatively short drying time of 4 hr for the case in which the dry layer is continuously removed. In laboratory tests on freeze drying of 1 in. thick slabs of beef, an actual drying time of slightly over 2 hr has been achieved, compared with about 15 hr for conventionally dried slabs [23].

In spite of these apparent advantages, the application of microwaves to freeze drying has not been successful. The major reasons for the failures are the following:

TABLE 11-4

Dielectric Properties of Thawed, Frozen, and Freeze-dried Beef at 3000 Mc/sec [a]

Temperature (°F)	Loss factor	
	Beef at normal water content	Freeze-dried beef
-40	0.083	0.0058
0	0.293	0.0079
+40	10.56	0.0122

[a] After Burke and Decareau [3].

1. Energy supplied in the form of microwaves is very expensive. A recent review estimates that it may cost 10-20 times more to supply 1 Btu from microwaves than it does from steam [17].
2. A major problem in application of microwaves is the tendency to glow discharge, which can cause ionization of gases in the chamber and deleterious changes in the food, as well as loss of useful power. The tendency to glow discharge is greatest in the pressure range of 0.1-5 torr and can be minimized by operating the freeze driers at pressures below 50 μm. Operation at these low pressures, however, has a double drawback: (a) it is quite expensive, primarily because of the need for condensers operating at a very low temperature, and (b) the drying rate at these low pressures is much slower.
3. Microwave freeze drying is a process which is very difficult to control. Since water has an inherently higher dielectric loss factor than ice, any localized melting produces a rapid chain reaction, which results in "runaway" overheating.
4. Economical microwave equipment suitable for the freeze drying of foods on a large continuous scale is not yet available.

In view of all of the above limitations, microwave freeze drying is at present only a potential development.

III. STRUCTURAL CHANGES DURING FREEZE DRYING OF FOOD MATERIALS AND THE PHENOMENON OF COLLAPSE

During freeze drying, a number of events occur which determine the structure of the freeze-dried matrix. Freezing causes a separation of the aqueous solutions present in foods into a two-phase mixture of ice crystals and concentrated aqueous solution. The properties of the concentrated aqueous solution depend on temperature, concentration, and composition [29]. If drying is conducted at a very low temperature, then mobility in the extremely viscous concentrated phase is so low that no structural changes occur during drying. The resultant structure consists of pores in the locations of former ice crystals surrounded by a dry matrix of insoluble components and precipitated compounds originally in solution. If, on the other hand, the temperature is above a critical level, mobility in the concentrated aqueous solution may be sufficiently high to result in flow and loss of the original structure and of the separation of solutes which exist immediately following freezing.

During freeze drying there exist both temperature and moisture gradients in the drying materials, and the mobility and therefore collapse of the concentrated solutions forming the matrix may vary from location to location. The collapse is more likely in locations having a high temperature and/or a high moisture content. However, since the gradients of these two parameters in freeze drying are opposite to each other, the collapse may actually occur in intermediate locations of a drying layer rather than at either extreme (since the coldest portion of the drying layer is also the one with the highest water content, and vice versa).

In food materials containing a variety of soluble and insoluble components, it is usually not possible to define a unique relation between moisture, temperature, and collapse. The phenomenon, however, has been extensively studied in freeze drying of solutions [28, 24] and it is possible to specify some "collapse temperatures." Some of the values reported by MacKenzie [28] are shown in Table 11-5.

Migration of soluble components during freezing can also result in formation of a surface layer of low permeability. Quast and Karel [36] have observed, for instance, that removal of a thin surface layer of frozen coffee concentrate, or freezing under conditions inhibiting the separation of a more concentrated phase at the surface, result in improved drying rates. Similar observations have led to some industrial patents.

The type of freezing and the rates of crystallization during freezing have important effects on structure of the freeze-dried materials. The factors involved are very complex, since the pore size distribution depends on the ice crystal size distribution. In general, very rapid freezing rates give smaller crystals and slower drying rates than those attainable with slower freezing rates, provided that surface skin formation is avoided. The effects on the various quality factors, such as color and cell membrane changes, however, may be improved by rapid freezing. An interesting consideration in freeze drying of coffee is the surface appearance. Rapidly frozen coffee has a pale surface due to the light scattering properties of small voids left by the crystals. This is considered an undesirable structural property and steps are often taken to avoid it by controlled surface melting or by a suitable choice of freezing conditions.

IV. RETENTION OF ORGANIC COMPOUNDS, INCLUDING FLAVORS, DURING FREEZE DRYING

The vapor pressures of many organic compounds occurring naturally in foods, including those compounds responsible for flavor, are relatively high. Many flavor compounds show vapor pressures which over the range of temperatures involved in freeze drying are above the vapor pressure of water. A natural consequence is that many of them are subject to evaporation and loss during drying. The laws governing their removal, however, are complex.

The simplest assumption, perhaps, is that an equilibrium exists at all times between organic compounds in the aqueous phase and the surrounding vapor space. Under these conditions and if no water is frozen out, the ratio of removal of water

TABLE 11-5

Collapse Temperatures for Some Freeze-drying Solutions[a]

Solute	Collapse temperature (°C)
Dextran, MW = 10,000	-9
Gelatin	-8
Glucose	-40
Maltose	-30 to -35
Sucrose	-32

[a] After MacKenzie [28].

to loss of the organic compound, hereafter referred to as "volatile," is given by:

$$\frac{W}{W_0} = \frac{C}{C_0}^{\alpha - 1} \tag{11-23}$$

where

W is the weight of water
C is the concentration of volatile in water in moles per unit weight of water
α is relative volatility
the subscript "0" refers to initial conditions

The relative volatility α is defined as follows:

$$\alpha = \frac{y_v}{y} \frac{N}{N_v} \tag{11-24}$$

where

y is mole fraction of water in vapor phase
N is mole fraction of water in liquid phase
y_v is mole fraction of volatile in vapor phase
N_v is mole fraction of volatile in liquid phase

Under ideal conditions, in the binary system of water-volatile and if Raoult's law is valid , we have:

$$\alpha = p^o{}_v / p^o \tag{11-25}$$

where

p° is vapor pressure of water
p°_{v} is vapor pressure of volatile

It is universally agreed that under conditions existing in foods ideality does not exist, and a number of modifications of the ideal equation have been suggested. Among those receiving some attention are the following:
Sivetz and Foote [44] have suggested:

$$\alpha = (p^{\circ}_{v} / p^{\circ}) \frac{18}{MW_{v}} \tag{11-26}$$

where MW is molecular weight. Buttery [4] has suggested that the partial pressure of organic compounds should be related to mole fraction at saturation.

$$\alpha = (p^{\circ}_{v} / p^{\circ}) N_{vs} \tag{11-27}$$

where N_{vs} is the mole fraction at saturation. Others have tried to relate the volatility to vapor pressures measured under specific limiting conditions [12, 46]. In all cases the correction can be expressed as an activity coefficient γ defined in Eq. (7-15), Chapter 7. Since this coefficient depends on composition as well as environmental factors, however, it usually must be determined experimentally.

It is interesting to note that all of these relationships underestimate the retention of volatiles in actual freeze-drying operations. Typical results are shown in Table 11-6. The high degree of retention in freeze-dried volatile-containing foods also has been noted by Rey and Bastien [39] and by Saravacos and Moyer [42]; the theoretical explanation of the phenomenon has been provided by the work of Flink and Karel in the United States [8] and by that of Thijssen and his co-workers in The Netherlands [47].

Early work tended to explain the retention of volatiles as an adsorption phenomenon [39]. Recent work has shown that this is not the major factor except in special cases, such as the adsorption of amines on acidic polysaccharides [15]. In most situations, the dominant mechanism is not adsorption on specific sites but localized entrapment of volatiles within the dry food matrix.

The Thijssen group has described this phenomenon of entrapment in terms of diffusional properties of water and volatiles. They have shown that as the water content of a dry carbohydrate "cake" decreases the diffusion coefficients for both water and for volatile decrease, but the coefficients for organic volatiles drop much more rapidly than that for water. Thus there is a preferential retention of organic compounds [47].

Flink and Karel [8, 9] studied retentions of numerous volatile-solute combinations and the effect of a number of processing variables on volatile retention in freeze-dried carbohydrate solutions. It was their further study of the freeze-dried system's properties in the dry and humidified state that led to the development of the "microregion permeability" concept. In several freeze-drying experiments, they showed that adsorption of volatile in the dry layer was not responsible for volatile retention. These experiments also demonstrated that volatile retained after freeze drying was quite strongly held in the dry material.

TABLE 11-6

Retention of Pyridine in Coffee Powder

Basis of Determination	Retention (%)
Predicted from theoretical relations	
Raoult's law	7.5
Volatility in infinitely dilute solutions, data of Thijssen [46]	0.01
Relation suggested by Sivetz and Foote [44]	5.0
Calculated using volatility determined experimentally for 0.01% aqueous solution at -20° C [12]	$<10^{-5}$
Observed retention	
Coffee powder [44]	50
Freeze-dried model (1% glucose) [12]	85
Freeze-dried model (1% starch) [12]	50

Freeze-dried material was evacuated for up to 12 hr at 125° F with no reduction of volatile content. Evacuation (at room temperature) of finely ground freeze-dried material for up to 13 hr also gave no reduction of retained volatile, showing that retention was occurring in areas of quite small dimension.

The localized nature of the retention units and relative openness of the freeze-dried cake was shown in two experiments in which the freeze-dried slab was sectioned perpendicular to the direction of drying. The essentially uniform volatile retention showed that the gross dry cake was relatively permeable to the flow of volatile. This was further illustrated by a "layering" experiment, in which samples were prepared by freezing alternate layers of volatile-containing and volatile-free solution. After freeze drying, analysis of each layer showed volatile retention only in those layers which originally contained volatile (Table 11-7). The results of these experiments indicated that the volatile was being retained in many small, localized, strongly bound units, distributed rather uniformly throughout a relatively open overall structure. The small units were called "microregions."

The structure of dry solids often has been investigated by water-vapor sorption isotherms, and the general method of humidification of the freeze-dried material has been utilized to further investigate the microregion concept. An interpretation of the standard BET water sorption isotherm has resulted in the conclusion that the amorphous freeze-dried carbohydrate cake is actually highly hydrogen bonded. The addition of water vapor to this hydrogen-bonded matrix can be expected to break carbohydrate-carbohydrate hydrogen bonds progressively from the weakest to the strongest. Microregion structure is dependent on the extent of

TABLE 11-7

Results of "Layering" Experiments [a]

	2-Propanol content (g/100 g maltose) after freezing and drying in layers			
	Before drying	After drying	Before drying	After drying
Top layer	0	0	4	2.52
Middle layer	0	0.05	0	0.05
Bottom layer	4	2.73	0	0.02

[a] From Flink and Karel [8].

hydrogen bonding, and thus the addition of water vapor results in a changing structure and loss of volatile. This has been observed for the rehumidified freeze-dried systems, where increasing the water content (and hence decreasing the microregion hydrogen bonding) has resulted in partial loss of volatile. Total loss of volatile occurs only when humidification is carried to the point of saturation or when other significant structural changes, such as crystallization, occur.

According to the microregion permeability concept, numerous, small carbohydrate-rich regions decrease in permeability to organic volatiles as water molecules are removed and carbohydrate-carbohydrate hydrogen bonding develops. Unstabilized microregions form during the freezing process when the concentrated solute phase develops. As drying proceeds and water-carbohydrate hydrogen bonds are replaced with carbohydrate-carbohydrate hydrogen bonds, the microregion becomes increasingly stabilized and the mobility of the individual carbohydrate molecules is reduced. This results in reduced loss of volatile from a microregion due to reduced permeability. Below a certain critical moisture content, no further loss of volatile can occur and the microregion is sealed.

The destruction of microregions is correlated with the appearance of "collapse" described in Section III. Thus, sugars which are susceptible to collapse (such as glucose) tend to be sensitive to high drying temperatures and may lose part of their capability to entrap volatiles.

In a recent study, Flink and Gejl-Hansen [7], by using scanning electron microscopy, have demonstrated the existence of entrapped hexanal droplets in freeze-dried tertiary solutions of maltodextrin and hexanal in water. Figure 11-11 shows typical maltodextrin "plates" in freeze-dried maltodextrin. Figure 11-12 shows a similar electron scanning microscope picture of maltodextrin containing entrapped droplets of hexanal. These droplets are about 0.5-2 μm in diameter.

Flink and his co-workers have shown that the size of microregion, amount retained, and microregion stability depend strongly on such physical properties as solubility of the solutes and on various environmental and process conditions [9, 10].

FIG. 11-11. Scanning electron micrograph of freeze-dried 20% maltodextrin solution (208x).

FIG. 11-12. Scanning electron micrograph of freeze-dried maltodextrin solution containing hexanal (266x).

Utilization of the above principles is likely to result in improvements in the heretofore empirical technology of "locking in" flavors.

V. PRINCIPLES OF INDUSTRIAL DEVELOPMENTS IN FREEZE DRYING

A. Status of Industrial Applications

During the 1960's a number of industrial organizations explored freeze drying for a variety of uses. Different modifications and improvements were described in patents and in scientific literature [32]. Practically all plants of that period operated on the batch principle, involving prefreezing of the material to be dried followed by vacuum drying with radiant heat transfer, breaking of vacuum with nitrogen, and packaging in rooms in which the humidity was maintained at a low level.

Of the various freeze-dried products which have been tried in markets of the United States, Europe, and some other areas, freeze-dried coffee has certainly been the most successful. It accounted for about 20% of the total instant coffee market in the United States in 1969 and 28% in 1970. In a report published in November 1970, the staff of the trade journal World Coffee and Tea has expressed the opinion that traditional spray-dried instant coffee is likely eventually be replaced entirely by coffee which is freeze dried or by coffee which is spray dried and then agglomerated into particles resembling in density and size those achieved by freeze drying [50].

The major barriers to large-scale applications of freeze drying have been cost, problems in storage stability of foods which are susceptible to oxidation, and loss of juiciness in foods derived from animal tissues. Another potential problem is loss of flavor components during drying. In the case of soluble coffee, the first three problems are relatively less important than in other foods. The raw material involved is sufficiently expensive to bear process costs and the consumer accepts the relatively slight loss of quality due to storage. Processing advances have been able to overcome the potential problem of flavor losses, and a superior freeze-dried instant coffee is now enjoying a spectacular rise in popularity.

This increase in popularity has occurred in spite of an increased cost, which can range up to 30% more per unit weight of dried coffee at the retail level based on data published by Consumer Reports in January of 1971 (p. 32).

Application of freeze drying to other products has been relatively slow. The following developments, however, merit recognition. Eggs are being freeze dried on a large scale by Affined Foods in Great Britain, and smaller quantities of vegetables, dairy products, and soup ingredients are freeze dried elsewhere in Europe. In Japan, three industrial installations dry various ingredients and elsewhere in the far east mushrooms are dried successfully. In Europe and Latin America, new plants are also on stream for production of orange juice.

In the United States, besides the expanding freeze-dried coffee markets, some activity is evident in the area of freeze drying components for the institutional market and also through contracts to government agencies, including the military. Although small in scale, a spectacular development in terms of resultant publicity has been the application of freeze-dried items to space diets. Of particular significance, in terms of laboratory research and pilot plant applications, are the following developments.

1. increased understanding of the heat and mass-transfer aspects of the freeze-dehydration process
2. increased understanding of processes which limit deterioration of freeze-dried foods in storage, and the incipient application of these principles to industrial practice
3. continuing search for continuous processes applicable to freeze drying on an industrial scale

B. Recent Developments in Continuous Processing of Liquid Foods

The fact that production of freeze-dried foods is now concentrated on a few items produced in relatively large volume has allowed the development of processes specifically tailored to the products being dried and has resulted in great improvements in process economy and product quality. A good review of problems of adjusting specific steps to specific products has been given by Warman and Reichel [49]. We shall limit ourselves here to only a few examples of process modifications for specific products.

1. Processing of Liquids Prior to Freezing

Solid food products may require various preparatory steps, such as cleaning, cutting, or grinding. Liquid products are equally demanding in prefreezing considerations. The first consideration is the concentration to be used. Since freeze drying is an expensive process, removal of a portion of the water by other methods may reduce the cost. On the other hand, the dry layers of concentrated solutions may have lower permeabilities and may also have a different final quality. In addition, the bulk density of the freeze-dried product is affected by the concentration of solids in the feed.

Where a low bulk density is considered desirable but a concentrated feed is desired for reasons of economy, foaming prior to freezing can be practiced. This method of preparing coffee concentrates for subsequent freeze drying is the subject of several patents issued to Nestlé Products Limited and to General Foods Corporation.

Preconcentration can be achieved by any of the methods discussed in Chapter 9, but where a very high quality is to be maintained, freeze concentration is the method of choice.

2. Prefreezing

Freezing of materials to be subsequently dried affects the quality very strongly. As we have discussed in Sec. III, permeability of the dry layer depends on freezing rate; in addition, formation of an impermeable film at the surface of frozen liquids can result from concentration of solute at the surface. This last effect can be eliminated either by slush freezing or by mechanical disruption of the surface layer prior to drying, and both practices are mentioned in patents and other industrial literature.

An increasingly important method of prefreezing of liquids is the production of frozen granulate. This granulate is prepared by freezing flakes or sheets of ice to a soft consistency followed by hardening in a belt, blast, or other kind of freezer. Subsequent reduction to the desired size is achieved by milling and sieving at a very low temperature.

3. Dehydration Process

As mentioned in Sec. V. A, the initial industrial experience has been based on batch driers, in which the freeze-dehydration cabinet is loaded with trays at the beginning of the cycle, the cabinet is closed and evacuated, and heating platens are maintained at a given temperature program for the duration of the cycle. At the end of the cycle, lasting usually several hours, the cabinet is filled with nitrogen and opened and the trays containing the dry material are removed for packaging, storage, and distribution.

The most recent development in industrial practice is a trend toward continuous methods of drying. These methods are labor saving and therefore are economically superiod; however, they are usually designed with a specific product in mind and require a high initial investment. They are therefore employed when a specific type of application has been firmly established.

There are two basic types of continuous driers: (a) tray driers, in which the product is stationary on a tray, and the trays are continuously moved through the drier and (b) dynamic or trayless driers, in which particles of the product are moved through the drier.

Figure 11-13 is a schematic diagram of a tray drier with semicontinuous operation. In this arrangement a bank of trays containing prefrozen product is placed in an entry lock where the initial evacuation takes place. The trays are then moved into the first of two drying tunnels. By means of several discrete movements the product is transported through the two tunnels and into the exit lock, where the vacuum is broken with inert gas. The entry lock and the two drying tunnels are equipped with heating platens and the temperature program in each section can be adjusted for the specific requirements of the product. The entry lock and the two drying tunnels are connected to two condensers each. Only one condenser of each pair is used at any one time, while the other is being defrosted during the same interval. This avoids shutting down the operation for defrosting.

Many different arrangements for continuous tray driers are possible and are in fact utilized. Only a few of the possible arrangements can be mentioned here.

a. A single drying tunnel can be used with entry and exit locks accommodating only a single tray. The semicontinuous operation then consists of a series of discrete movements in which individual trays are moved in a predetermined pattern.
b. Individual sections of a multi-section drying tunnel, as shown in Figure 11-13 can be operated at different pressures.
c. Condensers can be located in the drying tunnels, with some arrangements allowing continuous or periodic defrosting of some of the condensers. This arrangement is claimed to offer economical construction and operation.
d. The movement of the trays through the driers can be accomplished by any of a number of mechanical devices, including rail carts, belts, and chains.

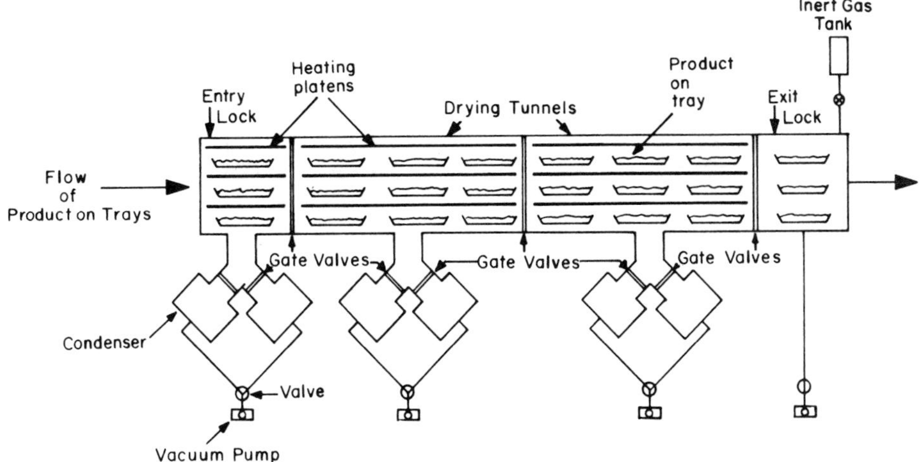

FIG. 11-13. Schematic diagram of one type of continuous tray freeze drier.

e. The continuous or semicontinuous operation may include the freezing steps and the postdrying packaging steps, and there may be an associated continuous tray cleaning cycle.
f. The trays can be of different designs and may be "ribbed," pocketed, or otherwise structured to optimize heat transfer and material handling.

Finally, it should be noted that the continuous tray driers are closely related in their design to batch driers and may, in fact, be composed simply of several modules, each of which is essentially equivalent to a batch drying unit.

The second type of continuous freeze drier is one in which the particles of the material to be dried are moved through the drier by some means other than trays. There are many types of such dynamic or trayless driers. Some types are mentioned below:

a. Belt freeze-driers (Fig. 11-14).
b. Circular plate driers. These driers contain circular heater plates arranged above each other with a rotating shaft in the center. The shaft has attached to it blades which gently move the product from one edge of the plate to the other edge, from which the product falls to the plate below. By the time the product traverses the stack of plates, it is dry (Fig. 11-15).
c. Vibratory driers. The product is moved from plate to plate by oscillations of these plates, which may be horizontal or somewhat inclined to facilitate transport. There are several variations in which the vibrating plates are replaced by a spiral chute or by disk-shaped heaters. Since the residence time, rate of heat and mass transfer, and the temperature and moisture profiles in the product are all highly dependent on the type of flow the product is experiencing, the design of vibratory driers is a very demanding task and is usually undertaken with a specific product in mind. Most specific designs appear to be custom tailored for specific industrial customers.

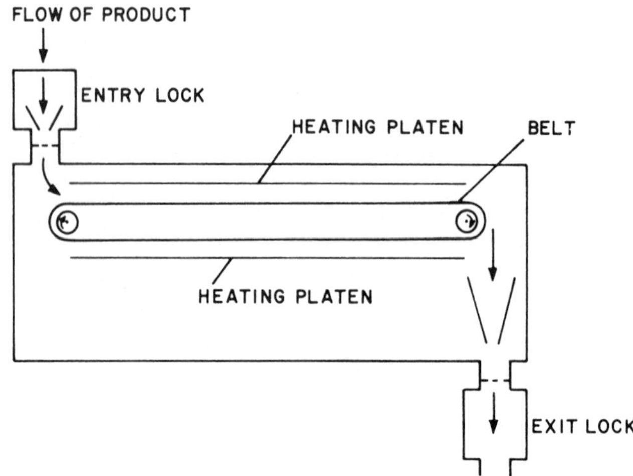

FIG. 11-14. Schematic diagram of continuous belt freeze drier.

d. Fluidized-bed driers and spray freeze driers. With particles of small size, it is possible to conduct the drying operation while particles are suspended in another fluid, which is usually air, nitrogen, and another gas. Industrial applications of these processes for freeze drying in vacuum seem less likely than for high-pressure dehydration.

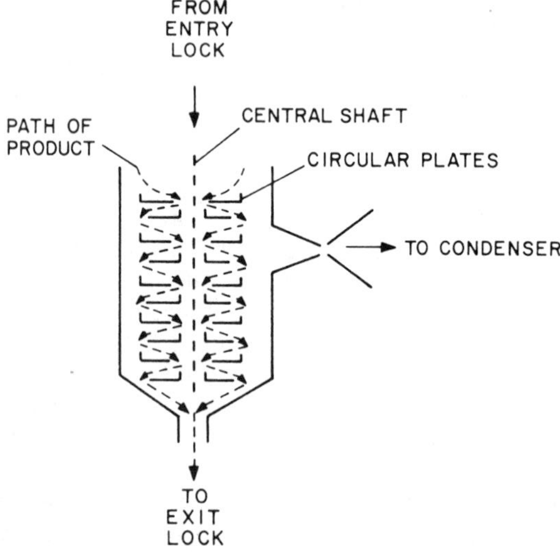

FIG. 11-15. Schematic diagram of continuous circular plate freeze drier.

In all of the continuous systems in which the particles of the product are subject to movement, excessive abrasion of the product must be avoided.

C. New Processes Suggested for Freeze Drying

Very extensive research in the field of freeze drying is still under way and suggestions for modified freeze-drying processes are continuously appearing in the literature. Some of these processes include the following:

1. A pressure-temperature cycle in which periods of evacuation are alternated with intervals of heating without vapor removal.
2. Improved heat transfer by use of sonic vibrations.
3. Use of beds in which a product layer is alternated with a desiccant layer. Both sets of layers are swept continuously with an inert gas, which carries water from food to desiccant and heat from desiccant to food.
4. Atmospheric freeze drying, in which air, nitrogen, or helium with an extremely low pressure of water is used as a water-removal and heating medium.
5. Azeotropic freeze drying utilizing various organic solvents, and complex freeze drying in which a structure is developed by consecutive drying from different solvents.

Additional information can be obtained in the current literature [19, 25, 30, 33-35, 38].

SYMBOLS

(Units are given for some quantities. These are intended as examples. Other consistent units in either British or metric systems may be used.)

A	area (ft^2)
C	concentration (moles/liter)
E	electric field strength (V/cm)
G	drying rate (lb/hr)
G_{max}	maximum drying rate defined in Eq. (11-1)
K_1	constant defined in Eq. (11-1)
L	thickness of slab (ft)
MW	molecular weight
N	mole fraction of water in liquid phase
N_v	mole fraction of volatile solute in liquid phase
N_{vs}	mole fraction of volatile solute in liquid phase at saturation

R_d	resistance of "dry" layer in the food
R_s	resistance of space between food and condenser
T	temperature
W	weight of water
b	permeability of dry layer (lb ft^{-1} hr^{-1} $torr^{-1}$)
$f(x_d)$	function defined in Eq. (11-20)
k	thermal conductivity (Btu ft^{-1} hr^{-1} $°F^{-1}$
m	moisture content (lb water/lb solids)
p	partial pressure (torr)
p_i	vapor pressure of ice (torr)
p°	vapor pressure of water (torr)
p°_v	vapor pressure of volatile (torr)
q	heat flux (Btu/hr)
t	time (hr)
x	distance (ft)
y	mole fraction of water in vapor phase
y_v	mole fraction of volatile in vapor phase
α	relative volatility
ΔH_s	latent heat of sublimation (Btu/lb)
ε''	loss factor
υ	frequency
ρ	bulk density of dry food (lb/ft^3)

Subscripts

c	condenser
d	dry layer
f	final
i	ice or frozen layer
0	initial
s	space in freeze drier
w	wall of heat-exchange plates

REFERENCES

1. R. Bralsford, J. Food Technol. (London), 2, 339 (1967).

2. R. Bralsford, J. Food Technol. (London), 2, 353 (1967).

3. R. F. Burke and R. V. Decareau, Advan. Food Res., 13, 1 (1966).

4. R. G. Buttery, J. L. Bomben, D. G. Guadagni, and L. C. Ling, J. Agr. Food Chem., 19, 1045 (1971).

5. D. A. Copson, Microwave Heating, AVI Publ. Co., Westport, Conn., 1962.

6. D. F. Dyer, D. K. Carpenter, and J. E. Sunderland, J. Food Sci., 31, 196 (1966).

7. J. Flink and F. Gejl-Hansen, J. Agr. Food Chem., 20, 691 (1972).

8. J. Flink and M. Karel, J. Agr. Food Chem., 18, 295 (1970).

9. J. Flink and M. Karel, J. Food Sci., 35, 444 (1970).

10. J. Flink and M. Karel, J. Food Technol. (London), 7, 199 (1972).

11. Y. H. Foda, M. G. E. Hamed, and M. A. Abd-Allah, Food Technol., 24, 1392 (1970).

12. R. Fritsch, W. Mohr, and R. Heiss, Chem. Ing. Tech., 43, 445 (1971).

13. S. A. Goldblith, Advan. Food Res., 15, 277 (1966).

14. S. A. Goldblith, M. Karel, and G. Lusk, Food Technol., 17, 258 (1963).

15. J. I. Gray and D. G. Roberts, J. Food Technol. (London), 5, 231 (1970).

16. R. I. N. Greaves, Ann. N.Y. Acad. Sci., 85(Art. 2), 682 (1960).

17. A. C. Grimm, RCA Review, Dec. 1969, p. 593.

18. J. C. Harper, AIChE J., 8, 298 (1962).

19. J. C. Harper and A. Tappel, Advan. Food Res., 7, 172 (1960).

20. J. C. Harper and A. E. El Sahrigi, I & EC Fundamentals, 3, 318 (1964).

21. J. D. Hatcher, D. W. Lyons, and J. E. Sunderland, J. Food Sci., 36, 33 (1971).

22. J. E. Hill, "Sublimation dehydration in the continuum, transition and free molecule flow regimes," Ph.D. Thesis, Georgia Institute of Technology, Sept. 1967.

23. M. W. Hoover, A. Markantonatos, and W. N. Parker, Food Technol., 20, 807 (1966).

24. K. Ito, Chem. Pharm. Bull. (Japan), 19 (6), 1095 (1971).

25. J. King, CRC Critical Reviews in Food Technology, 1, 379 (1970).

26. J. B. Lambert and W. R. Marshall, in Freeze-Drying of Foods (F. R. Fisher, ed.), National Academy of Sciences-National Research Council, Washington, D. C., 1962, p. 105.

27. G. Lusk, M. Karel, and S. A. Goldblith, Food Technol., 18(10), 121 (1964).

28. A. P. MacKenzie, Bull. Parenteral Drug Assoc., 20, 101 (1966).

29. A. P. MacKenzie, Ann. N. Y. Acad. Sci., 125(Part 2), 522 (1965).

30. J. F. Maguire, "Vacuum techniques in freeze-drying pilot plants," Preprint 54B, Fifty-Ninth Annual Meeting of the AIChE in Detroit, Mich., Dec. 4-8, 1966.

31. J. W. McCulloch and J. E. Sunderland, J. Food Sci., 35, 834 (1970).

32. R. Noyes, Freeze Drying of Foods and Biologicals, Noyes Development Corporation, Park Ridge, N. J., 1968.

33. G. W. Oetjen, Vakuum-Technik, 9/69, 203 (1969).

34. G. W. Oetjen, Vakuum-Technik, 10/69, 237 (1969).

35. H. A. A. Pyle and H. Eilenberg, Vacuum, 21(3/4), 103 (1971).

36. D. Quast and M. Karel, J. Food Sci., 33, 170 (1968).

37. L. Rey, Proceedings of the Eleventh International Congress of Refrigeration, Vol. 1, International Institute of Refrigeration, Paris, 1963, p. 67.

38. L. Rey, Advances in Freeze Drying, Hermann, Paris, 1966.

39. L. Rey and M. C. Bastien, in Freeze-Drying of Foods (F. R. Fisher, ed.), National Academy of Sciences-National Research Council, Washington, D. C., 1962, p. 25.

40. O. C. Sandall, C. J. King, and C. R. Wilke, AIChE J., 13, 428 (1967).

41. O. C. Sandall, C. J. King, and C. R. Wilke, Chem. Eng. Progr. Symp. Series, No. 86, 64, 43 (1968).

42. G. D. Saravacos and J. C. Moyer, Chem. Eng. Progr. Symp. Series, No. 86, 64, 37 (1970).

43. G. D. Saravacos and M. N. Pilsworth, J. Food Sci., 30, 773 (1965).

44. M. Sivetz and H. E. Foote, Coffee Processing Technology, Vol. 1, AVI Publ. Co., Westport, Conn., 1963.

45. R. M. Storey and G. Stainsby, J. Food Technol., (London), 5, 157 (1970).

46. H. A. C. Thijssen, J. Food Technol. (London), 5, 211 (1970).

47. H. A. C. Thijssen and W. H. Rulkens, Ingenieur, 80(47), 45 (1968).

48. T. A. Triebes and C. J. King, I & EC Process Des. Devel., 5(4), 430 (1966).

49. K. G. Warman and A. J. Reichel, The Chemical Engineer, May 1970, p. 134.

50. World Coffee and Tea, Special wrap-up report: The Coffee Plant, World Coffee and Tea, 1970, p. 13.

PROTECTION OF FOOD DURING STORAGE

Chapter 12

PROTECTIVE PACKAGING OF FOODS

Marcus Karel

CONTENTS

I. PROTECTION OF PRODUCT AS A MAJOR FUNCTION OF PACKAGING

A. Functions of Packaging

Most food is consumed far removed in time and space from the point of its production; hence the need for the preservation processes discussed in this book. A necessary aid for the storage and distribution of food is packaging. The functions of packaging are several.

Packaging serves as a material handling tool containing the desired unit amount of food within a single container and may facilitate the assembly of several such units into aggregates. For example, some fluids are packaged in bottles which may be placed in boxes, and these boxes in turn can be assembled into easily handled pallets.

The package may also serve as a processing aid. For instance, the metal can be used in heat sterilization of many food items serves not only a protective function but, by its dimensional stability, assures that when fully packed the food maintains a certain shape and location for which heat penetrations can be calculated. In the case of flexible heat-processable pouches, which do not provide such dimensional permanence, it is necessary to introduce guiding racks or separators, which in a sense constitute temporary packages.

The package is a convenience item for the consumer. The examples one could choose here are very numerous. A beer can, for instance, serves as the drinking utensil as well as a process, storage, and distribution container. A variety of packages aid in handling, preparation, and consumption of foods by the consumer. Unfortunately, numerous examples also exist of poorly designed containers providing a barrier to efficient utilization of foods.

The package is a marketing tool. The sales appeal and product identification aspects of packaging are particularly important to the sales and marketing branches of food companies, and since these branches often have a dominant role in business decisions it is these aspects that often dominate package design. A trip to any supermarket can provide any reader with literally thousands of examples of packaging directed at marketing. These include such obvious product identification aspects as keeping a constant shape and labeling pattern for a container (consider Campbell Soup cans and the Coca Cola bottle). Various advertizing-oriented features are also used (for instance, bottles of products advertized as delicate often have "hourglass" shapes, and "robust" products often are packaged in squarish, husky-appearing bottles). This aspect of packaging often has been the target of consumer protection associations and advocates and has also encouraged the formulation of regulations to prevent deceptive packaging.

Packaging, when properly used, can be a cost-saving device. This aspect is very complicated, however, because it requires the simultaneous consideration of numerous interactions. Certain packages have obvious economic benefits, such as prevention of spills, ease of transporting, prevention of contamination, reduction of labor cost, and so on. There are also the obvious inherent costs: the packaging material itself, the packaging machinery, shipping weight of the package, and so on. In addition, there may be social costs hidden from immediate view: cost of disposal, cost of avoiding ecological disturbances due to accumulation of wastes (e.g., pollution), and costs associated with changes in energy consumption patterns. With respect to the last point, modern packaging systems have tended to evolve to minimize consumption of muscle power at the expense of electrical power. These hidden costs and benefits sometimes favor one kind of packaging, whereas the immediately visible balance is in favor of another. Unfortunately, thorough cost-benefit analyses of food packaging systems have not yet appeared.

Protection of the product is the most important aspect of packaging and the only one we shall consider in this chapter. It should be remembered, however, that the other functions cannot be ignored by the food technologist.

B. Protective Packaging

In considering the protective function of packaging we have to consider what aspects of quality are being protected. The quality of products as they reach the consumer depends on the condition of the raw material, on method and severity of processing, and on conditions of storage. Figure 12-1 demonstrates this point graphically. In this figure we show quality (by which we mean the totality of desirable attributes, including eating quality, and absence of undesirable ones, such as toxic factors) decreasing in storage. (There are, of course, exceptions to this pattern, e.g., aging of wines and beef.) The rate and extent of this decrease depends on the conditions of the environment in which the food is stored (Fig. 12-2). This internal environment, $\overline{I}$ may be maintained at conditions different from the external environment, $\underline{EO}$ (and preferably more favorable than EO), by interposing a protective barrier, $\overline{B}$.

The chemical, physical, and biological mechanisms of food deterioration are sensitive to various environmental factors, and the most pertinent barrier property of the package varies with each product. Some of the environmental factors and the corresponding pertinent packaging properties are listed in Table 12-1.

II. NATURE OF PACKAGING MATERIALS

The protection offered by a package is determined by the nature of the packaging materials and by the type of package construction. A great variety of materials is used in packaging, as shown in Table 12-2.

A. Glass

Glass containers have been used for many centuries and still are among the important packaging media. Physically, glass is a supercooled liquid of very high

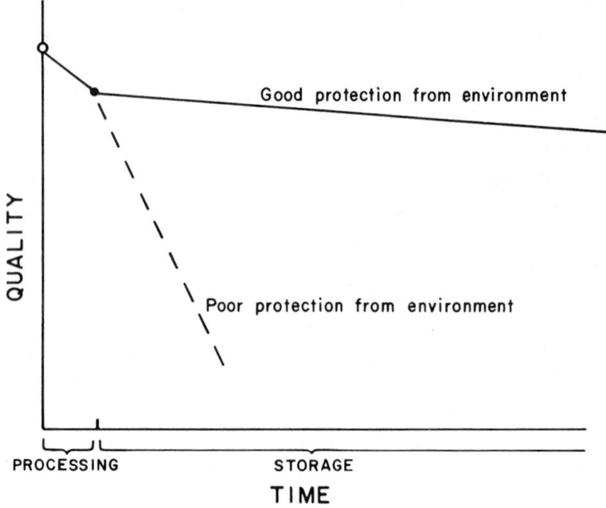

FIG. 12-1. Schematic representation of quality loss during processing and storage.

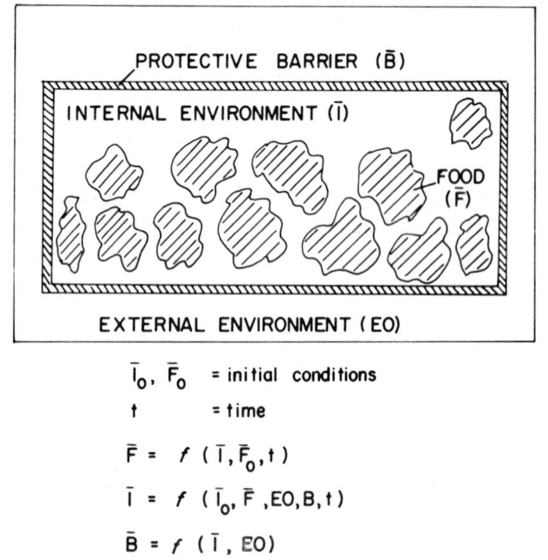

$\bar{I}_0$, $\bar{F}_0$ = initial conditions

t = time

$\bar{F} = f(\bar{I}, \bar{F}_0, t)$

$\bar{I} = f(\bar{I}_0, \bar{F}, EO, B, t)$

$\bar{B} = f(\bar{I}, EO)$

FIG. 12-2. Schematic representation of interaction between packaged food and the environment.

TABLE 12-1

Package-Environment Interactions

Environmental factors	Pertinent package properties
Mechanical shocks	Strength
Pressure of oxygen, water vapor, etc.	Permeability
Light intensity	Light transmission
Temperature	Thermal conductivity Porosity Reflectivity
Biological agents	Penetrability

viscosity. Chemically, it is a mixture of inorganic oxides of varying composition. Most container glass is of the soda-lime-silica type.

Container glass usually consists of 70-75% sodium and calcium silicate, 6-12% calcium and magnesium oxide, and small amounts of aluminum, barium, and other metallic oxides. The materials are mixed with "cullet," or crushed glass from

TABLE 12-2

Approximate Value of Packaging Materials Shipped Annually in the United States (1970)

Materials	Value in billions of dollars
Paper containers	7.8
Paper for flexible packaging	0.6
Plastic films	1.0
Metal containers	4.6
Metals for flexible packaging	0.1
Glass containers	2.0
Rigid plastic containers	0.75
Wooden containers	0.6

broken or defective containers, and the mix is melted in a furnace at temperatures in excess of 2700°F.

Bottles are formed in two steps. An exact amount of molten glass is metered into a mold, smaller than the final bottle, and compressed gas or a mechanical plunger presses the glass against the walls of the mold. The partially formed bottle is transferred to a second mold, where it is blown by compressed air into its final shape. The mouths of the container are fire polished and the bottles are annealed at high temperature. All containers are inspected; in most modern installations they are subjected to mechanical tests (squeezing by automatic equipment), and defective containers are rejected and used for cullet.

The most important packaging properties of glass, such as moldability, inertness, transparency, and strength, can be modified by relatively minor changes in the glass composition. Color, for instance, can be controlled by inclusion of small amounts of oxides of various metals, such as chromium, cobalt, and iron, and semiopacity can be imparted by addition of fluorine compounds.

The major packaging design problems involving glass arise from its mechanical properties. Glass is brittle and its tensile strength depends greatly on the condition of the surface. Commercial glass containers have only a small fraction of the potentially possible tensile strength because the surface conditions cannot be maintained at the perfect level required for maximum strength. As a consequence the containers must be made of fairly thick glass, and they therefore have substantial weight. Recent developments in glass packaging have included an attempt by manufacturers to improve mechanical properties and to reduce weight. Shatter-resistant bottles are said to be under development and one product is already on the market.

A recent development is the use of very thin glass encased in a plastic exterior. The resultant bottle has the advantage of inertness of interior surface with resistance to abrasion by the exterior surface.

The properties of containers also depend on type of construction. Types of containers commonly used include bottles, jars, jugs, and tumblers. A variety of closures is available, and the recent trend toward maximum convenience for the consumer has resulted in substantial efforts toward development of easily opened closures. Convenience as well as esthetic considerations are often paramount in package design, and satisfying these requirements demands substantial ingenuity by the design engineer.

Glass containers possessing the proper type of closure can be used for practically all types of packaging applications, including heat processing, gas packaging, and other types of preservation requiring hermetically closed containers. For more detailed discussion of glass in packaging the reader is referred to specialized publications [10, 13, 34].

B. Metals

1. Rigid Metal Containers

The most common container for packaging of foods other than beer and carbonated beverages is the so-called tin can. It is traditionally used for heat-sterilized products and is made of tinplate, which consists of a base sheet of steel with a coating of tin applied to it by "hot dipping" or by the electrolytic process. The amount of tin coating depends on the process and the type of can. A typical tinplate may be 0.01 in. thick, with the tin coating contribution only about 0.00005 in. to this thickness. To make the can more suitable for specific packaging applications enamels and linings can be applied to the tin or to the tinless steel sheets. The composition of these linings varies with intended use. Plastics, shellacs, resins, glass, and inorganic oxides are among materials used for this purpose.

Three different types of base steel are common:

1. Type "L". This steel is very corrosion resistant and therefore most suited to canning of acid products.
2. Type "MR". This steel is less restricted than "L" in contents of C, S, P, M, and Si and is suitable for mildly corrosive products.
3. Type "MC". This is least restricted in composition and is used for non-corrosive products.

In recent years very substantial changes have occurred in the can industry, and a great variety of cans different from the can made of standard tinplate has been developed. In particular, these developments have been aimed at beer and carbonated beverage applications but may have additional applications in other foods. Tin-free steel, utilizing direct chromium-containing coatings or such very thin thermoplastic coatings as nylon, has been used. The other innovation is the utilization of cans with cemented, or welded, rather than soldered side seams. This development is in part connected with the utilization of tin-free steel. However, in recent years there also has been increasing concern with the potential contamination of foods by lead from soldered side seams.

Coatings for cans can be of the following types: oleoresin, vinyl, epoxy, phenolic, and wax. The industry refers to all of these, except for waxes, as "lacquers" or "enamels."

The "C" and "R" enamels are of the oleoresin type and are most common in food industry. The "C" enamel contains zinc oxide, designed to react with sulfides evolved during heat processing. The sulfides otherwise tend to react with iron. Iron sulfide is of course black and is objectionable from an esthetic point of view. The "R" enamels contain no zinc oxide. The other enamels are used in specific applications, since each type has specific advantages and restrictions.

Tin cans are fabricated as follows (Figs. 12-3, 12-4). Sheets of tin plate with or without an enamel coating are cut into pieces of a size sufficient to form the body (sides) of a can. Each such piece or "body blank" is notched at the corners, hooked or turned back at the edges to be joined, formed into a cylinder, and joined by means of the hooks. The hooks are then flattened and soldered to form the "side seam." The side seam consists of four layers of tin plate except for the portion that engages the ends (top and bottom) of the cans. At these two intersections the body blanks are notched so that a two-layer lap seam is formed. The lap seam, being less bulky, folds more easily and securely into the seam which is formed when the ends are attached. Cutouts, tabs, or small incisions are sometimes incorporated in the side seam design. These modifications serve to increase the surface area of the soldered joint, which in turn strengthens the can.

A small flange is next formed on each rim of the cylindrical body to facilitate sealing the can ends. The can ends are formed by stamping metal disks from strips of tin plate. The various contours necessary for strengthening the can

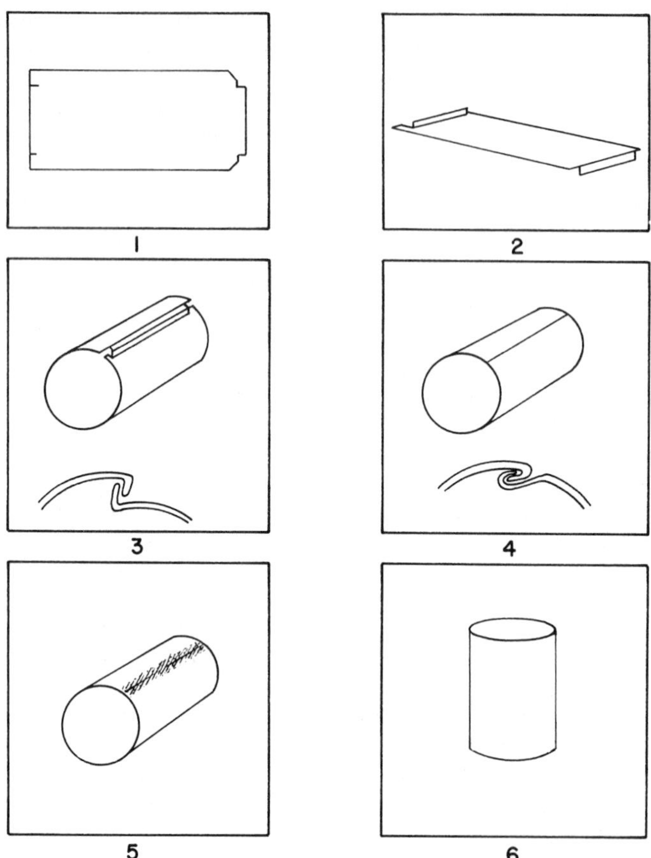

FIG. 12-3. Steps in formation of a can. (1) Body blank with notched corners; (2) Body blanked hooked; (3) Cylinder formed; (4) Hooks flattened; (5) Side seam formed; (6) Completed cylindrical body.

ends are formed during this operation. The edges of the can ends are curled and a rubberlike sealing compound is flowed into the curl to form a gasket. The bottom end of the can is attached by means of a "double seam" (Fig. 12-4). The double seam consists of five layers of tin plate plus the gasket. No solder is used for making the end seals (double seams).

The cans are tested individually to insure the absence of leaks, after which they are shipped to the canner. Generally, the canner can expect to find no more than one defective can in each 10,000. After the can is filled, the canner applies the top end, again by means of a "double seam." The double-seaming operation can be accomplished at speeds approaching 800 cans per minute per machine.

A simple convention is used by the canning industry to identify the size of a can. Such numbers as 211 x 300, 303 x 406, and 307 x 409 are common designations. In each case the first three-digit number indicates the diameter of a can and the second three-digit number indicates the height. The first digit of the number indicates inches and the last two digits indicate sixteenths of inches. For example, 211 x 400 indicates the can is 2-11/16 in. in diameter and exactly 4 in.

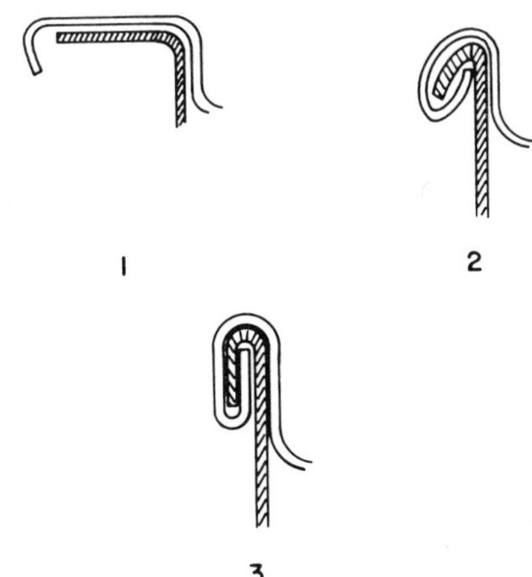

FIG. 12-4. Steps in formation of a double seam.

tall. In addition to this dimensional system, most sizes of cans have simple number designations. For example, a 307 x 409 can is also known as a No. 2 can.

Sanitary cans made of aluminum have in recent years moved into a position of commercial importance. Although cans fabricated from aluminum are presently more expensive than cans fabricated from tin plate, there are certain applications where the initial cost disadvantage is offset by a functional advantage or by a saving in shipping cost resulting from the lightness of aluminum.

Recognized advantages of the aluminum can as compared to the tin plate can are lighter weight (lower shipping cost), ease of salvageability, absence of sulfide discoloration, increased resistance to certain types of corrosion, and ease of opening (pull tab). On the other hand, aluminum cans possess two major disadvantages as compared to tin plate cans: less strength and higher cost. Pure aluminum is quite weak. It can be strengthened by the addition of small quantities of magnesium, but this has an adverse effect on its resistance to corrosion. Even the magnesium-aluminum alloys posses less strength than tin plate of the same thickness. As a consequence, aluminum must be of heavier gage than tin plate in order to achieve cans of equal durability.

Cans made of aluminum have made a very substantial impact on the beverage market but have not been used extensively for heat-sterilized food products. It is known, however, that successful sterilization and a satisfactory shelf life can be attained with properly coated all-aluminum cans.

In terms of rigid metal construction the variety of container types is very large and all cannot be discussed here. The reader is referred to Paine [34] and to Brody [3].

2. Metal Foils

The most important food-packaging foils consist essentially of pure aluminum. They are used in thicknesses ranging from 0.00025 to 0.006 in. The thickness of uncoated aluminum foil largely determines its protective properties. Foils of low thickness have microscopic discontinuities (pinholes) and allow limited diffusion of gases and vapors. According to Cooke [6], foil 0.0015 in. thick or thicker is conceded to have essentially zero water-vapor permeability, indicating a complete absence of pinholes. The properties of thinner foils can be improved by combination with one or more plastics in the form of coatings or laminations.

The use of foils made of metals other than aluminum is very limited. Lead, tin, and zinc foils are mainly of historic interest, and the recently developed steel foils have yet to find major packaging uses.

C. Paper

A very considerable proportion of packaged foods is stored and distributed in packages made of paper or paper-based materials. Because of its low cost, ready availability, and great versatility paper is likely to retain its predominant packaging position for some time to come.

Paper consists primarily of cellulose fibers which are obtained by one of several pulping processes. The discussion of paper manufacture is beyond the scope of this chapter, and the interested reader is referred to standard works on paper [4].

The packaging properties of paper vary considerably depending on the manufacturing processes and on additional treatments to which the finished paper sheet can be subjected. Its strength and mechanical properties depend on mechanical treatment of the fibers and on the inclusion of fillers and binding materials. The physicochemical properties of paper, such as permeability to liquids, vapors, and gases, can be modified by impregnating, coating, or laminating. The materials used for this purpose include waxes, plastics, resins, gums, adhesives, asphalt, and other substances.

The quality of materials made by the above converting processes varies considerably. Some laminated, waxed, and coated papers deserve the designation of protective materials. Certain grades of converted papers, however, offer little more than protection from light and mechanical damage.

Papers can be used as flexible packaging materials or as materials for constructing rigid paper containers. Flexible packaging applications include the use of papers for wrapping materials as well as for manufacture of bags, envelopes, liners, and overwraps. Some of the more important types of paper used for these purposes include Kraft paper, greaseproof papers, glassines, and waxed papers, as well as papers prepared by conversion of these basic types.

Rigid paper containers include a great variety of types, varying in materials and methods of construction. They may include cartons, boxes, fiber cans, drums, liquid-tight cups, tetrahedral packs, and many others. They can be constructed from paperboard, laminated papers, corrugated board, and the various types of specially treated boards and papers. Frequently, paper containers have liners or overwraps, made usually of protective grades of paper, plastics, or metal foil. The packaging applications of paper are more adequately discussed in several publications on packaging [10, 30].

D. Plastics

Plastics are organic polymers of varying structure, chemical composition, and physical properties. A detailed description of the chemical nature and physicochemical properties of plastics is beyond the scope of this chapter, and the reader is referred to modern texts on high polymers [1]. A brief review of the types of plastics used in food packaging is given below. Some properties of importance to maintenance of nutritional value are also discussed in later sections of this chapter.

1. Cellophanes

Cellophanes are frequently classified as plastics, and this usage is followed here, in spite of the fact that the major constituent of cellophanes — cellulose — is a natural rather than a synthetic polymer. In addition to cellulose, all cellophanes contain a plasticizer, such as glycerol or ethylene glycol. (Plasticizers are materials, usually solvents of low volatility, added to plastics to reduce the attractive forces between the polymer chains. This results in plastics of better flexibility.) Plasticized cellulose, called plain cellophane, offers no protection to the diffusion of water vapor. It is usually coated with protective agents, such as nitrocellulose, waxes, resins, and synthetic polymers. The packaging properties of cellophanes are determined primarily by the nature of these coatings.

2. Cellulosics

Cellophane provides the base material for the production of cellulose acetate, ethyl cellulose, and cellulose nitrate. The cellulosics have properties similar to those of cellophanes but have fewer applications in the packaging of foods.

3. Polyolefins

Polyethylene is one of the most important packaging materials of the present time. It is a hydrocarbon polymer with the nominal formula $-(CH_2-CH_2)-_n$. Commercial polyethylenes are produced with a variable amount of branching within this nominally linear polymer. High-density polyethylene has the least branching and as a result the greatest thermal stability and the lowest permeability. High-density polyethylene is used in film form as well as for production of rigid plastic containers, including milk bottles. Low-density polyethylene has the advantage of maximum flexibility and low cost and is widely used for packaging of foods in bags or as an overwrap. Polyethylenes are also widely used in laminations, where they provide the inner layer requiring good heat sealability. Polypropylene is closely related to polyethylene and its properties and uses are similar.

4. Vinyl Derivatives

Vinyl derivatives have the general formula $-(CH_2-CXY)_n-$, where X and Y are either hydrogen atoms or other substituents, such as chlorine, benzene, methyl and hydroxyl. The properties of vinyl polymers are dependent on the nature of substituents, on molecular weight, on the spatial arrangement of groups within the chains, and especially on orientation and crystallinity.

The following generalizations about the effects of polymer structure on functional properties are of particular importance in packaging.

1. Strength, resistance to high temperatures, resistance to action of solvents, and resistance to diffusion increase with increasing degree of crystallization.
2. The presence of polar groups decreases resistance to diffusion of polar molecules.
3. Linearity of chains and orientation of crystallites improve strength and impermeability.
4. Addition of plasticizers increases permeability and solubility.

The important food-packaging materials based on vinyl polymers and their chemical formulas are listed below.

Saran, copolymer of vinyl chloride $-(CH_2-CHCl)_n-$ and vinylidene chloride $-(CH_2-CCl_2)_n-$
Polyvinyl alcohol $-(CH_2-CHOH)_n-$
Polyvinyl acetate $-(CH_2-CHOCOCH_3)_n-$
Polystyrene $-(CH_2-CH-C_6H_5)_n-$
Polyvinyl chloride $-(CHCl-CH_2)_n-$

5. Polyesters

The polyester of most importance in food packaging is polyethylene terephthalate, a condensation product of ethylene glycol and terephthalic acid, which is better known in the United States under the Du Pont trade name Mylar. Mylar is a crystalline linear polymer of excellent strength and inertness. Mylar has been widely used in lamination in particular as the outer, abrasion-resistant layer of laminations to be used for food pouches requiring good protective properties.

6. Pliofilm

Rubber hydrochloride, or Pliofilm, is used for the packaging of certain types of food. Its properties are determined by the type and amount of plasticizer, which is always added to this type of film.

7. Polyfluorocarbons

Several polymers with the backbone composed of carbon and fluorine, or of hydrogen and chlorine atoms in addition to C and F, form the basis for several packaging films with remarkable properties. These materials are very costly but have unique properties. Teflon and other fluorinated hydrocarbons which are the equivalents of polyethylenes with hydrogen replaced by fluorine have remarkable inertness and high permeability to oxygen. In contrast, trifluorochloroethylene has an extremely low permeability to all gases and is in particular effective as a water vapor barrier. Polyvinyl fluoride is intermediate in permeability but has excellent resistance to sunlight.

8. Polyamides

A number of polyamide films, which are condensation polymers of diamines with diacids, are available on the market and have found extensive applications. Nylons, as these polymers are usually called in the trade, are inert and heat resistant and have excellent mechanical properties. They are usually coated or used in combination with other materials to produce packaging materials of food inertness as well as low permeability.

9. Water-soluble and Edible Films

There has been a recent increase in interest in edible and other water-soluble films. In part, this interest stems from ecological considerations, and is aimed at minimizing solid wastes. The other driving force for this type of package development is the increased trend toward convenience foods. Water-soluble films include polyvinyl alcohol and some cellulose derivatives, as well as other polysaccharides, including those derived from the linear starch component, amylose. Some proteins, in particular collagen, are also suitable materials [25, 31].

10. Other Films

Recent developments in packaging films include the utilization of polymeric films containing ionized groups contributing to film strength through ionic crosslinks. These films, known as "Ionomers," are particularly resistant to effects of low temperatures. The other major area of packaging material research has been on heat-resistant materials suitable for producing sterilizable packages [10]. Polymer alloys are another likely future development.

E. Wood

Containers made of wood are used extensively for shipment and storage of food packages. Their use as direct packages for food products is of little importance.

Shipping containers are manufactured from different types of wood and they vary widely in construction. The types of construction commonly encountered in food packages include cases, boxes, crates, and barrels.

F. Combinations

For many packaging applications it is necessary to combine two or more different materials in order to obtain a package with satisfactory properties. Some of the combinations used in food packaging have been mentioned in the preceding sections. The following list is intended to show the great variety of materials that can be used for this purpose:

a. plastic combinations containing two or more plastic films laminated with various types of synthetic and natural adhesives
b. plastic, silicone, and wax coatings on packages made of metals, paper, wood, and other materials
c. rigid cans made of paper, with metal bottoms and tops
d. plastic films with thin metallic coatings
e. metal foils with thin plastic coatings
f. barrier materials constructed of as many as six or seven materials including paper, textiles, metals, plastics, waxes, and resins.

III. EFFECT OF ENVIRONMENT ON FOOD STABILITY AND THE NEED FOR PROTECTIVE PACKAGING

The package affects the quality of foods by controlling the degree to which factors connected with processing, storage, and handling can act on components of foods. The processing and storage factors amenable to control by packaging include light, oxygen concentration, moisture concentration, heat transfer, contamination, and attack by biological agents.

A. Light

Food can be adversely affected by prolonged exposure to light. The chemistry of light-catalyzed changes can be quite complex. For instance, light promotes the following chemical reactions in food: oxidation of fats and oils to produce the complex of changes known as oxidative rancidity; oxidation of milk leading to formation of volatile and unpleasant mercaptans; and changes in various pigments, including the pink pigment astaxathin (in salmon and shrimp) and the green, yellow, and red pigments of plant products, as well as the red pigment of meats (myoglobin). Riboflavin is especially photosensitive and when exposed to sunlight it loses its vitamin value and also activates or sensitizes other components to photodegradation. Ascorbic acid, or vitamin C, is also quite sensitive to light and it interacts with other components during light exposure.

The catalytic effects of light are most pronounced with light of the highest quantum energy, i.e., light in the lower wavelengths of the visible spectrum and in the ultraviolet spectrum. Specific reactions involved in deterioration of foods, however, may have specific wavelength optima; in particular the presence of sensitizers may shift the effective spectrum very substantially. Sensitizers present in food include riboflavin, β-carotene, vitamin A, and peroxidized fatty acids.

Sensitivity of a given class of foods to various wavelengths of light can vary with the method of processing. Fresh meats are known to discolor rapidly when exposed to ultraviolet light but are resistant to visible light. Meats preserved with nitrite, however, undergo discoloration due to visible as well as ultraviolet light.

In addition to wavelength the intensity of light and duration of exposure are important. In this respect we again must distinguish between the sensitivity of different products. As in the case of ionizing radiation, the penetration of light into food follows an exponential equation:

$$I_X = I_o e^{-\mu X} \tag{12-1}$$

where

I_X is the intensity of light at depth X in food
I_o is the intensity of light at the surface of food
μ is a characteristic constant (absorbance) for the given wavelength of light and a given food material

The most penetrating quanta of light are those with wavelengths which are least effective per quantum of energy. Thus, ultraviolet rays penetrate less deeply into food than the red light components. In some cases the relatively shallow penetration of light into food products is an adequate protection in itself. Consider, for instance, the potential for loss of riboflavin in bread. In breads with well-developed crusts, 95% of the short-wavelength light is absorbed in the outer 2 mm layer of bread. If we make an unrealistically harsh assumption that all of the vitamin in this layer is destroyed and that the bread is exposed to light from all sides; then the volume in which the riboflavin is lost (V_i) is equal to the surface area of bread (A_i) multiplied by 2 mm. Assume a brick-shaped loaf of 10 cm x 20 cm. A_i is then equal to $[2(10 \times 10) + 4(10 \times 20)] = 1000\ cm^2$, and $V_i = 1000\ cm^2 \times 0.2\ cm = 200\ cm^3$. Since total volume of the loaf is $10 \times 10 \times 20\ cm^3 = 2000\ cm^3$, the maximum loss under these exaggerated conditions is only 10%.

The loss of riboflavin in a similarly shaped bottle of milk during exposure to light cannot be dismissed so lightly. This is true even when it is assumed light penetration in milk and bread are identical. The reason is that while the light cannot reach the vitamin located in the depth of the bottle, a diffusion process can occur in this liquid product, and an exchange can occur between riboflavin in the surface and riboflavin in the interior, thus continuously supplying some more vitamin to the irradiation surface. Similarly, free radicals produced by light at the surface of the liquid can migrate inward and react to damage the vitamin molecules in the interior. Even in a solid product with limited diffusion, surface reactions often suffice to produce damage resulting in unsaleability of the product. This is obviously the case with sensitive colors or when small amounts of potent off flavors or toxic compounds are produced.

B. Oxygen

Oxidation reactions are often the cause of undesirable changes in foods. One such reaction has already been mentioned, namely oxidative rancidity due to peroxidation of fats and oils in various foods. In addition, many vitamins, pigments, and some amino acids and proteins are oxygen sensitive. Packaging can control two variables with respect to oxygen, and these have different effects on rates of oxidation reactions in foods. The first variable is the total amount of oxygen available. For instance, by packaging in a hermetically closed container which is constructed of an impermeable material, it can be assured that the total amount ofoxygen available to react with food is finite. Under these circumstances, no matter how fast the rate of reaction, the extent of the reaction cannot exceed that which corresponds to complete depletion of oxygen. If the extent of reaction is of no consequence from the point of view of food quality, then the rate of reaction is irrelevant to shelf life.

On the other hand, situations do exist where the above is not true, either because the amount of oxygen in a sealed container is in fact significant with respect to deterioration potential or because the container allows some permeation. In these instances another variable is important, namely, the concentration of oxygen in the food, and this in turn depends on oxygen pressure.

In Chapter 7, we have considered equilibrium and rate with respect to sorption of water in foods. Similar considerations can be applied to the distribution of oxygen between the food and the gas phase surrounding it. In many cases, for the purposes of assessing the influence of packaging measures on oxidation-caused deterioration in foods we can proceed by establishing relations between the rates of reaction in the food and the oxygen pressure in the space surrounding the food, without worrying about diffusion effects in food. Where the food itself is very resistant to diffusion of oxygen this correlation is not useful, unless the size and surface area of the material for which the correlation has been determined is specified.

Quast and Karel [36] have studied the diffusion of oxygen in foods under conditions that can occur during processing, handling, packaging, and storage. They have found that oxygen diffusion can affect rates of oxidation in two types of situations (Table 12-3): (1) when oxidation occurs very rapidly (such as in fish meal during curing in bulk bins), and (2) in a very dense product (such as butter). Oxygen diffusion also influences meat color [32].

The relation between rate of oxidation and oxygen pressures varies with type of reaction, type of product, and temperature. We have seen previously (Fig. 8-15) how oxidation of potato chips varies with oxygen. The relationship is nonlinear and above some level oxygen pressure has no effect on oxidation rate. This behavior is typical for oxidizing fats and oils in products with large surface to volume ratios and little internal diffusion resistance. When the surface to volume ratio decreases, oxidation of lipids can approach a linear dependence on oxygen pressure (Fig. 12-5).

Oxidation of meat pigment to an undesirable brown color has a complicated response to oxygen pressure. In fresh meats discoloration rates show a maximum at about 24 mm oxygen pressure corresponding to about 2-3% oxygen at atmospheric pressure (Fig. 12-6). The discoloration of cured meat pigments is almost directly proportional to oxygen pressure. Of course these differences in discoloration behavior dictate different requirements for package properties.

Another complicated oxygen permeation requirement is that of fresh fruits and vegetables. The respiration of these commodities continues after harvest and, as shown in Table 12-4, some vegetables and fruits have very high demands for oxygen and produce an essentially equimolar amount of carbon dioxide. The rate

TABLE 12-3

Effect of Diffusion on Oxidation of Some Food Products in Air[a]

Product	Temperature (°C)	Effective oxygen diffusivity (cm^2/sec)	Product thickness (cm)	Reduction of oxidation due to diffusion resistance (%)
Potato chips (past induction)	37	2×10^{-1}	100	0
Fish meal	37	8×10^{-2}	10	0
Fish meal	37	8×10^{-2}	100	50
Butter	~ 5	1.6×10^{-8}	10	60

[a] From Quast and Karel [36].

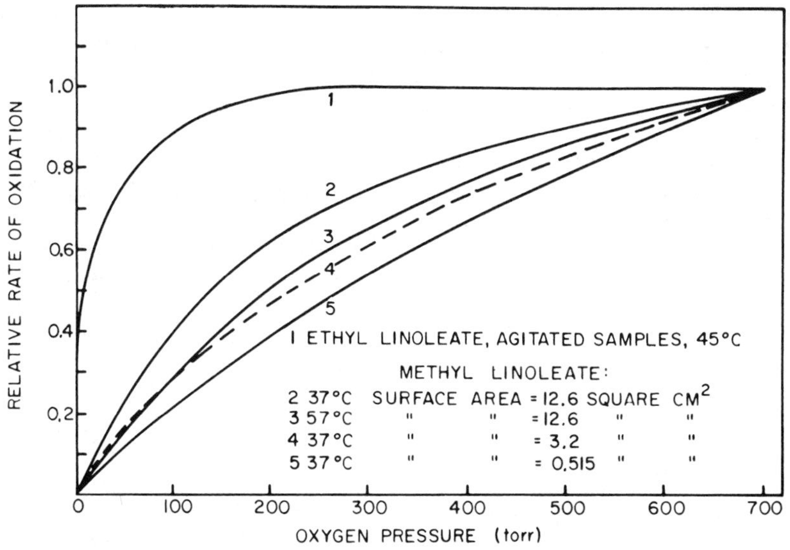

FIG. 12-5. Dependence of oxidation of lipids on oxygen pressure for different surface areas per unit volume.

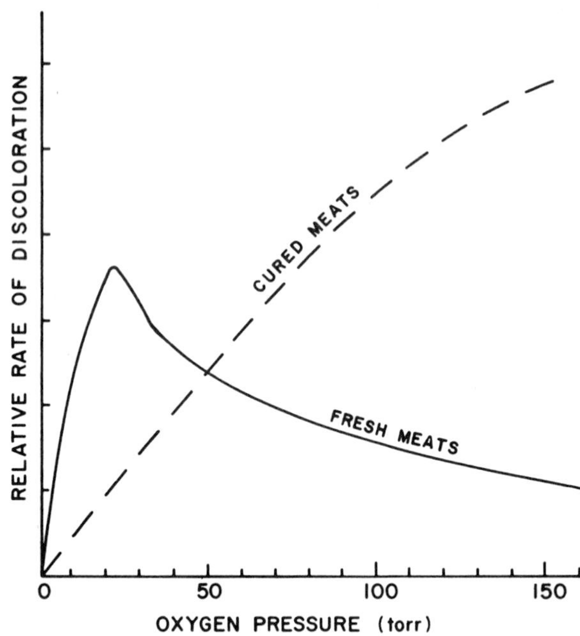

FIG. 12-6. Rates of discoloration of meats as a function of oxygen pressure.

of respiration can be reduced by reducing the partial pressure of oxygen, as shown in Figure 12-7. As can be seen in the same figure, however, a reduction of oxygen concentration to a very low level produces a sharp rise in the respiratory quotient, defined as the ratio of carbon dioxide evolved to oxygen consumed. This is due to the fact that there exists a critical oxygen value which varies with the commodity and with temperature and is often called the "extinction point of aerobic respiration." Below this value the plant tissues produce alcohol and carbon dioxide without absorbing oxygen (anaerobic respiration). If anaerobic respiration continues the fruit or vegetable spoils rapidly. Although it is desirable to slow rates of aerobic respiration of fruits and vegetables by lowering oxygen pressure, the oxygen level must be maintained above the critical value at which anaerobic respiration starts.

The above examples show that the oxygen response of foods is varied and must be known before package selection is undertaken.

C. Water

The water content of foods is dependent on the relative humidity of the immediate environment. In Chapter 8 we have reviewed in considerable detail the water relations of foods and need only to reemphasize here that these relations strongly affect packaging requirements.

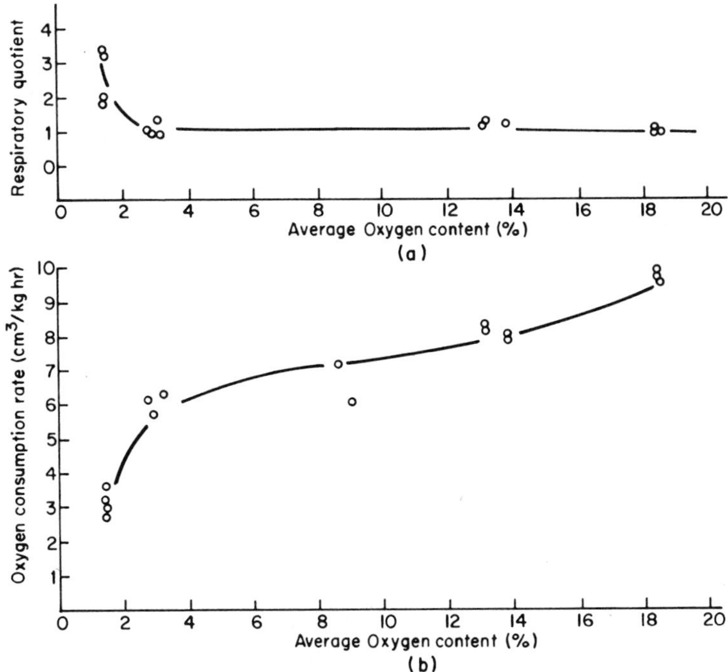

FIG. 12-7. Respiratory quotient (a) and respiration rate (b) for apples, as a function of oxygen concentration.

D. Temperature

In practically every preceding chapter we have had occasion to mention the effects of temperature on rates of biological, chemical, and physical events in foods. Some degree of temperature control can be attained by packaging measures, but other control measures are often more feasible.

TABLE 12-4

Rates of Respiration of Some Fruits and Vegetables

Fruit or vegetable	CO_2 production (mg/kg hr) at		
	5°C	15°C	20°C
Peas	100	300	400
Asparagus	50	200	300
Banana	—	60	120
Tomato	20	30	40
Grapefruit	5	10	15

E. Sensitivity to Mechanical Damage

One of the functions of the package is to protect the food from various mechanical disturbances, such as spillage, or various mechanical deformations, such as breakage, abrasion, compression, or bending. A complete discussion of food rheology would be necessary to assess the differences in sensitivity among different products. Conversion of laboratory tests on food rheology to specification of requirements for mechanical properties of containers is in any case still quite primitive. For present purposes, it is sufficient to say that most if not all containers require adequate strength to maintain their own integrity (and therefore maintain separation between the food and the external environment). An additional requirement for cushioning (or protection of food from mechanical shocks) may also arise.

F. Sensitivity to Attack by Biological Agents

All food is subject to attack by some biological agents. The sensitivities of foods to microorganisms and viruses vary greatly with environmental conditions, and the corresponding requirements for packaging protection vary accordingly. Thus, foods relying on destruction of microorganisms prior to or immediately after packaging (sterilized products) require absolute package protection from recontamination. Other kinds of products, for instance those preserved by addition of chemicals, are much less sensitive and may tolerate some contamination.

Similar considerations apply, albeit to a lesser degree, to attack by insects, a potential danger for practically all types of foods. Other biological agents, such as rodents, can be more effectively controlled by proper warehouse and truck sanitation and by proper design of shipping containers than by packages that protect individual consumer portions of the product.

IV. PROPERTIES OF PACKAGES AND OF PACKAGING MATERIALS WHICH DETERMINE DEGREE OF PROTECTION AGAINST INDIVIDUAL ENVIRONMENTAL FACTORS

A. Light-protection Characteristics

The total amount of light absorbed by the food is given by the following formula:

$$I_{abs} = I_o Tr_p \frac{1 - R_f}{(1 - R_f)R_p} \tag{12-2}$$

where

I_{abs} is the intensity of light absorbed by the food
I_o is the intensity of incident light
Tr_p is the fractional transmission by the packaging material

R_p is the fraction reflected by the packaging material
R_f is the fraction reflected by the food

The fraction of the incident light transmitted by any given material can be considered to follow the Beer-Lambert law:

$$I_x = I_o e^{-\mu X} \tag{12-1}$$

The absorbance, μ, varies not only with the nature of the material but also with the wavelength. The light transmitted through a given material therefore gives a characteristic spectrum of transmitted light dependent on incident light and the properties of the package. Figure 12-8 shows light-transmission curves for several food-packaging materials. It is readily apparent that several of the

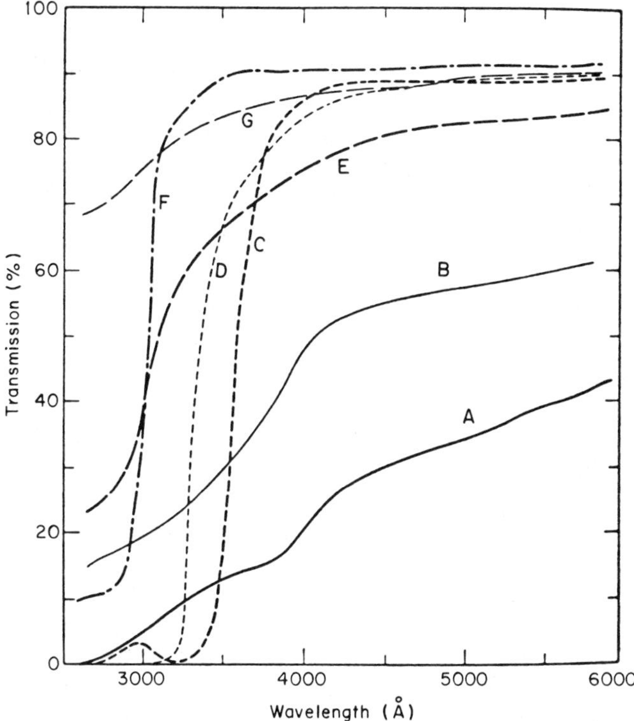

FIG. 12-8. Light transmission through various flexible packaging materials. (a) Low-pressure polyethylene, 0.0035 in. thick; (b) translucent wax paper, 0.0035 in. thick; (c) Saran, 0.0011 in. thick, (d) pliofilm, 0.0014 in. thick; (e) Pliofilm, 0.0013 in. thick; (f) cellulose acetate, 0.001 in. thick; (g) conventional polyethylene, 0.0015 in. thick.

plastics, while transmitting similar amounts of light in the visible range, offer varying degrees of protection against the damaging ultraviolet wavelengths. The protection against uv light by different packaging materials often can be characterized by a "cutoff" wavelength below which transmission of light becomes negligible, and these wavelengths are sometimes listed in compilation of package properties.

The barrier properties with respect to light of a packaging material can be improved by special treatments. Glass is frequently modified by inclusion of color-producing agents or by application of coatings. Figure 12-9 shows the differences in light transmission between transparent and colored bottle glass. The light transmission curve of window glass of comparable thickness is included for comparison.

Modification of plastic materials may be achieved by incorporation of dyes or by application of coatings. Figure 12-10 shows the effect of such treatments on cellophanes.

As a consequence of the successive absorption of light in both the packaging material and in the food, the total energy delivered to different points in the food varies and the spectral characteristics of the light vary as well [46]. Figure 12-11 shows schematically how relative intensity of light at all wavelengths and the relative intensity of short wavelengths may vary in a packaged product. Consider the situation as indicated in Figure 12-11 with the thickness of the packaging

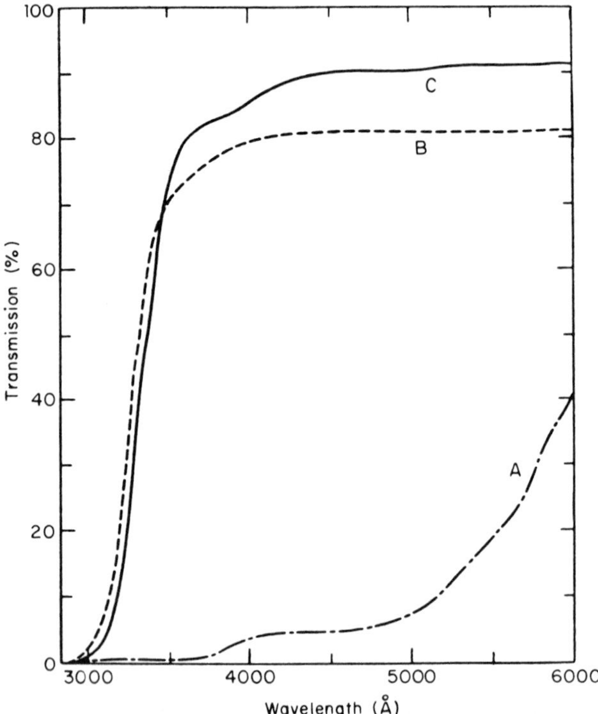

FIG. 12-9. Transmission of light through three types of glass. (a) Amber bottle glass, 0.118 in. thick; (b) transparent milk bottle glass, 0.118 in. thick; (c) window glass, 0.119 in. thick.

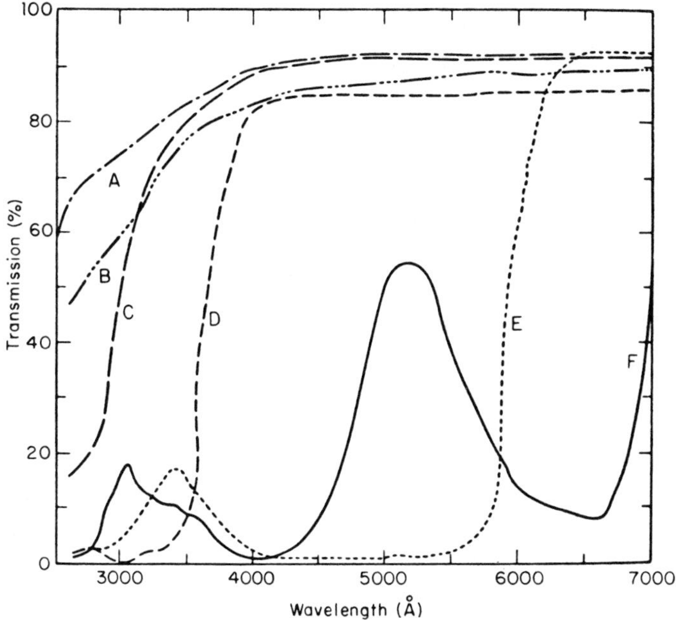

FIG. 12-10. Transmission of light through various types of cellophane. (a) Plain cellophane, 0.0009 in. thick; (b) Saran-coated cellophane, 0.0015 in. thick; (c) nitrocellulose-coated cellophane (conventional moistureproof cellophane), 0.001 in. thick; (d) moistureproof cellophane coated with a special ultraviolet absorbing composition, 0.001 in. thick; (e) dyed cellophane (red), 0.001 in. thick; (f) dyed cellophane (dark green), 0.001 in. thick.

material being X_p and its absorbance for any one wavelength μ_p. If a food has absorbance μ, then we can calculate the intensity of light of a given wavelength at plane X in the food as follows (assuming that reflection can be ignored):

$$I_i = I_o e^{-\mu_p X_p} \tag{12-3}$$

$$I_x = I_i e^{-\mu X} \tag{12-4}$$

where I_i refers to intensity at the surface of the food, and I_x to intensity at plane X in the food. Combining Eqs. (12-3) and (12-4) we obtain:

$$I_x = I_o \; e^{-(\mu_p X_p + \mu X)} \tag{12-5}$$

and therefore

$$\ln (I_o/I_x) = \mu_p X_p + \mu X \tag{12-6}$$

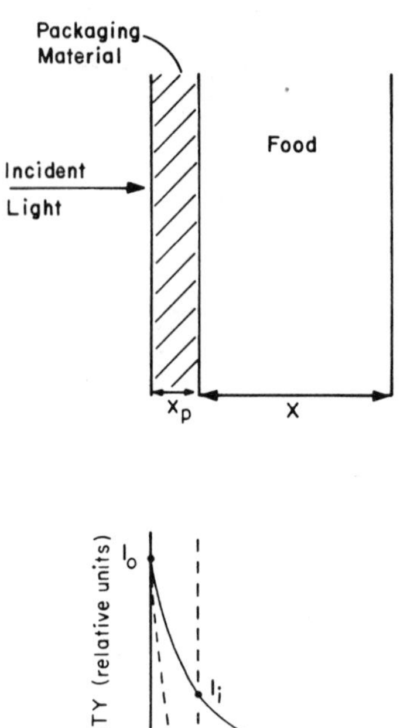

FIG. 12-11. Schematic representation of light absorption in a packaged food product.

B. Permeability to Gases and Vapors

1. Definitions

The protection of foods from gas and vapor exchange with the environment depends on the integrity of packages, including the seals and closures, and on the permeability of the packaging materials themselves. Gases and vapors can permeate through materials by macroscopic or microscopic pores and pinholes or they may diffuse by a molecular mechanism, known as "activated diffusion." In activated diffusion the gas is considered to dissolve in the packaging material at one surface, to diffuse through the packaging material by virtue of a concentration gradient, and to reevaporate at the other surface of the packaging material. The equilibrium and kinetic considerations governing mass transfer have been discussed in Chapter 7 and found applicable to dehydration and freeze drying. They apply here as well.

In particular, for the case of gas transport in one direction (from the atmosphere into a package) the diffusion law applies:

$$J = -DA \frac{dc}{dX} \tag{12-7}$$

where

J is the flux of gas in appropriate units (e.g., moles/sec)
A is the area (cm^2)
D is the diffusion coefficient for the gas in the membrane (cm^2/sec)
c is concentration of the gas in the membrane ($moles/cm^3$)
X is distance in membrane measured in direction of flow (cm)

If D is a constant and steady-state conditions exist, then (as shown in Fig. 12-12):

$$J = DA \frac{c_1 - c_2}{\Delta X} \tag{12-8}$$

However, c_1 and c_2 are difficult to measure within the membrane. If Henry's law applies, we have:

$$c = Sp \tag{12-9}$$

where:

S is solubility ($moles/cm^3/atm$)
p is partial pressure of gas (atm)

Then we can combine Eqs. (12-8) and (12-9) and obtain

$$J = DSA \frac{p_1 - p_2}{\Delta X} \tag{12-10}$$

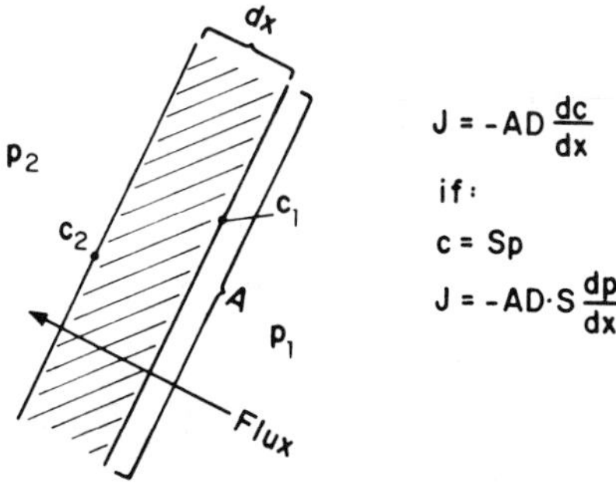

FIG. 12-12. Schematic representation of steady-state "activated diffusion" across a packaging film.

The quantity DS is known as the permeability coefficient. We shall use for it the symbols B or b. The permeability constant is defined in terms of quantities defined below:

$$B \equiv \frac{\text{(amount of gas) (thickness)}}{\text{(area) (time) (pressure difference)}}$$

There is no single convention as to units to be used for the permeability constant and Table 12-5 shows that many different units are in fact being used in the literature. In addition to permeability coefficients as defined above, some authors use permeance which is not corrected to unit thickness. Furthermore permeance to water vapor is sometimes reported in units which are neither corrected to unit thickness nor to unit pressure, but this value always must be reported with specified thickness, humidity, and temperature. Such a unit is WVP ("water vapor permeability"), used extensively in practical packaging technology and usually defined as grams of water per day per 100 in.^{2} of package surface, for a specified thickness and temperature and for a relative humidity on one side of approximately 0% and on the other side of 95%. In our discussion we shall use primarily the following two units for describing permeability behavior: either $B^* = (cm^3)(mm)(cm^{-2})(sec^{-1})(cm\ Hg^{-1})$ or $B = (cm^3)(mil)(m^{-2})(days^{-1})(atm^{-1})$. In each case cm^3 refers to volume at standard conditions of temperature and pressure and 1 mil = 0.001 in. The two units are easily convertible: to convert from B* to B, multiply by 2.58 x 10^{12}; to convert from B to B* multiply by 2.86 x 10^{-13}. With respect to water vapor we shall also use $b = g\ mil\ m^{-2}\ days^{-1}\ torr^{-1}$ and the WVP unit defined previously.

TABLE 12-5

Some Units Used in Reporting Permeability

Amount of gas	Thickness	Area	Time	Pressure difference
cm (STP)	mm	cm^2	sec	cm Hg
moles	cm	cm^2	sec	cm Hg
cm^3 (STP)	cm	cm^2	sec	cm Hg
grams	cm	cm^2	sec	mm Hg
cm^3 (STP)	mil	100 in^2	24 hr	atm
grams	cm	cm^2	24 hr	0.21 atm
cm^3 (STP)	mil	m^2	24 hr	atm
cm^3 (STP)	cm	cm^2	min	atm
cm^3 (STP)	100 μm	100 cm^2	24 hr	atm
grams	mil	100 m^2	hr	atm
cm^3 (STP)	mil	100 in^2	100 hr	atm

2. Measuring Permeability

There are many methods for measuring permeability and it is not possible to review them in detail here. Two major types of measurement are shown schematically in Figure 12-13. In the pressure-increase method a membrane is mounted between the "high-pressure" and the "low-pressure" sides of a permeability cell. Both sides are evacuated and the membrane is degassed. Then, at "zero" time a known consant pressure p_H of the test gas is introduced on the high side, and p_L ("low-side" pressure) is measured as a function of time. If the measurement is continued only as long as p_H remains much larger than p_L, the Δp remains essentially constant and the permeability constant can be calculated as shown in Figure 12-14.

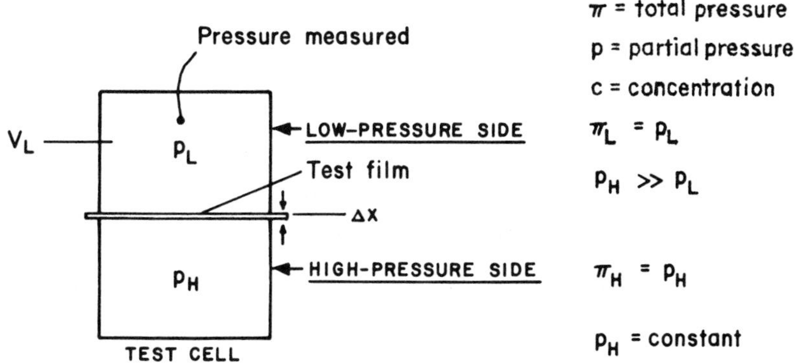

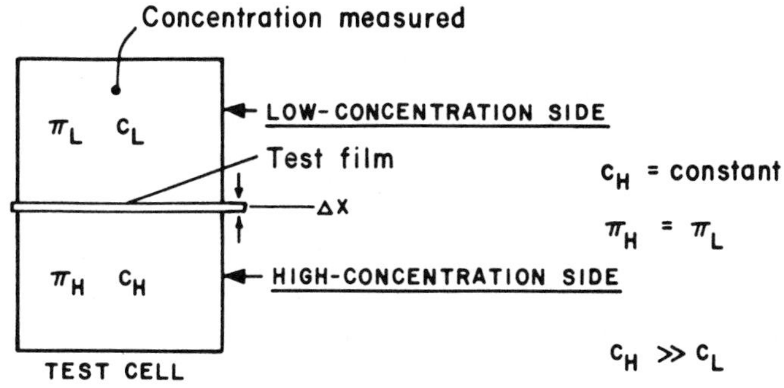

FIG. 12-13. Principles of two types of permeability measurement.

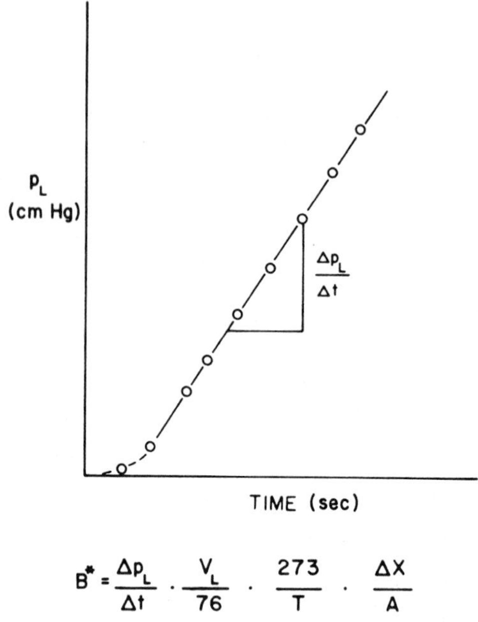

FIG. 12-14. Calculation of permeability from the slope of pressure-increase curve.

In the "concentration-increase" method, both sides are at the same total pressure but a partial pressure difference is maintained by sweeping one side continuously with the test gas and by maintaining on the other side an inert gas into which the test gas is diffusing. Figure 12-15 shows an experimental arrangement for measuring gas permeability by concentration increase using gas chromatographic analysis for oxygen or carbon dioxide [21]. More modern instrumental methods allow continuous direct recording of oxygen permeation using an oxygen electrode [37].

Water vapor permeability is usually measured gravimetrically. A desiccant maintaining a low water-vapor pressure is sealed in an aluminum cup (WVP cup) by a film sample as shown in Figure 12-16. The cup is placed in a cabinet at constant temperature and constant humidity and the permeability is determined from the rate of weight increase due to water adsorbed by the desiccant. The desiccant must be of a type maintaining a low vapor pressure of water in spite of water sorption. Calcium sulfate, magnesium perchlorate, and calcium chloride have been used for the purpose. Silica gel also can be used for comparative tests but the partial pressure of water sorbed on it increases with coverage. Sometimes it is advantageous to conduct such permeability tests in sealed pouches rather than in the WVP cups [8].

Another way of measuring water-vapor permeability is to measure relative humidity in a compartment separated by the test sample from a source of water. A simple arrangement for this purpose, involving an electric hygrometer has

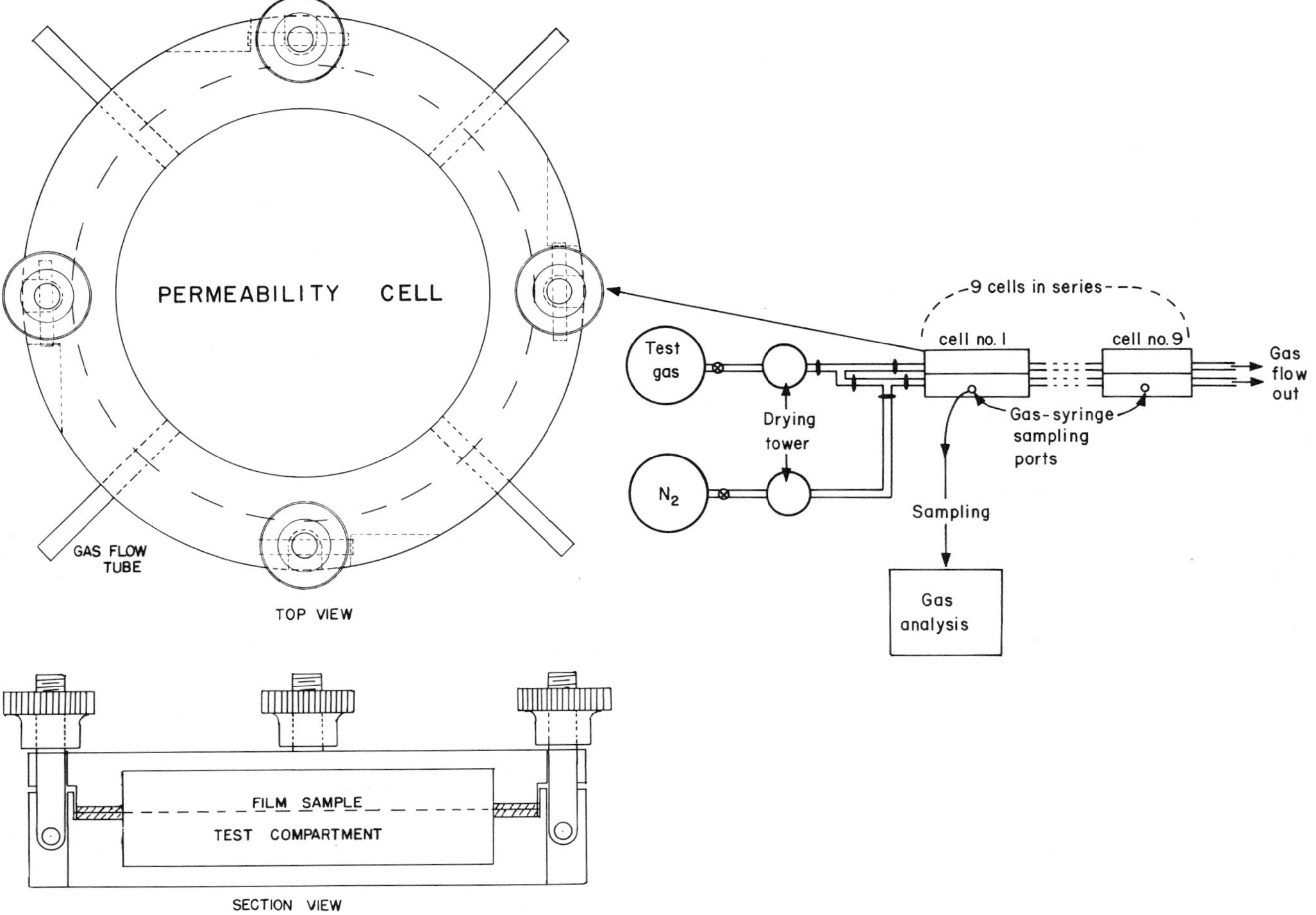

FIG. 12-15. System for permeability measurement using a concentration-increase method.

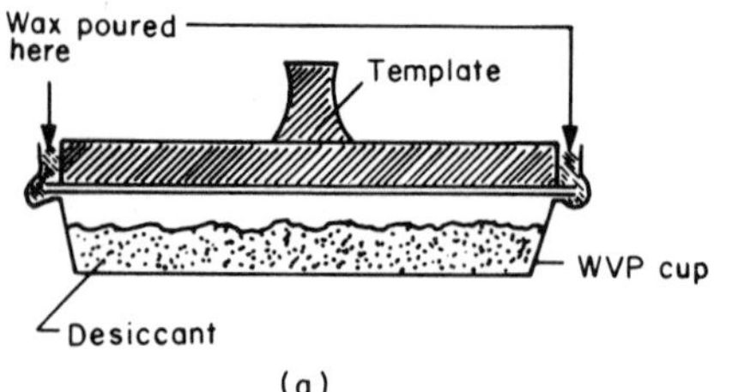

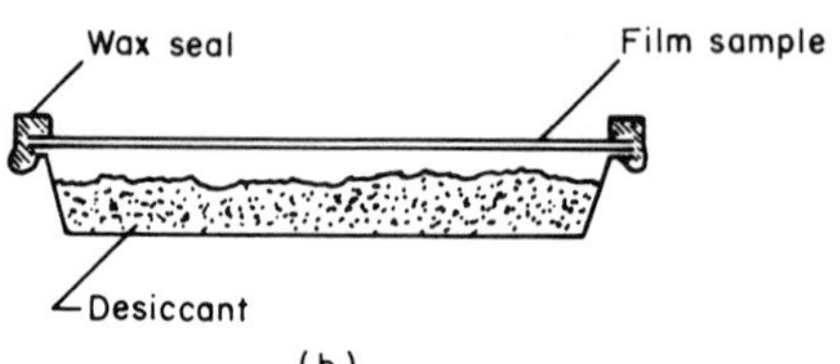

FIG. 12-16. Cup used for gravimetric measurement of water vapor permeability: (a) formation of wax seal; (b) cup with wax seal in place.

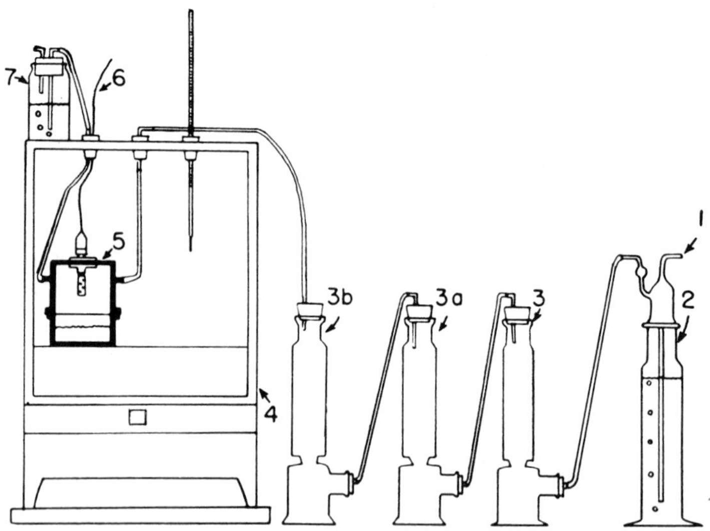

FIG. 12-17. System for water vapor permeability measurement using an electric hygrometer. (1) Connection with compressed-air line, (2) gas-washing bottle, (3, 3a, 3b) the three gas-drying towers, (4) constant-temperature cabinet, (5) the permeability cell, (6) connection with the electric indicator, and (7) the back-pressure bottle.

TABLE 12-6

Permeability of Some Materials to Hydrogen and Helium

Material	Gas	Temperature (°C)	Permeability (cm^3 cm cm^{-2} sec^{-1} cm Hg^{-1})
Quartz	H_2	300	0.5×10^{-9}
	H_2	700	4×10^{-9}
	He	300	10^{-11}
	He	700	2.5×10^{-9}
Pyrex glass	He	0	4×10^{-12}
	He	300	4×10^{-10}
Palladium	H_2	350	6.5×10^{-5}
Rubber	H_2	18	8×10^{-9}
	H_2	60	5×10^{-8}

been designed by the author and his co-workers and is shown in Figure 12-17. More elaborate and expensive instrumentation for the same purpose, based on the same principle, is also available commercially (Honeywell Co.). Measurement of permeability of organic vapors can be achieved by gravimetric [35], manometric [44], radiotracer [14], other analytical methods [19], or organoleptic analysis [17].

3. Permeabilities of Packaging Materials to "Fixed Gases"

Oxygen, nitrogen, hydrogen, carbon dioxide, and other gases with low boiling points are known as "fixed gases." We group them together because they show similar "ideal" behavior with respect to permeability through packaging materials. Table 12-6 shows the permeability of helium and hydrogen through several materials. Metals and glass in thicknesses used in packaging have permeabilities too low to be significant, so that only polymeric materials need be considered. Table 12-7 shows permeabilities of a number of packaging materials to O_2, CO_2, and N_2 at room temperature. Table 12-8 shows some oxygen permeability values. Table 12-9 shows typical ratios for several materials.

These values illustrate the following points:

First, for any given gas there exist materials with widely differing permeabilities. Thus polyvinylidene chloride (saran) is 100,000 times less permeable to oxygen than is silicone rubber.

TABLE 12-7

Permeability (B) for Some Plastic Films at Room Temperature

Film	B^a		
	Nitrogen	Oxygen	Carbon dioxide
Saran	3	13	75
Nylon 6	25	100	400
Mylar (polyester)	20	80	260
Trithene or Kel-F	40	150	1,000
High-density polyethylene	700	2,000	10,000
Low-density polyethylene	3,500	12,000	70,000
Natural rubber	20,000	60,000	350,000
Silicone rubber	—	10^6	6×10^6

[a] $B = cm^3 \text{ mil } m^{-2} \text{ day}^{-1} \text{ atm}^{-1}$.

Second, there are regularities in transmission of different gases through the same material. Normally, carbon dioxide permeates four to six times faster than oxygen, and oxygen four to six times faster than nitrogen. Since carbon dioxide is the largest of the three gas molecules, one expects its diffusion coefficient to be the lowest, and so it is. Its permeability coefficient, however, is the highest, because its solubility, S, in polymers is much greater than that for other gases [27].

"Fixed" gases also show the following "ideal" behaviors: (1) the permeabilities can be considered independent of concentration and (2) the permeabilities change with temperature in accordance with the following relation:

$$B = B_o e^{-E_p/RT} \tag{12-11}$$

where

E_p is the activation energy for permeability (kcal/mole)
B_o is a constant

Table 12-10 shows some values of E_p and Figure 12-18 shows how permeability varies with the reciprocal of absolute temperature. For some materials there is a definite break in the permeability-temperature curve and above a critical temp-

TABLE 12-8

Permeability to Oxygen of Various Packaging Materials at 25°-30°C

Material	Permeability range (cm^3 mil day^{-1} m^{-2} atm^{-1})
Conventional polyethylene	6,000-15,000
High density polyethylene	1,500-3,000
Pliofilm	200-5,000[a]
Saran	10-350
Plain cellophane	20-5,000[b]
Polyvinyl fluoride	25-100
Mylar (polyester)	50-100
(Poly)trifluorochloroethylene	50-1,000[a]
Cellulose acetate	1,000-3,000[b]
Silicone rubber	over 250,000
Coated and waxed papers	100-15,000[c]
Cellulose acetate	2,000-5,000
Foil laminations	0[d]
Plastic laminations	10-400[e]

[a] Depends on type and amount of plasticizer.

[b] Depends on humidity.

[c] Depends on type, quality, and amount of coating; humidity; conditioning; and other factors.

[d] For undamaged samples

[e] Primarily for cellophane-polyethylene and mylar-polyethylene laminations.

erature (or temperature range) the material is much more permeable. Such effects (for instance, in polyvinyl acetate around 30°C and in polystyrene around 80°C) are due to a glass transition at a temperature, T_g. Below T_g, the material is "glassy" and above it, rubbery."

At this point some discussion of structure and morphology of polymeric packaging materials is necessary. Polymeric solids can be crystalline,

TABLE 12-9

Typical Ratios of Permeabilities for Various Polymers (Molar Units)

Material	Permeability ratios H_2O/O_2	CO_2/O_2
Polyfluorocarbon (FEP)	10	1
Silicone rubber	20	6
Polypropylene	30	4
Polystyrene	100	5
Polyethylene terephthalate	5,000	3
Cellulose acetate	10,000	6
Polyvinyl fluoride	15,000	5

TABLE 12-10

Some Approximate Activation Energies for Permeation of Gases

Material	Activation energy for specified gas (kcal/mole) O_2	CO_2
Saran	14	12
Mylar (polyester)	6.5	6
Low-density polyethylene	10	9
Natural rubber	7	5
Nylon	10	9

amorphous, or semicrystalline. Packaging materials are either semicrystalline or amorphous. In semicrystalline materials part of the volume is occupied by ordered aggregates of polymer chains, "crystallites." These crystallites are considered impermeable to gases and vapors. The rest of the volume is occupied by amorphous regions, i.e., polymer chains in a substantially disordered state. Amorphous polymers exist almost totally in a disordered state. Below T_g the "holes," "pores," and other voids between chains remain fixed because the mobility of the disordered chains is severely limited. Above T_g the chains have greater freedom of motion and the voids between them assume an amoebalike quality of constantly changing shape, and location. Obviously these changes have

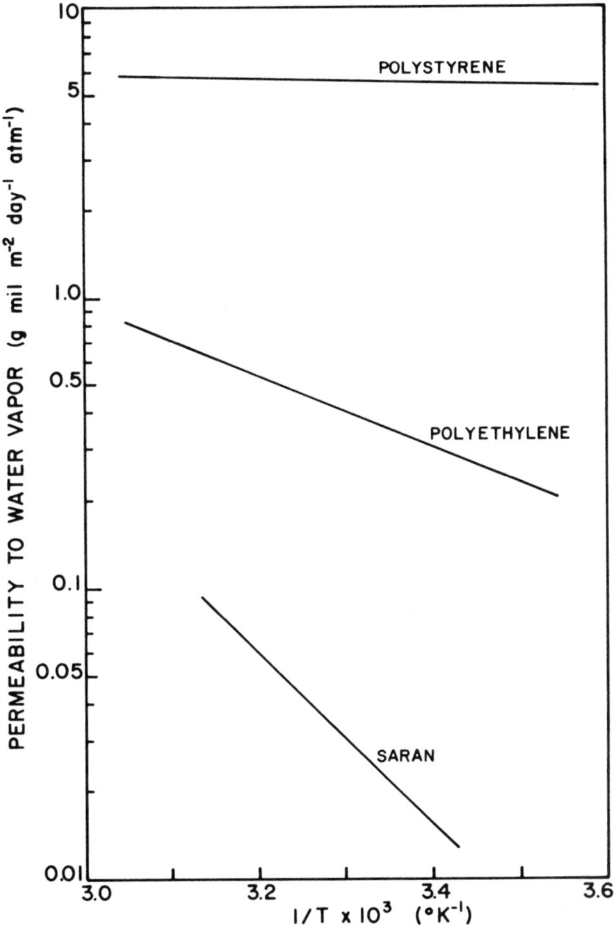

FIG. 12-18. Dependence of permeability on absolute temperature.

an important effect on the ability of molecules to diffuse through the material, especially fairly large molecules.

The differences in the ability of different polymer films to transmit gases arise in part from differences in crystallinity, in part from mobility differences between different types of polymeric chains, and finally from specific influences of functional groups of the polymers on the solubility of gases and vapors in the amorphous portion of the polymer. (The crystalline portions are considered nonsorbing). The crystalline content of a polymer is of primary importance in determining its permeability. Ideally one expects the permeability to increase linearly with the square of the noncrystalline volume fraction. This relation, however, holds only for such small molecules as hydrogen and it deviates very substantially for very large molecules, such as sulfur hexafluoride. These deviations occur because the presence of crystalline regions exerts some restraining influence on the mobility of polymer chains in noncrystalline regions, and such restrictions are particularly impeding to the flow of large molecules.

The permeability of polymers also can be modified by crosslinking polymers (for instance, vulcanizing, which is a process of crosslinking natural rubber). Crosslinking restricts mobility of chains and reduces permeability, provided that crystallinity is not impaired. Orientation of polymer chains by stretching the films increases crystallinity and decreases permeability.

Ideal behavior (D independent of concentration, log B proportional to 1/T) of films does not occur (1) when polar gases, especially water, diffuse through a hydrophilic polymer and (2) when low boiling organic compounds, including solvents and flavors, diffuse through any polymer. The theory of transport of organic volatiles through polymeric films is not considered here. However, further attention is given to the permeability of polymeric films to water.

TABLE 12-11

Permeability of Various Packaging Materials to Water Vapor at 100 °F and 95% versus 0% Relative Humidity

Material	Permeability range (g mil 24 hr^{-1} 100 $in.^{-2}$
Plain cellophane	20-100
Nitrocellulose-coated cellophane	0.2-2.0
Saran-coated cellophane	0.1-0.5
Polyethylene, conventional	0.8-1.5
Polyethylene, low pressure	0.3-0.5
Saran	0.1-0.5
Vinyl-chloride-based films	0.5-8.0
Aluminum foil, 0.00035 in. thick	0.1-1.0
Aluminum foil, 0.0014 in. thick	< 0.1
Plastic paper foil laminations	< 0.1
Waxed papers	0.2-15.0
Coated papers	0.2-5.0
Mylar	0.8-1.5
(Poly)trifluorochloroethylene	0.01-0.1
Silicone rubber	> 200
Polypropylene	0.2-0.4

4. Permeability to Water

As indicated above, with water vapor and hydrophilic polymers we encounter deviations from "ideality." The permeability "constant" varies with water concentration as well as with temperature, and the temperature dependence may not be as simple as that shown in Eq. (12-11). Most data are reported for 100 °F and for a humidity (at the "high-pressure side") of 95% RH. Some data are shown in Table 12-11.

The permeability of films to water vapor usually increases rapidly at high relative humidities due to sorption of water and concomitant swelling of the material. Figure 12-19 shows how the rate of water transfer through a nylon film varies with water activity at the surface of the "high-pressure" side of the film. Table 12-12 shows how the permeability constant varies when the "high-pressure" side is at 100% RH and the humidity at the "low-pressure" side is varied.

The calculation of water transport through such materials as nylon, cellulose, polyvinyl alcohol, and others which show a humidity-dependent permeability can be made only if this dependence can be characterized by a suitable mathematical function. A number of approaches to obtaining such information are possible and techniques have been reported for this purpose [11, 18, 23].

Temperature is another factor that has a bearing on permeability. For the system water-polar polymers, Eq. (12-11) can be used only over narrow temperature ranges [18].

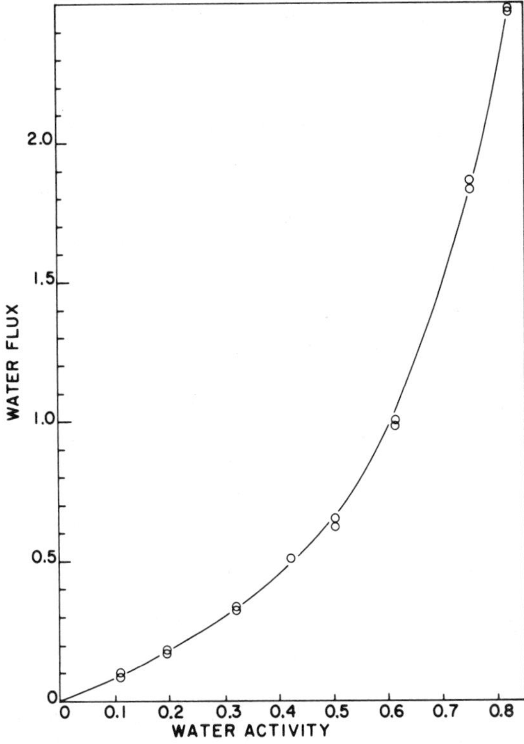

FIG. 12-19. Dependence of water-vapor transfer rate through nylon film on water activity at the "high-humidity" side of film.

TABLE 12-12

Permeability (b)[a] of Several Films to Water Vapor when "High-pressure Side" is at 100% RH and the Other Side is at Various RH's

Material	RH 31%	RH 56%	RH 80%
Polyethylene	0.18	0.19	0.17
Polyvinyl alcohol	48.5	67	97
Cellulose acetate	39	48.5	73
Nylon	7.3	12.1	21

[a] $b = g \text{ mil } m^{-2} \text{ day}^{-1} \text{ torr}^{-1}$.

[b] From Hauser and McLaren [11].

5. Permeability to Humidified Gases

Exposure of hydrophilic films to high humidity results in sorption of water, swelling, and greatly increased mobility of polymer chains. Water has effects similar to other "plasticizers" or solvents of relatively low volatility which enhance flexibility of polymer chains. Plasticizers tend to increase the permeability to all gases, and sorbed water is no exception. The increase in permeability is particularly dramatic for transport of gases through materials which are excellent gas barriers when dry. This situation is illustrated in Table 12-13, which shows the permeabilities of (1) cellophane, (2) polyethylene, and (3) a laminate of polyethylene and cellophane. The cellophane permeability depends on humidity, and when it is kept dry or protected from high humidity by a layer of polyethylene it is a good barrier to carbon dioxide.

6. Permeability of Multi-layer Materials

The following relationship applies to the flux of gas or vapor through several materials arranged in series (steady state is assumed; see the hypothetical three-layer material shown in Fig. 12-20).

$$J_1 = J_2 = J_n \tag{12-12}$$

$$J_1 = \frac{B_1 A_1}{\Delta X_1} (p_1 - p_2) \tag{12-13a}$$

$$J_2 = \frac{B_2 A_2}{\Delta X_2} (p_2 - p_3) \tag{12-13b}$$

$$J_3 = \frac{B_3 A_3}{\Delta X_3} (p_3 - p_4) \tag{12-13c}$$

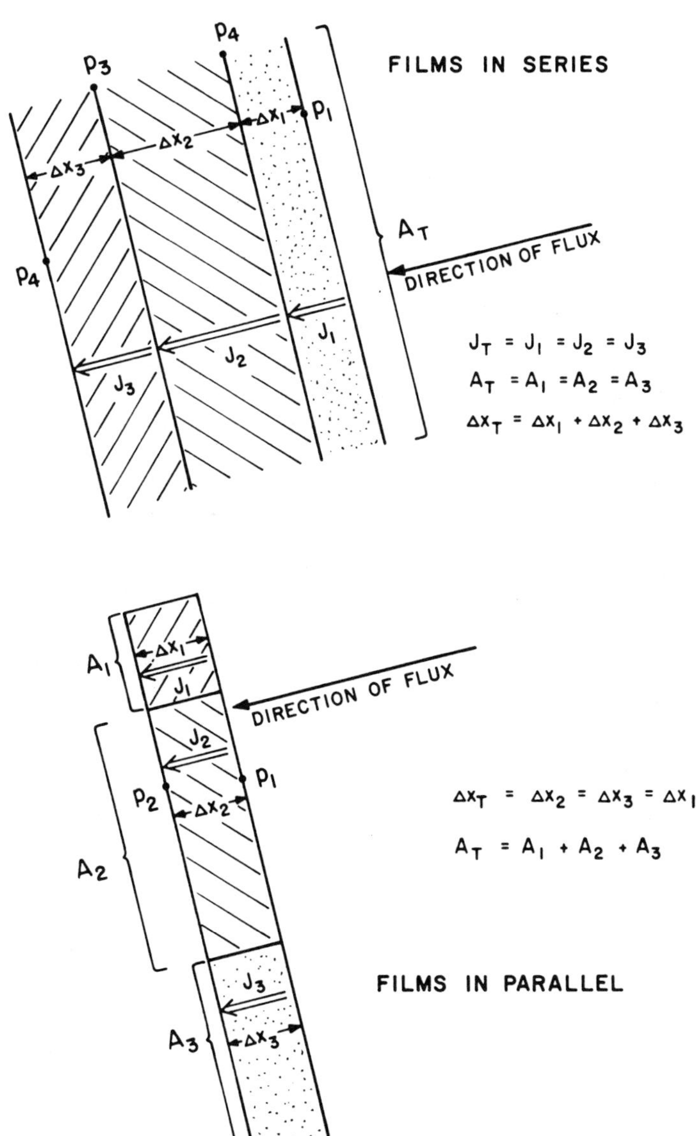

FIG. 12-20. Schematic representation of permeation through several materials in series and in parallel.

$$J_T = \frac{B_T A_T}{\Delta X_T} (p_1 - p_4) \tag{12-14}$$

Since

$$A_1 = A_2 = A_3 = A_T \tag{12-15}$$

$$\frac{\Delta X_T}{B_T} = \frac{\Delta X_1}{B_1} + \frac{\Delta X_2}{B_2} + \frac{\Delta X_3}{B_3} \tag{12-16}$$

$$B_T = \frac{\Delta X_T}{(\Delta X_1/B_1) + (\Delta X_2/B_2) + (\Delta X_3/B_3)} \tag{12-17}$$

If all the permeabilities are independent of concentration, then Eq. (12-17) can be used to calculate the permeability constant for any multilayer material. Knowledge of the individual thickness and permeabilities is needed, of course. In some situations, Eq. (12-17) is not useful. We have already encountered such a situation in Table 12-13. Films or papers with a high permeability are often coated with resistant coatings. Usually the permeability of such coated materials cannot be readily calculated using Eq. (12-17). This is because the permeability constant for coatings is often a function of coating thickness. Furthermore, coated materials often have permeabilities which vary with water vapor concentration (Fig. 12-21). Calculation of water-vapor fluxes through such composite materials, however, can be achieved using slightly more sophisticated mathematical techniques [23].

TABLE 12-13

Effect of Relative Humidity on Transmission of CO_2 Through Polyethylene and Cellophane

Material	Position	RH of "high-humidity side" (%)	CO_2 transmission ($cm^3\ m^{-2}\ days^{-1}\ atm^{-1}$)
C	—	0	10
C	—	100	3,000
P	—	0	50,000
P	—	100	47,000
C + P	P faces high RH	0	18
C + P	P faces high RH	100	18
C + P	C faces high RH	0	18
C + P	C faces high RH	100	3,500

[a] P = 1 mil polyethylene, C = 1 mil cellophane.

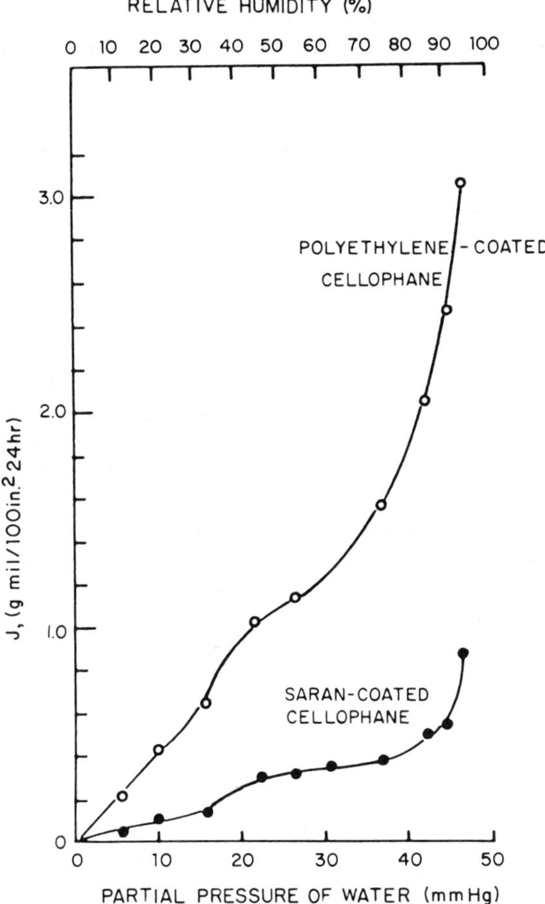

FIG. 12-21. Water-vapor transfer through coated films as influenced by humidity at the "high-humidity" side of the film.

The above relations for films in series hold only for flat membranes. When cylinders, spheres, or other shapes are considered one must consider changes in cross-sectional area for flow. Under these conditions one uses approximations or exact solutions of the type we have considered for flow of heat in paths with changing cross-sectional area (mean transfer areas).

When several materials are arranged in parallel the following equations hold given equal thicknesses (Fig. 12-20):

$$J_T = J_1 + J_2 + J_3 \tag{12-18}$$

$$B_T A_T = B_1 A_1 + B_2 A_2 + B_n A_n \tag{12-19}$$

$$B_T = \frac{\Sigma B_i A_i}{A_T} \tag{12-20}$$

When several materials in parallel have different thicknesses or greatly different permeabilities the idealized equations do not hold because the flow is no longer unidirectional. The total flux, however, can be approximated by simple relations derived from Eq. (12-18).

7. Flow-through Pinholes, Cracks, and Seal Imperfections (Convective Flow)

In cases where pores or pinholes provide the major means of transport, molecular interactions between the packaging material and the gas or vapor become unimportant and the flow depends on properties of the gas and on size and shape of the pores. Depending on the total pressure at which the transport occurs we can have: (1) molecular diffusion, (2) laminar flow, (3) slip flow, or (4) surface flow.

The flow of gases through fine pores and capillaries occurs by one of these four regimes. Molecular flow occurs at very low pressures when intermolecular collisions are relatively infrequent and the flow is affected mainly by molecules colliding with walls of the capillaries. Laminar flow occurs when pressure is sufficiently high so that intermolecular collisions predominate. Slip flow occurs at pressures intermediate between the two extremes. Surface flow occurs when molecules adsorbed onto the surface of the walls of the capillaries can migrate from one localized adsorption site to another.

For the present purposes it is sufficient to characterize any material in terms of an overall convective flow constant, β

$$J = \beta A \frac{dp}{dx} \tag{12-21}$$

where β has the same dimensions as B. The major differences between behavior of β and B are as follows.

a. Effect of Different Packaging Materials. B depends primarily on molecular structure of the packaging material, β on pore size distribution. Hence a paper sieve and one made of a metal give similar values of β when their pore sizes are similar.

b. Effect of Gas Properties. The important properties of a gas in "activated diffusion" are molecular size and solubility in the polymer. In convective flow the important properties are molecular size and viscosity. As a consequence, B usually varies as follows:

$$CO_2 > O_2 > N_2$$

On the other hand β for different gases varies only slightly and larger molecules tend to permeate less readily.

c. Effect of Total Pressure. Total pressure as well as total pressure difference have a very significant effect on β since they determine which of the several flow regimes is controlling. B, on the other hand, is ideally completely independent of total pressure.

d. Effect of Temperature. There is a drastic difference in the effects of temperature on B and β. Gas viscosity increases with increasing temperature, hence one of the factors controlling β causes a decrease in β as temperature increases. On the other hand, B is a product of D and S, both of which increase exponentially with temperature. In general, therefore, B increases rapidly with increasing T but β either increases slightly, stays constant, or actually decreases (depending on packaging material, pore size distribution, and pressure).

For most packaging materials there is also a large difference in magnitude of the fluxes involved. A pinhole in a metal can or an imperfection in the seal of a saran pouch usually transmits gas orders of magnitude faster than the intact surface of these containers.

C. Temperature Control by Packaging

Packaging affects the rate of heat transfer to and from the enclosed food products. The heat transfer may take place by conduction, convection, or radiation; and is accordingly affected by the packaging material thickness and thermal conductivity, its porosity, and its reflectivity, and these properties collectively determine the insulating value of the package.

The temperature rise in refrigerated food products exposed for a short period of time to elevated temperatures can be greatly retarded by the use of insulating containers. Ratzlaff [42] studied the temperature changes in pasteurized milk during bottling, cooling, storage, and distribution. He found that the temperature of milk depended on the type of the container and on the type of outer packing case. Precooled (40°-43°F) milk exposed to ambient temperatures of 77°- 79°F reached a temperature of 70°F in 6 hr when packaged in glass bottles housed in wire or wood packing cases, and in 10.5 hr when packaged in waxed paper cartons housed in wire packing cases. Milk in waxed paper cartons and wood packing cases had a temperature of less than 70°F after 12 hr of storage at ambient temperature.

Insulating containers that slow the warming of cool food products also retard the cooling of warm food products. In the study mentioned above, the time necessary to cool milk from 60° to 50°F varies from 2.5 to 17 hr, depending on the type of packaging.

The insulating properties of packages are of even greater importance in the handling and distribution of frozen foods. Thawing rates of frozen food depend on the type of container. Single containers of food defrost more rapidly than containers packaged in shipping cases. Other packaging factors of importance are size of package and location in the transporting vehicle.

A review of controlled-temperature shipping containers has been presented by Bond [2]. He has noted that two elements are necessary to provide simple systems for control of product temperature by packaging. The packaging must be a good insulator, and an energy source for temperature adjustment must be available. Among the available packaging materials, the synthetic foams of polystyrene and of polyurethane have excellent insulation properties. Their thermal conductivities are in the range of 0.11-0.24 Btu in ft^{-2} hr^{-1} $°F^{-1}$. Bond [2] also discusses the potential for eutectic temperature-control units. In these units substances undergoing phase transition are included within the container to provide controlled energy adsorption and release. Utilization of salt-ice mixtures, of course, is well established but other eutectic substances can be utilized, including such organic compounds as glycols, hydrocarbons, and waxes. Bond's article provides excellent illustrations of available eutectic substances and their possible applications. According to Bond [2], a commercial unit utilizing a 3 x 6 x 12 in. module and a commercially available eutectic compound can be maintained at 40°F for 120 hr when the ambient temperature is 0 °F, for 101 hr when the ambient temperature is 70 °F, or for 49 hr when the ambient temperature is 110 °F. The potential for utilizing eutectic coolants and "warmers" other than ice and salt is only beginning to be realized.

An interesting problem in insulative packaging is the control of temperature changes in food packages placed in refrigerated display cabinets. In this instance, most of the cooling takes place by conduction and convection, while simultaneously there is heat input by radiation from fluorescent lamps used for lighting. Under these conditions the high reflectivity and high conductivity of aluminum foil offers advantages, particularly with such products as icecream.

D. Control of Biological Attack by Packaging

1. Microorganisms

The protection of package content from attack by microorganisms depends on mechanical integrity of the package (absence of breaks and seal imperfections) and on resistance of the package to penetration by microorganisms. Mechanical integrity is treated in Section IV. E.

Extensive studies on a variety of plastic films and metal sheets have shown that microorganisms, including molds, yeast, and bacteria cannot penetrate these sheets in the absence of pinholes. However, in practice, thin sheets of packaging materials, including aluminum foil and plastic, do contain pinholes. Fortunately there are several safeguards against the passage of microorganisms through pinholes in films:

a. Because of surface tension effects, microorganisms cannot pass through very small pinholes unless the microorganisms are suspended in solutions containing wetting agents and the pressure outside the package is greater than within.
b. Materials of packaging are generally used in thicknesses such that pinholes are very infrequent and small.
c. For applications in which package integrity is essential (such as sterilization of food in pouches) adequate test methods are available to assure freedom from bacterial recontamination [26].

The potential for growth of microorganisms, in particular molds, on package surfaces is greatest when the surfaces are subjected to very high humidities. Under tropical conditions this danger becomes very significant and may require the treatment of package surfaces with antimicrobial agents.

Under more moderate conditions packages remain free of mold growth without such measures provided recommended practices with respect to sanitation and housekeeping are followed.

2. Resistance to Insect Infestation

In order to avoid insect infestation one must assure that the package contents are free of viable insect eggs or larvae, and that the package cannot be penetrated from the outside by adults or larvae. (Eggs and pupae of insects obviously have no penetration power of their own but can penetrate by virtue of gross contamination with dust and soil.) In storage and distribution the most desirable condition is one in which the environment external to the packages (truck interiors, warehouses, railroad cars) is insect free. Such conditions are not always possible and the package is expected to provide a defense against insect penetration. The following are some of the decisive factors in such a defense:

a. Condition of Package Surface. Highly polished and slippery surfaces, free of debris and dust, are most desirable. In order to bore into the package the insects require a "foothold" on the surface. It is believed that wax-dipped packages used in World War II to distribute rations in insect-infested areas were effective partly because of surface conditioning rather than because of increased package barrier properties.

b. Resistance to Emission of Odors. Insects do not read labels; they rely on chemical senses in their quest for food. A perfect odor barrier combined with cleanliness of surfaces helps avoid insect attack.

c. Paper Package Construction. Packaging materials that are resistant to insect penetration and have well-formed seals and closures aid greatly in assuring protection.

d. Use of Repellants. The surfaces of packaging materials can be further protected against insect attack by incorporation of insect repellants. In such cases it is essential to assure that these chemicals do not enter the food supply in levels presenting potential health or esthetic problems. Various chemicals have been proposed for this purpose, including pyrethrin, malathion, DDT, and Lindane, and these can be applied in waxes, coatings, or surface sprays. A particularly suitable method of application appears to be to incorporate such coatings in multi-wall packages, since under these conditions the danger of transfer of the chemicals to the food is minimized.

Different insects have different penetrating powers and also different sensitivities to various environmental conditions. Beetles and moths present most problems. Table 12-14 shows the main foods attacked and the environmental requirements of several insects attacking stored food products. Table 12-15 shows the ability of insects to penetrate different types of packaging materials.

TABLE 12-14

Characteristics of Some Insects Attacking Stored Food Products

Insect name		Main food attacked	Comments
Common	Scientific		
Confused flour beetle	*Tribolium confusum*	Cereals, flour, dried fruit	Optimum 20°-30° C
Cadelle	*Tenebroides mauritanicus*	Grains, cereals	Very cold resistant, very penetrating
Mediterranean flour moth	*Ephestia kuhniella*	Most cereal products, dried fruit	
Lesser grain borer	*Rhyzopertha dominica*	Grain	Requires warm climate, optimum over 30° C
Yellow meal worm	*Tenebrio molitor*	Cereal products, meat scraps	Requires high humidity, prefers dampness

TABLE 12-15

Penetration of Packages by Insects

Material	Average number of weeks before penetration by a mixed population of insects
Paper and jute bags	< 1
Polyethylene	3-4
Mylar	6-8
Laminates of plastic and paper	6-8
Aluminum-containing laminates	> 12

E. Protection from Mechanical Damage

1. Strength of Packaging Materials

Maintenance of the integrity of the package depends first of all on the mechanical properties of the packaging materials. The strength of materials is a complicated matter and we present here only a generalized picture intended to acquaint the reader with fundamentals. There are different types of tests which are usually conducted on materials and they measure strength and deformability under various controlled conditions. Figure 12-22 shows schematically how force is applied to sheets of materials in five such tests: tensile, compression, impact, tear, and bursting strength. In all five tests one measures the force (or energy) required to achieve the specific failure. The tensile test is capable of providing additional information about the mechanical properties of the material tested, i.e., tensile strength, modulus of elasticity, maximum elongation, and resilience. Figure 12-23 show graphically what each of these represents in the stress vs strain diagram. Each test must be conducted under controlled conditions of temperature, material precondition, sample preparation, and rate of application of force (rate of loading).

Mechanical properties of some packaging materials as measured in such standardized tests are shown in Table 12-16. The significance of values obtained in such standardized tests to practical package performance must be assessed by packaging engineers in accordance with intended use. Tests of bursting strength are routinely performed on paperboards which are designed to contain heavy loads. High tensile strength is clearly required for applications in which tension exists in the plane of the package wall, e.g., in shopping bags.

Additional specialized tests can be performed. Among these are folding endurance for carton materials, puncture resistance, abrasion resistance, and wet strength. In many situations, however, results of mechanical tests on packaging

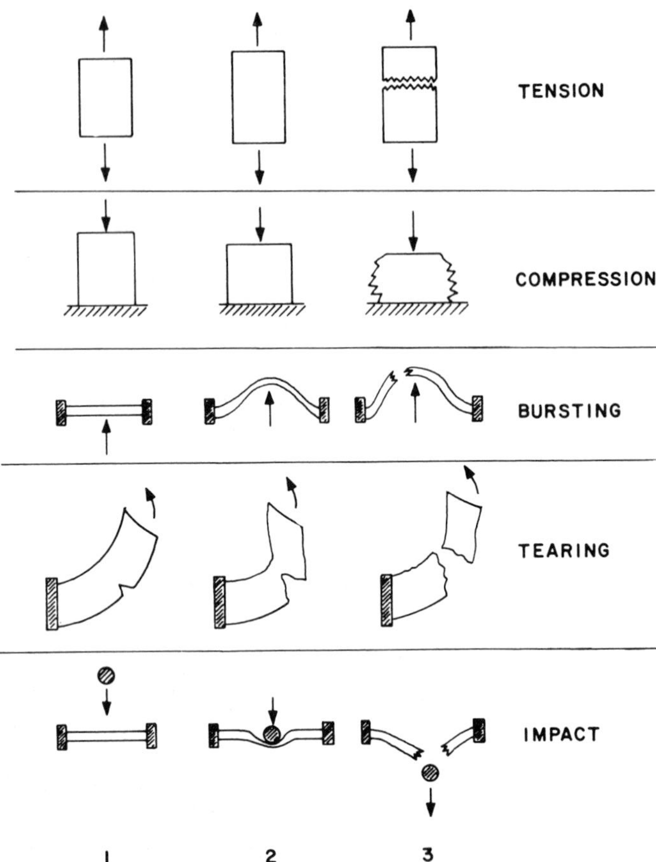

FIG. 12-22. Schematic representation of forces applied to materials in tests of mechanical properties. Arrows represent direction of applied forces. Three states are shown; (1) material at beginning of test, (2) deformation due to applied forces, and (3) failure of material.

materials are difficult to relate to package performance and additional tests on completed containers, with or without contents, may be necessary. It may also be necessary to conduct accelerated aging tests to determine whether mechanical strength of the materials changes during storage in a given environment.

"Wet" strength of paper and paperboard is particularly important, since cellulose-based materials tend to have a much reduced strength when they are "plasticized" by water. Organic plasticizers have similar effects on nonpolar polymers and are often added to increase flexibility, provided the associated decrease in ultimate strength is not detrimental.

2. Tests on Packages to Determine Adequacy of Mechanical Strength and Integrity

These test procedures vary with the container and the intended use.

Metal cans are used often for sterilized foods and require testing to assure absence of leaks. Freedom from pinholes is tested for by manufacturers, before lids are sealed on, by applying overpressure. Correct sealing procedures for lids

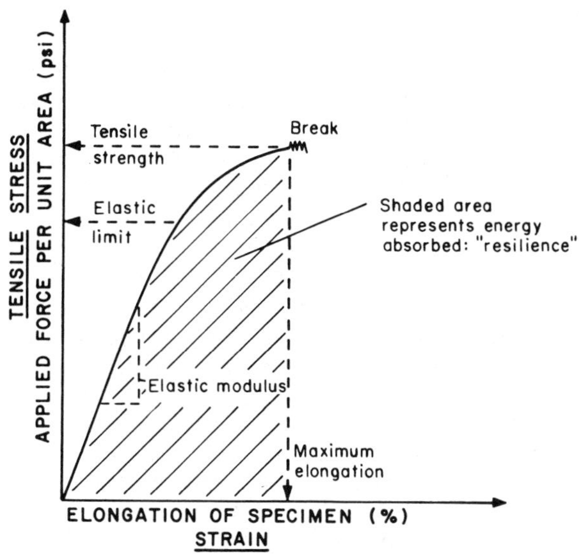

FIG. 12-23. Stress-strain diagram for a hypothetical material.

TABLE 12-16

Mechanical Properties of Selected Packaging Materials

Material	Tensile strength (10^3 lb/in.^{2})	Elongation (%)
Cellophanes	4-12	5-50
Cellulose acetate	5-10	5-50
Polyethylene (LD)	2-3	100-800
Mylar	17-22	40-100
Saran	4-9	20-150
Aluminum foil	5-30	20-50[a]
Packaging papers[b]	5-15	2-50
Aluminum-plastic laminations	10-20	10-40
Steel foil	30-100	1-15

[a] Depends strongly on temperature.

[b] Depends strongly on additives to paper and on relative humidity.

are checked by the processors. Processors also conduct checks on sealed cans containing no products or on cans from which the product has been removed. This test is usually done by attaching a tube to the side of the can away from the seam, submerging the can in water, and applying gas pressure through the tube. Rising gas bubbles are indicative of a leak. Similar internal pressures also can be generated by sealing in the can an amount of dry ice (solid CO_2) calculated to give the desired test pressure (1.5-2 atm) upon sublimation.

In addition, it is possible to check for inadequate vacuum (whether due to leaks, poor gas exhaustion, or microbial growth) by checking the sound made when the lid is tapped or by any mechanical or optical method capable of detecting the deviation of the can lid from the shape assumed when the can is sealed under proper vacuum. Some of these methods can be used effectively and automatically for 100% inspection on the line. Internal pressure of a can can be readily and nondestructively measured by any of several contact gages which depend on measuring can lid deflection when vacuum is applied to the top surface of the can lid.

Glass containers require close control of such dimensions as height and mouth diameter because of demands of automatic filling and capping equipment. Another test conducted routinely on the line is that for resistance to temperature fluctuations. This done by subjecting glass containers to temperature drops under controlled conditions. Because glass strength is extremely dependent on surface conditions of the container, visual inspection of the surface is an important part of testing. Some manufacturers also routinely test glass containers for resistance to internal pressure and compression.

Plastic pouches and laminated pouches require tests on strength and integrity of heat seals. Tests similar to tensile strength tests on flat sheets can be used to measure strength of heat seals, and the internal pressure tests mentioned above for cans can also be applied to pouches. One such simple test designed by Davis et al. [7] is shown in Figure 12-24.

Shipping containers and other containers undergoing hazards of mechanical shocks during distribution should be subjected to tests simulating such mechanical abuses. Two major types of test are usually employed: (1) impact tests and (2) compression tests. Impact tests usually take the form of drop tests of various types. Drop-testing equipment has been designed to simulate various types of hazards in shipping. Compression tests also simulate shipping hazards as well as the behavior of containers stacked or palletized in warehouses.

3. Cushioning

Cushioning means protecting the product from shock. Mechanical shock is usually caused by rapid deceleration, a quantity usually measured in multiples of gravitational acceleration, g, and the sensitivity of articles to mechanical shocks is called the G factor. Thus an item with a G factor of 5 means that it will withstand deceleration five times greater than g, or $5 \times 32\ \text{ft/sec}^2$. Cushioning acts by damping the shock imparted to the articles which are in the package. A simple model of this action is shown in Figure 12-25, which considers the cushion as a compressible spring that on impact temporarily absorbs energy which otherwise can be imparted to the product over a short period of time with potentially damaging effects. Pertinent characteristics of the cushioning materials are indicated in Figure 12-26. The packaged articles undergo maximum deceleration at the time the cushion is compressed to the "stopping distance," and this maximum deceleration is defined as follows:

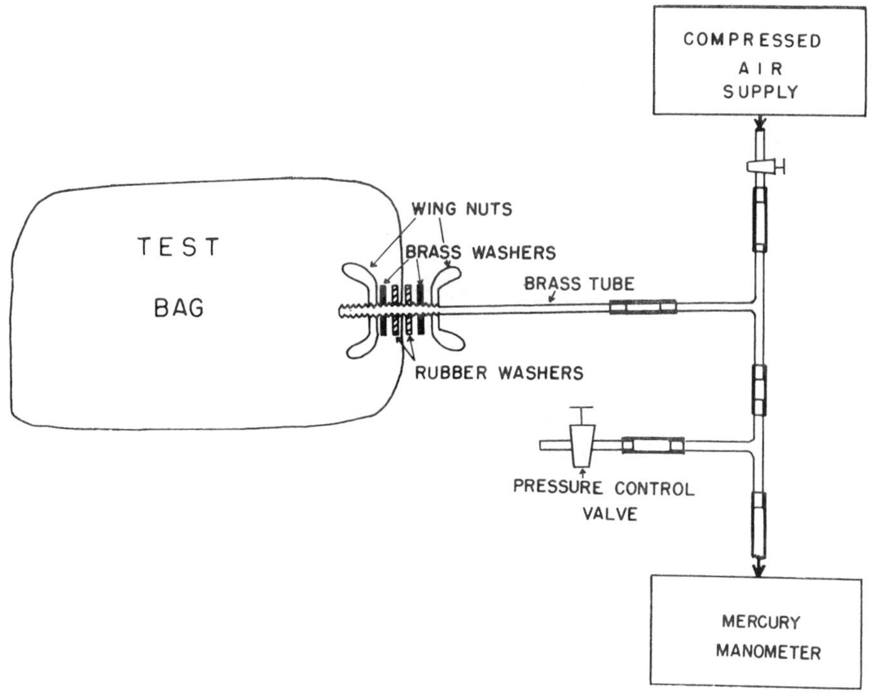

FIG. 12-24. Apparatus for testing "bursting strength" of bags.

$$\text{Maximum deceleration} = \frac{\text{energy absorbed by cushioning}}{\text{stopping distance x mass of product}}$$

Cushioning design, however, is complicated by the differences in load absorption characteristics of different materials. These characteristics are shown schematically in Figure 12-27, from which it is evident that cushion response is often time-dependent. Thus, two cushioning materials with the same load-compression curves under static loads applied relatively slowly may have different responses in dynamic testing in which the vibrations or shocks are applied very rapidly.

Most of the work on cushioning of packaged articles has been done on nonfood items, such as electronic components. This kind of information is applicable to some items that are brittle, such as eggs, freeze-dried foods, and some bakery

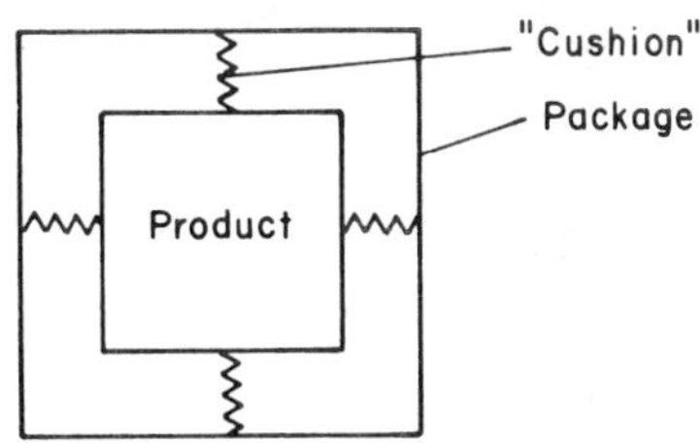

FIG. 12-25. Schematic representation of cushioning of packaged foods.

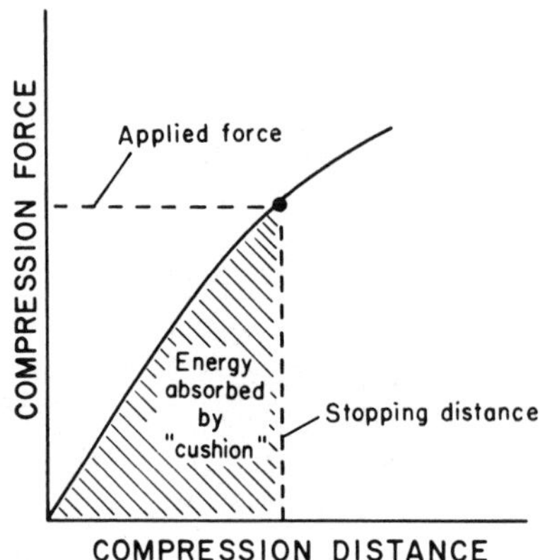

FIG. 12-26. Schematic representation of properties of a hypothetical cushioning material.

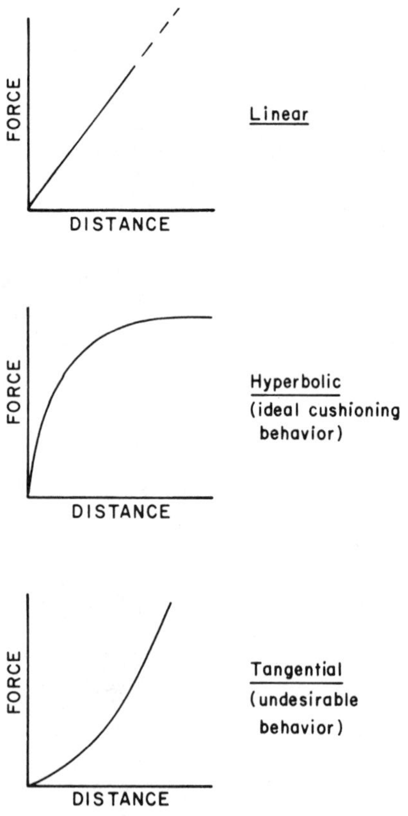

FIG. 12-27. Mechanical behavior of three types of cushioning materials.

items (biscuits and cookies). In freeze-dried foods the amount of damage is often measured in terms of total weight of "fines." The amount of fines can be reduced in such products by packaging them in skin-tight pouches, which are then spot glued to the interior of paperboard cartons. The skin-tight pouch prevents piece-to-piece impacts and the carton minimizes the effect of impacts from outside.

Other types of foods have their own internal cushioning (mainly water-swollen tissues, such as fresh meats of fruit; fatty tissues; or air entrapping structures, such as in lettuce or broccoli), and in these instances the problem is not with breakage but with surface bruising. The analysis of damage sensitivity in such items is usually done on a statistical basis, from consideration of total bruise area or similar quality defects.

V. CALCULATION OF SHELF LIFE AND REQUIREMENTS FOR PACKAGING

A. General Approach

Given the great complexity of factors involved in package-product interactions, the choice of proper packaging methods for a given product and the prediction of effectiveness of a given packaging system have been determined either by extended storage tests or by judicious guessing. This approach, however, has substantial economic penalties and severely limits freedom of choice, resulting usually in overpackaging. This is not only wasteful but severely impedes the development of entirely new packaging concepts which are currently needed to reduce problems of environmental pollution [33].

Recently it has become possible to apply knowledge of mechanisms of nutrient degradation to predict shelf life of a particular food-package combination or, conversely, to determine protective packaging requirements for a given food if the needed shelf life is known. These techniques are based on kinetic parameters of reactions causing deterioration of foods and on mass-transfer properties of the packaging materials. Analytical or numerical solutions useful for predicting the desired quantities can be obtained either with or without computer assistance.

The work on prediction of storage life as a function of efficiency of packaging measures and of environmental storage conditions was initiated on a large scale following World War II. Initial work was stymied by the complexity of the factors involved in this problem. The general opinion often voiced was that the complexity of these factors precluded laboratory prediction of shelf life. Initial laboratory evaluation of packaging problems can, however, greatly simplify the overall task of eliminating obvious failures and pinpointing likely successes. Such an evaluation should answer the following questions.

1. What are the optimal conditions for storage of the particular food product in terms of the important environmental factors?
2. What are the external environmental conditions to which the package is likely to be exposed?
3. What barrier properties are required in order to maintain an optimal internal environment?

In a series of studies reviewed recently by Karel [20], an approach was taken based on the following five assumptions (Fig. 12-2):

First, properties of food which determine quality, $\overline{F}$, depend on the initial condition of the food, $\overline{F}_0$, and on reactions which change these properties with time, t. These reactions, in turn, depend on the internal environment of the package, I.

It is assumed that deteriorative mechanisms limiting shelf life and their dependence on environmental parameters (oxygen pressure, water activity, temperature) can be described by a mathematical, although not necessarily analytical, function. As an example consider Eq. (12-22), which describes a deterioration dependent on the three above-mentioned parameters.

$$\frac{d\overline{D}}{dt} = f\ (RH,\ PO_2,\ T) \tag{12-22}$$

where

$\overline{D}$ is the deteriorative index
t is time
RH is equilibrium relative humidity
PO_2 is oxygen pressure
T is temperature

Second, maximum acceptable deterioration level $\overline{D}_{max}$ can be determined by correlating objective tests of deterioration with organoleptic or toxicological parameters.

Third, the internal environment, I, depends on the conditions of the food, F; on package properties, B; and on external environment, EO. It is assumed that changes in environmental parameters can be related to food and package properties. For instance,

$$a = f\ (a_0,\ t,\ RH,\ k_1 \ldots k_n,\ T \ldots) \tag{12-23}$$

where

a is water activity in the food at any time
a_0 is initial a
$k_1 \ldots k_n$ are constants characterizing sorptive and diffusional properties of the food and package.

Fourth, barrier properties of the package can in turn be related to internal and external environments (I and EO).

Fifth, the various equations can be combined and solved with or without the aid of a digital computer. The solutions predict storage life or required package properties for a given storage.

B. Analysis of Storage Requirement of Moisture-sensitive Foods

Many of the reactions which cause quality deterioration depend on moisture content. As mentioned in Chapter 8, we can distinguish two types of dependence: (1) those reactions for which a definite critical moisture content exists and below which the rate of spoilage is insignificant; and (2) reactions which proceed at all moisture contents, at rates which depend strongly on moisture content. The prediction of moisture protection required for the first case (that of a single critical moisture content) is relatively easy if some simplifying assumptions are accepted.

1. Moisture Increase in a Simplified Situation

Consider, for instance, the case of a product requiring a shelf life of t_c and in which the water activity should never exceed 0.7. Let us assume that the average storage conditions are 90°F and 85% RH, and that the sorption isotherm of the food is linear as represented by Eq. (12-24).

$$m = \sigma a \tag{12-24}$$

where

m is moisture content (g H_2O/g solids)
a is water activity
σ is a constant

Since $a = p/P^\circ$ (where P° is vapor pressure of water at 90°F) we have

$$p = \frac{m}{\sigma} P^\circ \tag{12-25}$$

Since $RH = (p/P^\circ)\,100$, we also have

$$P_{outside} = 0.85P^\circ$$

The increase in moisture content (m) of the food in the package is equal to the amount of water diffusing in, divided by the weight of solids (assuming no internal resistance to diffusion in the food) and is given by:

$$\frac{dm}{dt} = \frac{bA}{\Delta X}\frac{1}{s}(0.85P^\circ - p) \tag{12-26}$$

where

t is time (days)
b is permeability (g mil m^{-2} day^{-1} $torr^{-1}$)
ΔX is thickness of the packaging material (mil)
A is package area (m^2)
s is weight of solids (g)
$0.85P^\circ$ is the constant partial pressure of water outside the package

Combining Eqs. (12-26) and (12-25)

$$\frac{dm}{dt} = \frac{bAP^\circ}{\Delta X s \sigma}(m_e - m) \tag{12-27}$$

where m_e is the moisture content in equilibrium with outside RH, in our case = $85\sigma/100$. Integrating, we obtain:

$$\ln\left(\frac{m_e - m_i}{m_e - m_c}\right) = \frac{bAP^\circ}{\Delta X s \sigma} t_c \tag{12-28}$$

where m_i is initial moisture and $m_c = 0.7\sigma$. We can now calculate the required permeability:

$$b = \ln\left(\frac{m_e - m_i}{m_e - m_c}\right) \frac{\Delta X s \sigma}{AP^\circ t_c} \tag{12-29}$$

2. Loss of Vitamin C During In-package Desiccation

Another situation in which it is relatively easy to determine package requirements is one in which a deteriorative reaction depends only on moisture content and time. (Temperature is assumed to be constant at the expected average level.) In many cases, for instance in dehydrated powders, vitamin C loss can be approximated by [22]:

$$-\frac{dC}{dt} = km \tag{12-30}$$

where C is vitamin C content.

Let us now assume that the powder is placed in a hermetically closed can containing a package desiccant in order to achieve slow drying to the desired moisture content. A question that may be asked is what must be the minimum area of the internal package of desiccant to assure that no more than, say, 5% of the vitamin is lost before the moisture content is brought down to a very low level (when $m = 0.05$ assume that the rate of vitamin C loss is negligible). If we assume that the partial pressure of water at the desiccant surface is close to zero, we have $m_e = 0$ and the loss of water from the food is given by:

$$-\frac{dm}{dt} = \frac{-AbP^\circ}{\sigma \Delta X s}(m) \tag{12-31}$$

Combining Eqs. (12-31) and (12-30) we obtain

$$-\frac{dm}{dt} = \frac{-AbP^\circ}{\sigma \Delta X s k} \frac{dC}{dt} \tag{12-32}$$

Hence, in this very simple case, loss of the vitamin is proportional to loss of water, and the required area for our specific example is given by

$$A = \frac{(m_i - 0.05)(\sigma \Delta X s k)}{bP^\circ (C_i - 0.95C_i)} \tag{12-33}$$

where C_i is the initial vitamin C content.

3. Nonenzymatic Browning in a Dehydrated Product

A more complicated but very common situation is one in which a dehydrated product is limited in shelf life by nonenzymatic browning, but neither the isotherm nor the dependence of reaction rate on moisture are linear. Such a situation was analyzed by Mizrahi et al. [29], using dehydrated cabbage as an example. In this product, at 37° C, the isotherm is expressed by Eq. (12-34):

$$a = \frac{k_1 + m}{k_2 + m} \tag{12-34}$$

The rate of browning related to moisture content is expressed in Eq. (12-35):

$$\bar{D}' \equiv \frac{d\bar{D}}{dt} = k_3 \left[1 + \sin\left(\frac{-\pi}{2} + \frac{m\pi}{k_4}\right)\right]^{k_5} \tag{12-35}$$

where

$\bar{D}'$ is the rate of browning
$\bar{D}$ is the extent of browning
k_3, k_4, k_5 are constants

It was found, furthermore, that the cabbage did not offer important internal resistance to equilibration with infiltrating water, and that browning did result in production of additional water through chemical reactions in the product.

The information was then combined with the already familiar permeability equation, and a computer was used to calculate the moisture changes and extent of browning with time. Table 12-17 shows a simplified iteration scheme. (We shall ignore the water generated by the chemical reaction itself.) In Table 12-17 and in our discussion the subscripts 0, t, and t + Δt refer, respectively, to zero time and to time at the beginning and end of a time interval Δt. The iteration is as follows:

a. Rate of browning ($\bar{D}'$) at the beginning of the time interval is determined by Eq. (12-35) and the increase in browning $\Delta\bar{D}$ is added to obtain $\bar{D}_{t+\Delta t}$.
b. The water activity, a, of the food is calculated for the beginning of Δt.
c. The amount of moisture increase due to permeation into the package is calculated using Eq. (12-36).

$$\Delta m = K(a_{out} - a)\,\Delta t \tag{12-36}$$

where a_{out} is the water activity outside the package and

$$k = \frac{(b)(A)(P^{\circ})}{(\Delta X)(s)} \tag{12-37}$$

TABLE 12-17

Iteration Procedure

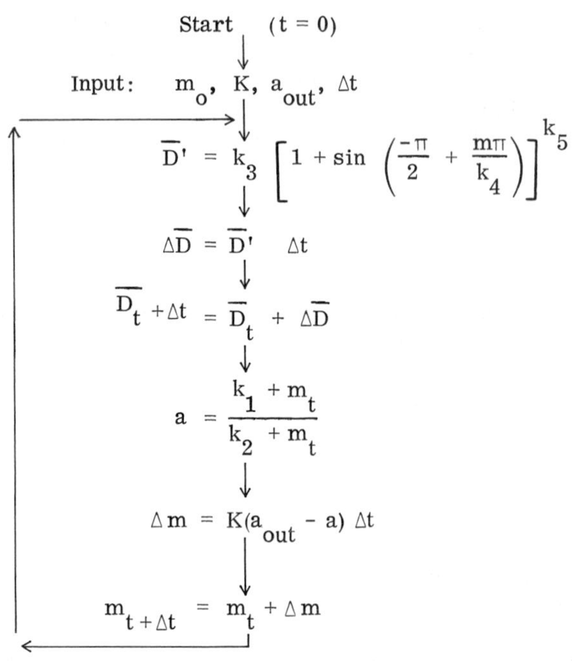

The process is then repeated to calculate changes occurring in the next time interval.

Predicted values of the increase of moisture content due to permeation, and the predicted extent of browning for samples packaged in two types of materials studied by Mizrahi et al. [29] are presented in Figures 12-28 and 12-29. The estimated uncertainties of the calculations are expressed as 2 times the standard deviation and are shown as broken lines. The actual increases in moisture content, obtained by weighing the samples during storage, are scattered around the predicted values with maximum deviation less than 3%. In the case of the extent of browning, the deviations between actual and predicted values are less than 10%.

C. Accelerated Testing

Accelerated tests are necessary in order to allow the development of valid storage life predictions within a reasonable time span. Acceleration is usually within a reasonable time span. Acceleration is usually achieved by raising temperatures and by increasing moisture content. Most calculations are based on the assumption that the Arrhenius equation is applicable (see Chapter 3).

Charm et al. [5] used the Arrhenius equation and organoleptic data to estimate the shelf life of fish at refrigerator temperatures. Wanninger [45] used the

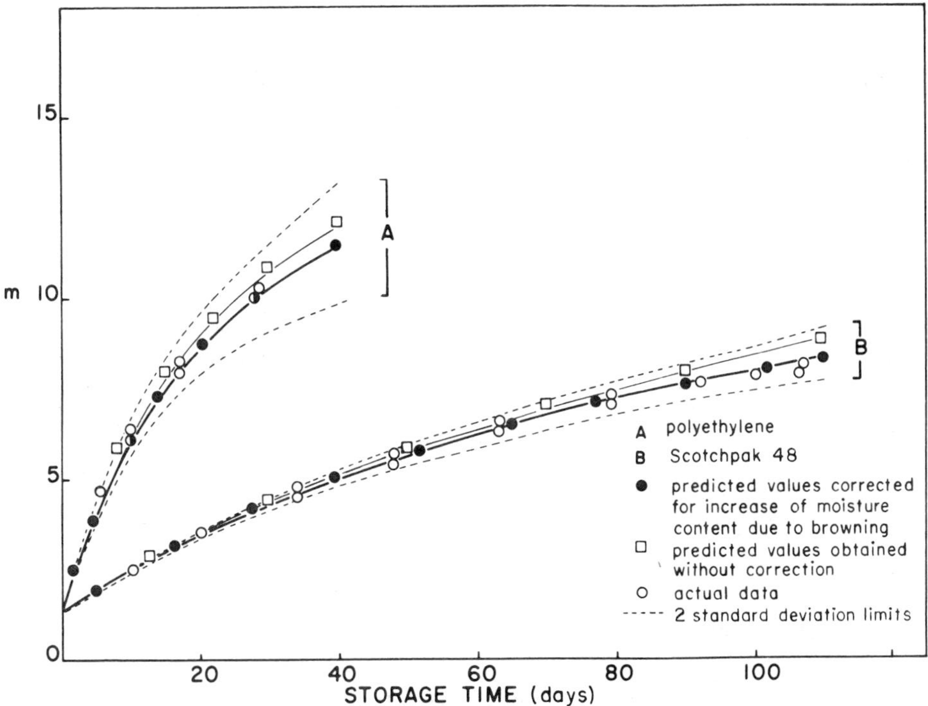

FIG. 12-28. Predicted and observed increase in moisture content of dehydrated cabbage packaged in two types of packaging materials.

dependence of ascorbic acid decomposition on moisture and temperature to design a simplified model predicting stability of this vitamin.

There also have been some good studies on accelerated tests for pharmaceuticals based on temperature exaggeration. Researchers have developed techniques for nonisothermal stability studies, in which temperature is changed during the accelerated test in some predicted manner (often linearly with time) and the kinetic constants are extracted from the test results with the aid of suitable mathematical techniques.

D. Packaging Requirements of Fresh Fruits and Vegetables

It is desirable to slow respiration rates of fruits and vegetables by lowering oxygen pressure, but the oxygen level must be sufficient to forestall significant anaerobic respiration. In Figure 12-7 we have noted the relationship between respiration rate and oxygen pressure. The curve represents a function:

$$-\frac{dO_2}{dt} = f(p) \tag{12-38}$$

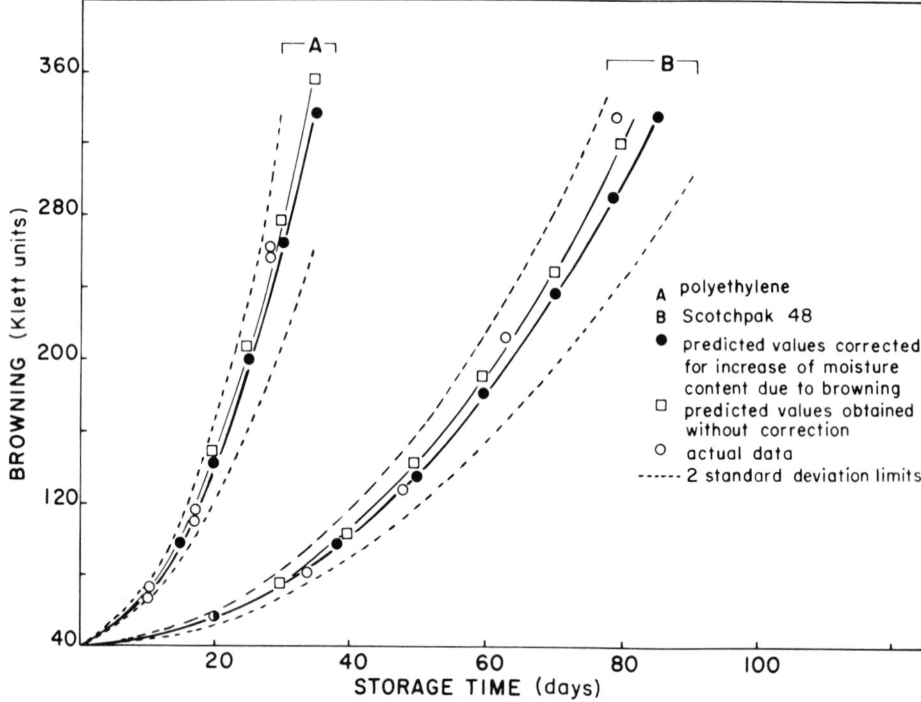

FIG. 12-29. Predicted and observed browning of dehydrated cabbage packaged in two types of packaging materials.

where p is partial pressure of oxygen in the package in atmospheres. The minus sign indicates that oxygen is being removed from the package atmosphere by the fruit. Simultaneously, however, oxygen is being supplied through the package wall, and the rate of input is also a function of p.

$$\frac{dO_2}{dt} = \frac{AB}{\Delta X}(0.21 - p) \tag{12-39}$$

If we set the two functions equal we can calculate the oxygen pressure at which the consumption and supply balance exactly (Fig. 12-30). As long as this pressure is above the critical lower limit of oxygen pressure, and the RQ $\simeq$ 1, the consumption of oxygen equals evolution of CO_2 and we can calculate the steady-state concentration of CO_2 as well. Figure 12-31 shows that such steady-state concentrations are rapidly attained in a polyethylene bag of bananas. By using several materials in parallel it is also possible to independently control both O_2 and CO_2 in a package for considerable periods of time (until the natural aging process of the fruit changes its response to the environment). Control of gas exchange by semipermeable packages has practical value as well as theoretical interest [20].

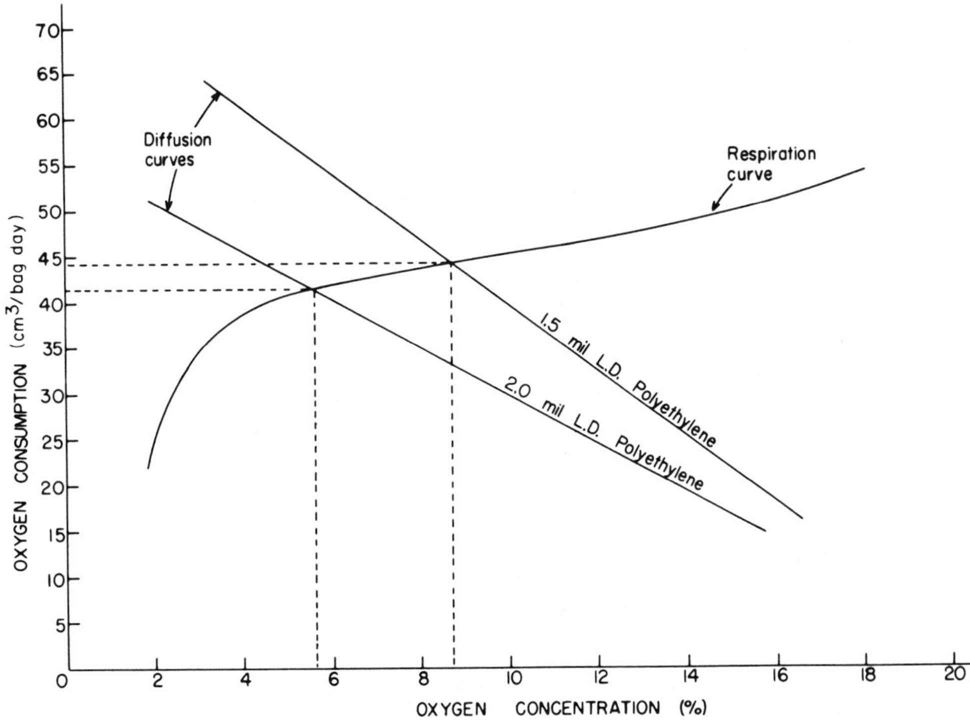

FIG. 12-30. Graphical estimation of steady-state oxygen concentration in packages containing apples.

E. Oxygen-sensitive Foods

Another packaging problem is the exclusion of oxygen to protect foods containing unsaturated fatty acids. These foods undergo oxidative reactions at rates which depend on oxygen concentration in the package. As mentioned previously, among the oxygen-sensitive nutrients are several vitamins, all of the essential fatty acids, and many amino acid residues.

Past measurements of storage lives of foods deteriorating through oxidation have involved packaging the foods in materials with different permeabilities to oxygen and to water. The foods have been stored at various external conditions and tested periodically either for chemical deterioration, for organoleptic acceptability, or for nutritional value. Recently we have been able to improve this trial-and-error method significantly through the use of mathematical models and computers.

The most complicated storage problem studied so far is that of dehydrated foods, which deteriorate either because of absorption of water or because of oxidation. Shelf life is defined here as the time to reach a critical humidity (which corresponds to a maximum moisture content) or a critical extent of oxidation. If one of these limits is reached before the other, the product is overprotected with respect to the other. Since very impermeable materials are expensive, this overprotection is unnecessarily costly. Therefore, we have

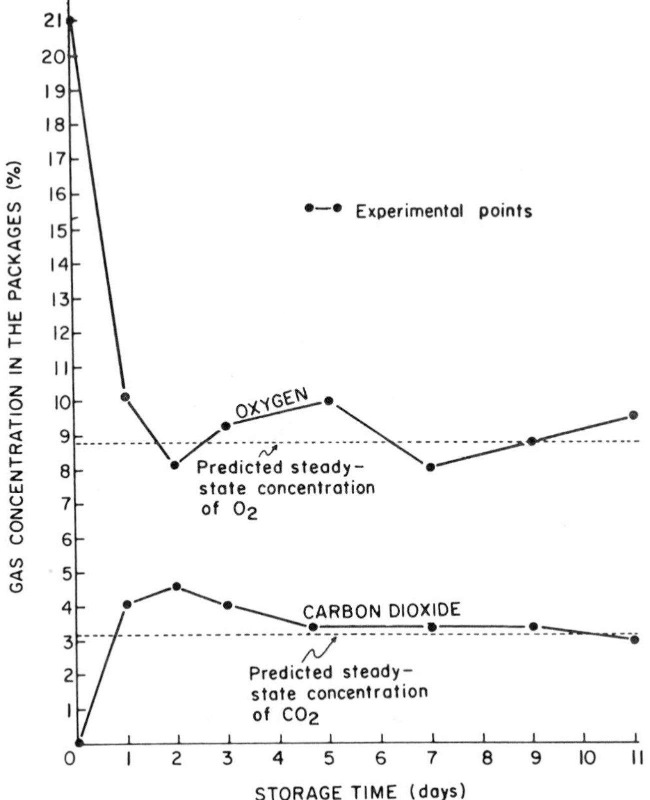

FIG. 12-31. Predicted and observed gas concentrations in packages containing apples.

developed a computer program to calculate optimal permeabilities so that deterioration caused by water and oxygen reaches the critical point at the end of the desired shelf life [40].

The significance of optimal permeabilities is best illustrated by the example of a film laminate. If a single film has high oxygen permeability and low water permeability, it must be very thick to prevent oxygen transmission. However, the thickness and oxygen transmission can be greatly reduced if a second film with high water permeability and low oxygen permeability is laminated to the first film. This lamination can significantly reduce cost.

The changes in packaged potato chips during storage can be described by a system of three equations [37, 38]. When solved with a computer, these equations predict changes with time of (1) oxygen pressure within a package, (2) extent of oxidation expressed as amount of oxygen consumed per unit weight of product, and (3) equilibrium relative humidity. These variables are interrelated because rate of oxidation depends on oxygen pressure and on the equilibrium relative humidity in the package.

Optimization can be calculated for a specified set of conditions: (1) initial environment in the package, (2) weight of product, (3) storage conditions, (4) film area and (5) film thickness.

Mathematical methods for optimization are not simple because of the complex interrelationships among variables. A major advantage of computer simulation, however, is that it allows rapid assessment of the influence of changes in the product or package conditions, without the need of multiple storage studies [39].

We have mentioned the possibility of inclusion in the package of a desiccant to allow the in-package removal of water from food. This process is called "in-package desiccation," An analogous system involves removal of oxygen within the package using "oxygen scavengers." Among the important systems are:

1. Use of the enzyme glucose oxidase, along with the substrate glucose. The oxidation of glucose catalyzed by this enzyme removes oxygen from the package. This system is suitable only for situations in which substantial amounts of water are permissible in the container, since some water is required for enzyme activity.
2. For dehydrated foods a more suitable system is the catalyzed oxidation of hydrogen gas to water. The amount of water produced in removing oxygen is not significant. (Removal of oxygen from a 100 cm^3 headspace consisting of air at 1 atm pressure produces only 40 mg of water.) The catalysts used for this purpose include palladium and platinum.

The American Can Company has developed a scavenger system incorporated in the packaging material itself, thus avoiding direct contact between the catalyst and the food. In this system the scavenger layer is present in a laminated packaging material separated from the inside of the package by a relatively permeable polyethylene layer. This allows the hydrogen and the oxygen to reach the catalyst system. Toward the outside there are several layers of essentially impermeable materials, thus assuring that oxygen from the outside does not reach the scavenger. This system has been proposed for packaging dry whole milk, which is very oxygen sensitive.

VI. INTERACTION BETWEEN PRODUCT AND PACKAGE

Packages are made from many components having varying chemical and physical characteristics. Some of these components can react with the package contents The toxicological consequences and potential hazards of such interactions have received a great deal of attention. The 1958 Food Additive Amendment specifically included in its definition of food additives substances which become a component of food or affect characteristics of food as a consequence of packaging. Passage of this amendment has stimulated research into food-package interactions. The problem is a complex and difficult one. Determination of the transfer of individual substances from package to food is difficult and regulations have been based on a less difficult procedure involving extraction of constituents of packaging materials into food-simulating solvents, including water, 3% aqueous NaCl, 3% aqueous $NaHCO_3$, 3% acetic acid, 3% lactic acid, 20% aqueous sucrose, and Lard or vegetable oil [10]. Other countries use different food-stimulating solutions [16].

The problem of extracting packaging constituents into fatty substances is particularly difficult, because some components of packaging materials are difficult to differentiate from normal components of fatty foods [24].

Accordingly, fat-simulating solvents, such as heptane, are often used to test the inertness of packaging materials. An even more imposing problem is to correlate the type and amount of extractive with physiological effects in animals [9].

Recent concern about potential toxicity of various heavy metals has resulted in research on transfer of contaminants from various metallic containers to food.

Apart from safety considerations, the problem of acceleration of nutritive deterioration by package components has also received some recognition. This problem can be solved by the application of proper packaging methods.

In metal containers, application of inert coatings (enamels and lacquers) guards against reaction between food and package. When foods are unusually corrosive special coatings are required and these must be tested for effectiveness.

Glass has excellent inertness, and when inert liners are used for closures there is essentially no interaction between food and package. Most of the flexible packages also possess adequate inertness.

A related problem concerns penetration of the package by free oil from packaged food products. Free oil is defined as the lipid available for staining the package. Penetration of packaging materials by free oil sets up a focal point for rancidity development that can affect the nutritional value of the packaged food. The importance of using greaseproof materials to control fat and oil penetration is obvious.

The effects of specific processing methods on the total package-food system must be considered. In the case of heat processing and radiation processing, FDA approval has been sought and obtained by the United States Army Natick Laboratory to utilize specific packages for specified foods.

VII. PACKAGING FOR SPECIALIZED SITUATIONS

In this chapter we have discussed the general principles of protective packaging of foods. There are additional special problems associated with certain food processing or distribution situations which require specialized applications of the scientific principles discussed here. Since these problems cannot be discussed here, the reader is referred to the following articles:

1. Packaging of heat-sterilized foods in rigid and flexible containers, which has been studied for the past 20 years [15], is reviewed by Brody [3].
2. Packaging of space foods for astronauts presented many difficult and unique challenges and has been reviewed aptly by Smith et al. [43]; Rambaut et al. [41]; and Heidelbaugh et al. [12].
3. Protection of food from deterioration in underdeveloped countries is a particularly challenging problem and has been reviewed by Majumder [28].

SYMBOLS

A	area (cm^2)
B	permeability constant (cm^3 mil m^{-2} day^{-1} atm^{-1})
B*	permeability constant (cm^3 mm cm^{-2} sec^{-1} cm Hg^{-1})

B_o	constant
$\overline{B}$	barrier properties of package
C	vitamin content (mg/100 g)
D	diffusion coefficient (cm^2/sec)
$\overline{D}$	deterioration extent (for instance; browning) of food
E_p	activation energy for permeability (kcal/mole)
EO	external environment
$\overline{F}$	food properties
F_o	initial food properties
$\overline{I}$	internal environment
I_o	light intensity at package surface
I_i	light intensity at food surface
I_x	light intensity at distance X in food
I_{abs}	total light absorbed by food
J	flux of gas or vapor (moles/sec)
P°	vapor pressure of water (torr)
PO_2	oxygen pressure
R_p	fraction of light reflected by packaging material
R_f	fraction of light reflected by the food
R	gas constant
RH	relative humidity
S	solubility of gas in membrane ($moles/cm^3$ atm)
T	absolute temperature
T_g	glass transition temperature
Tr_p	fraction of light transmitted by packaging material
a	water activity
b	permeability to water vapor (g mil m^{-2} day^{-1} $torr^{-1}$)
c	concentration of gas in membrane ($moles/cm^3$)
k_1-k_n	constants
m	moisture content (g H_2O/g solids)
p	partial pressure
s	weight of food solids (g)
t	time

β convective flow constant (g mil m^{-2} day^{-1} $torr^{-1}$)

μ absorbance of light in food

μ_p absorbance of light in packaging material

σ constant

REFERENCES

1. F. W. Billmeyer, Jr., Textbook of Polymer Science, 2nd ed., Wiley-Interscience, New York, 1971.

2. T. Bond, Mod. Packag., 46(7), 47 (1973).

3. A. Brody, CRC Crit. Revs. Food Technol., 2(2), 187 (1971).

4. J. P. Casey, Pulp and Paper-Chemistry and Chemical Technology. 2 vols. Interscience, New York, 1952.

5. S. E. Charm, R. J. Learson, L. R. Ronsivalli, and M. Schwartz, Food Technol., 26(7), 65 (1972).

6. N. A. Cooke, Know your packaging materials. American Management Association of Packaging, Series No. 46(3), 1955.

7. E. G. Davis, M. Karel, and B. E. Proctor, Mod. Packag., 33(4), 135 (1959).

8. E. G. Davis, M. Karel, and B. E. Proctor, Mod. Packag., 33(7), 208 (1960).

9. J. P. Frawley, Food Cosmet. Toxicol., 5, 293 (1967).

10. R. C. Griffin and S. Sacharow, Principles of Package Development. AVI Publ. Co., Westport, Conn., 1972.

11. P. M. Hauser and A. D. McLaren, Ind. Eng. Chem., 40, 112 (1948).

12. N. D. Heidelbaugh, M. C. Smith, P. C. Rambaut, L. Lutwak, C. S. Huber, C. Stadler, and B. M. Rouse, J. Amer. Dietet. Assoc., 62(4), 383 (1973).

13. R. Heiss, Principles of Food Packaging. An International Guide. P. Keppler Verlag KG, Heusenstamm, Germany, 1970.

14. W. Hoffman, H. Krämer, and V. Linowitzki, Chem. Ing. Technol., 37(1), 34 (1965).

15. K. H. Hu, A. I. Nelson, R. R. Lesgault, and M. P. Steinberg, Food Technol., 9, 236 (1955).

16. Institute of Packaging (London), Food Packaging and Health: Migration and Legislation, The Institute of Packaging, London, 1971.

17. G. Kaess, Z. Lebensm. Unters. Forsch., 90, 101 (1950).

18. M. Karel, B. E. Proctor and G. Wiseman, Food Technol., 13(1), 69 (1959).

19. M. Karel, B. E. Proctor and A. Cornell, Food Technol., 11(3), 141 (1957).

20. M. Karel, AIChE Symp. Ser., 69(132), 107 (1973).

21. M. Karel, P. Issenberg, L. Ronsivalli, and V. Jurin, Food Technol., 17(3), 91 (1963).

22. M. Karel and J. T. R. Nickerson, Food Technol., 18, 104 (1964).

23. M. Karel, S. Mizrahi and T. P. Labuza, Mod. Packag., 44(8), 54 (1971).

24. M. Karel and G. Wogan, Migration of Substances from Flexible Containers for Heat-Processed Foods, Report on Contract Research with U. S. Army Natick Laboratories, Contract DA 19-129- M-2080, MIT, Cambridge, Mass., 1963.

25. M. Kroger and R. X. Igoe, Food Prod. Dev., 5(7), 74 (1971).

26. R. A. Lampi, Activ. Report., 19(1), 51 (1967).

27. A. Lebovits, Mod. Plast., March, 139 (1966).

28. S. K. Majumder, in Proceedings of the Third International Congress on Food Science and Technology, Institute of Food Technology, Chicago, Ill., 1970, p. 518.

29. S. Mizrahi, T. P. Labuza, and M. Karel, J. Food Sci., 35, 799 (1970).

30. Modern Packaging Encyclopedia, McGraw-Hill, New York, 1971.

31. B. H. Morgan, Food Prod. Devel., 5(4), 75 (1971).

32. M. J. Morley, J. Food Technol., 6, 371 (1971).

33. F. A. Paine, Fundamentals of Packaging, Blackie and Sons, Ltd., London, 1962.

34. F. A. Paine, Packaging Materials and Containers, Blackie and Sons, Ltd., London, 1967.

35. J. Pinsky, Mod. Plast., April, 145 (1957).

36. D. Quast and M. Karel, J. Food Technol., 6, 95 (1971).

37. D. Quast and M. Karel, J. Food Sci., 37, 490 (1972).

38. D. Quast and M. Karel, J. Food Sci., 37, 679 (1972).

39. D. Quast and M. Karel, Mod. Packag., 46(3), 50 (1973).

40. D. Quast, M. Karel and W. M. Rand, J. Food Sci., 37, 673 (1972).

41. P. C. Rambaut, C. T. Bourland, N. D. Heidelbaugh, C. S. Huber and M. C. Smith, Food Technol., 26(1), 58 (1972).

42. A. Ratzlaff, J. Milk Food Technol., 18, 195 (1955).

43. M. C. Smith, C. S. Huber, and N. D. Heidelbaugh, Aerospace Med., 42(11), 1185 (1971).

44. V. Stannett and M. Szwarc, Permeability of Plastic Films and Coated Papers to Gases and Vapors, Monograph Series No. 23, Technical Association of the Pulp and Paper Industry, New York, 1962.

45. L. A. Wanninger, Food Technol., 26(6), 42 (1972).

46. G. Wildbrett, Fette-Seifen-Anstrichm, 69, 781 (1967).

SUBJECT INDEX

P

Q